ANALYSIS, CONTROL AND OPTIMIZATION OF COMPLEX DYNAMIC SYSTEMS

GERAD 25th Anniversary Series

- **Essays and Surveys in Global Optimization**
  Charles Audet, Pierre Hansen, and Gilles Savard, editors
- **Graph Theory and Combinatorial Optimization**
  David Avis, Alain Hertz, and Odile Marcotte, editors
- **Numerical Methods in Finance**
  Hatem Ben-Ameur and Michèle Breton, editors
- **Analysis, Control and Optimization of Complex Dynamic Systems**
  El-Kébir Boukas and Roland Malhamé, editors
- **Column Generation**
  Guy Desaulniers, Jacques Desrosiers, and Marius M. Solomon, editors
- **Statistical Modeling and Analysis for Complex Data Problems**
  Pierre Duchesne and Bruno Rémillard, editors
- **Performance Evaluation and Planning Methods for the Next Generation Internet**
  André Girard, Brunilde Sansò, and Félisa Vázquez-Abad, editors
- **Dynamic Games: Theory and Applications**
  Alain Haurie and Georges Zaccour, editors
- **Logistics Systems: Design and Optimization**
  André Langevin and Diane Riopel, editors
- **Energy and Environment**
  Richard Loulou, Jean-Philippe Waaub, and Georges Zaccour, editors

ANALYSIS, CONTROL AND OPTIMIZATION OF COMPLEX DYNAMIC SYSTEMS

Edited by

EL KÉBIR BOUKAS
GERAD and École Polytechnique de Montréal

ROLAND P. MALHAMÉ
GERAD and École Polytechnique de Montréal

Springer

El Kébir Boukas
GERAD & École Polytechnique de Montréal
Montréal, Canada

Roland P. Malhamé
GERAD & École Polytechnique deMontréal
Montréal, Canada

Library of Congress Cataloging-in-Publication Data

A C.I.P. Catalogue record for this book is available
from the Library of Congress.

ISBN-10: 0-387-25475-7 ISBN 0-387-25477-3 (e-book) Printed on acid-free paper.
ISBN-13: 978-0387-25475-3

Printed in the United States of America.

9 8 7 6 5 4 3 2 1 SPIN 11053064

springeronline.com

Foreword

GERAD celebrates this year its 25th anniversary. The Center was created in 1980 by a small group of professors and researchers of HEC Montréal, McGill University and of the École Polytechnique de Montréal. GERAD's activities achieved sufficient scope to justify its conversion in June 1988 into a Joint Research Centre of HEC Montréal, the École Polytechnique de Montréal and McGill University. In 1996, the Université du Québec à Montréal joined these three institutions. GERAD has fifty members (professors), more than twenty research associates and post doctoral students and more than two hundreds master and Ph.D. students.

GERAD is a multi-university center and a vital forum for the development of operations research. Its mission is defined around the following four complementarily objectives:

- The original and expert contribution to all research fields in GERAD's area of expertise;
- The dissemination of research results in the best scientific outlets as well as in the society in general;
- The training of graduate students and post doctoral researchers;
- The contribution to the economic community by solving important problems and providing transferable tools.

GERAD's research thrusts and fields of expertise are as follows:

- Development of mathematical analysis tools and techniques to solve the complex problems that arise in management sciences and engineering;
- Development of algorithms to resolve such problems efficiently;
- Application of these techniques and tools to problems posed in related disciplines, such as statistics, financial engineering, game theory and artificial intelligence;
- Application of advanced tools to optimization and planning of large technical and economic systems, such as energy systems, transportation/communication networks, and production systems;
- Integration of scientific findings into software, expert systems and decision-support systems that can be used by industry.

One of the marking events of the celebrations of the 25th anniversary of GERAD is the publication of ten volumes covering most of the Center's research areas of expertise. The list follows: **Essays and Surveys in Global Optimization**, edited by C. Audet, P. Hansen and G. Savard; **Graph Theory and Combinatorial Optimization**, edited by D. Avis, A. Hertz and O. Marcotte; **Numerical Methods in Finance**, edited by H. Ben-Ameur and M. Breton; **Analysis, Control and Optimization of Complex Dynamic Systems**, edited by E.K. Boukas and R. Malhamé; **Column Generation**, edited by G. Desaulniers, J. Desrosiers and M.M. Solomon; **Statistical Modeling and Analysis for Complex Data Problems**, edited by P. Duchesne and B. Rémillard; **Performance Evaluation and Planning Methods for the Next Generation Internet**, edited by A. Girard, B. Sansò and F. Vázquez-Abad; **Dynamic Games: Theory and Applications**, edited by A. Haurie and G. Zaccour; **Logistics Systems: Design and Optimization**, edited by A. Langevin and D. Riopel; **Energy and Environment**, edited by R. Loulou, J.-P. Waaub and G. Zaccour.

I would like to express my gratitude to the Editors of the ten volumes, to the authors who accepted with great enthusiasm to submit their work and to the reviewers for their benevolent work and timely response. I would also like to thank Mrs. Nicole Paradis, Francine Benoît and Louise Letendre and Mr. André Montpetit for their excellent editing work.

The GERAD group has earned its reputation as a worldwide leader in its field. This is certainly due to the enthusiasm and motivation of GERAD's researchers and students, but also to the funding and the infrastructures available. I would like to seize the opportunity to thank the organizations that, from the beginning, believed in the potential and the value of GERAD and have supported it over the years. These are HEC Montréal, École Polytechnique de Montréal, McGill University, Université du Québec à Montréal and, of course, the Natural Sciences and Engineering Research Council of Canada (NSERC) and the Fonds québécois de la recherche sur la nature et les technologies (FQRNT).

Georges Zaccour
Director of GERAD

Avant-propos

Le Groupe d'études et de recherche en analyse des décisions (GERAD) fête cette année son vingt-cinquième anniversaire. Fondé en 1980 par une poignée de professeurs et chercheurs de HEC Montréal engagés dans des recherches en équipe avec des collègues de l'Université McGill et de l'École Polytechnique de Montréal, le Centre comporte maintenant une cinquantaine de membres, plus d'une vingtaine de professionnels de recherche et stagiaires post-doctoraux et plus de 200 étudiants des cycles supérieurs. Les activités du GERAD ont pris suffisamment d'ampleur pour justifier en juin 1988 sa transformation en un Centre de recherche conjoint de HEC Montréal, de l'École Polytechnique de Montréal et de l'Université McGill. En 1996, l'Université du Québec à Montréal s'est jointe à ces institutions pour parrainer le GERAD.

Le GERAD est un regroupement de chercheurs autour de la discipline de la recherche opérationnelle. Sa mission s'articule autour des objectifs complémentaires suivants :

- la contribution originale et experte dans tous les axes de recherche de ses champs de compétence ;
- la diffusion des résultats dans les plus grandes revues du domaine ainsi qu'auprès des différents publics qui forment l'environnement du Centre ;
- la formation d'étudiants des cycles supérieurs et de stagiaires post-doctoraux ;
- la contribution à la communauté économique à travers la résolution de problèmes et le développement de coffres d'outils transférables.

Les principaux axes de recherche du GERAD, en allant du plus théorique au plus appliqué, sont les suivants :

- le développement d'outils et de techniques d'analyse mathématiques de la recherche opérationnelle pour la résolution de problèmes complexes qui se posent dans les sciences de la gestion et du génie ;
- la confection d'algorithmes permettant la résolution efficace de ces problèmes ;
- l'application de ces outils à des problèmes posés dans des disciplines connexes à la recherche opérationnelle telles que la statistique, l'ingénierie financière, la théorie des jeux et l'intelligence artificielle ;
- l'application de ces outils à l'optimisation et à la planification de grands systèmes technico-économiques comme les systèmes énergétiques, les réseaux de télécommunication et de transport, la logistique et la distributique dans les industries manufacturières et de service ;

- l'intégration des résultats scientifiques dans des logiciels, des systèmes experts et dans des systèmes d'aide à la décision transférables à l'industrie.

Le fait marquant des célébrations du 25e du GERAD est la publication de dix volumes couvrant les champs d'expertise du Centre. La liste suit : **Essays and Surveys in Global Optimization**, édité par C. Audet, P. Hansen et G. Savard ; **Graph Theory and Combinatorial Optimization**, édité par D. Avis, A. Hertz et O. Marcotte ; **Numerical Methods in Finance**, édité par H. Ben-Ameur et M. Breton ; **Analysis, Control and Optimization of Complex Dynamic Systems**, édité par E.K. Boukas et R. Malhamé ; **Column Generation**, édité par G. Desaulniers, J. Desrosiers et M.M. Solomon ; **Statistical Modeling and Analysis for Complex Data Problems**, édité par P. Duchesne et B. Rémillard ; **Performance Evaluation and Planning Methods for the Next Generation Internet**, édité par A. Girard, B. Sansò et F. Vázquez-Abad ; **Dynamic Games : Theory and Applications**, édité par A. Haurie et G. Zaccour ; **Logistics Systems : Design and Optimization**, édité par A. Langevin et D. Riopel ; **Energy and Environment**, édité par R. Loulou, J.-P. Waaub et G. Zaccour.

Je voudrais remercier très sincèrement les éditeurs de ces volumes, les nombreux auteurs qui ont très volontiers répondu à l'invitation des éditeurs à soumettre leurs travaux, et les évaluateurs pour leur bénévolat et ponctualité. Je voudrais aussi remercier Mmes Nicole Paradis, Francine Benoît et Louise Letendre ainsi que M. André Montpetit pour leur travail expert d'édition.

La place de premier plan qu'occupe le GERAD sur l'échiquier mondial est certes due à la passion qui anime ses chercheurs et ses étudiants, mais aussi au financement et à l'infrastructure disponibles. Je voudrais profiter de cette occasion pour remercier les organisations qui ont cru dès le départ au potentiel et la valeur du GERAD et nous ont soutenus durant ces années. Il s'agit de HEC Montréal, l'École Polytechnique de Montréal, l'Université McGill, l'Université du Québec à Montréal et, bien sûr, le Conseil de recherche en sciences naturelles et en génie du Canada (CRSNG) et le Fonds québécois de la recherche sur la nature et les technologies (FQRNT).

Georges Zaccour
Directeur du GERAD

Contents

Foreword v

Avant-propos vii

Contributing Authors xi

Preface xiii

Part I Production and Inventory Systems

1

Average Cost Optimality in Inventory Models with Markovian Demands and Lost Sales 3

Dirk Beyer and Suresh P. Sethi

2

Inventory Control of Switched Production Systems: LMI Approach 25

El Kébir Boukas and Luis Rodrigues

3

A Two-time-scale Approach for Production Planning in Discrete Time 43

Qing Zhang and George Yin

4

Evaluation of Throughput in Serial Production Lines with Non-Exponential Machines 61

Jingshan Li and Semyon M. Meerkov

5

Supply Chain Production Planning Modeling Facility Lead Time and Quality of Service 91

Osmon M. Anli, Michael C. Caramanis, and Ioannis Ch. Paschalidis

Part II Large Scale or Multi Agent System Problems

6

Providing QoS in Large Networks: Statistical Multiplexing and Admission Control 137

Nikolay B. Likhanov, Ravi R. Mazumdar, and François Théberge

7

Combined Competitive Flow Control and Routing in Networks with Hard Side Constraints 169

Rachid El Azouzi, Mohamed El Kamili, Eitan Altman, Mohammed Abbad, and Tamer Başar

8
Generalized Uplifts in Pool-Based Electricity Markets 193
François Bouffard and Francisco D. Galiana

9
Nash Equilibria for Large-Population Linear Stochastic Systems of Weakly Coupled Agents 215
Minyi Huang, Roland P. Malhamé, and Peter E. Caines

Contributing Authors

MOHAMMED ABBAD
Faculté des sciences, Maroc
abbad@fsr.ac.ma

EITAN ALTMAN
INRIA, France
eitan.altman@sophia.inria.fr

OSMAN M. ANLI
Boston University, USA
anli@bu.edu

RACHID EL AZOUZI
Universrity of Avignon, France
rachid.elazouzi@lia.univ-avignon.fr

TAMER BAŞAR
University of Illinois, USA
tbasar@control.cls.uiuc.edu

DIRK BEYER
Hewlett-Packard Laboratorie, USA
beyer@hpl.hp.com

FRANÇOIS BOUFFARD
McGill University, Canada
francois.bouffard@mail.mcgill.ca

EL KÉBIR BOUKAS
École Polytechnique de Montréal, Canada
el-kebir.boukas@polymtl.ca

PETER E. CAINES
McGill University, Canada
peterc@cim.mcgill

MICHAEL C. CARAMANIS
Boston University, USA
mcaraman@bu.edu

FRANCISCO D. GALIANA
McGill University, Canada
galiana@ece.mcgill.ca

MINYI HUANG
University of Melbourne, Australia
m.huang@ee.mu.oz.au

MOHAMED EL KAMILI
Faculté des sciences, Maroc
melkamili@yahoo.fr

JINGSHAN LI
GM Research and Development Center, USA
jingshan.li@gm.com

NIKOLAY B. LIKHANOV
Institute for Problems of Information Transmission, Moscow, Russia
likh1@online.ru

ROLAND P. MALHAMÉ
École Polytechnique de Montréal, Canada
roland.malhame@polymtl.ca

RAVI R. MAZUMDAR
Purdue University, West Lafayette, USA
mazum@purdue.edu

SEMYON M. MEERKOV
University of Michigan, USA
smm@eecs.umich.edu

IOANNIS CH. PASCHALIDIS
Boston University, USA
yannisp@bu.edu

LUIS RODRIGUES
Concordia University, Canada
el-kebir.boukas@polymtl.ca

SURESH P. SETHI
The University of Texas at Dallas, USA
sethi@utdallas.edu

FRANÇOIS THÉBERGE
University of Ottawa, Canada
theberge@ieee.org

GEORGE YIN
Wayne State University, USA
gyin@math.wayne.edu

QING ZHANG
The University of Georgia, USA
qingz@math.uga.edu

Preface

Recent years have witnessed important developments in the deployment of complex dynamic systems possibly involving multiple actors, whether in the areas of telecommunications, manufacturing, transportation or power networks under the current deregulation paradigm. A rapidly emerging research area is that of understanding and analyzing biological systems using the tools of systems theory. In addition, countries economies are increasingly dependent in the ongoing globalization of economic exchanges. This is in large part made possible by the massive development of information technologies, and the ubiquity of communication systems which permit the existence of global marketplaces.

Complex dynamical systems can be viewed as systems involving a great number of interconnections of simpler subsystems, and where, of necessity, details about the interconnected components survive in the modeling process. The interest in complex systems can be traced back to the sixties where efforts were directed at extending state space modern control techniques to systems involving a large number of variables (see [1],[2] for example), via state space aggregation and extended numerical techniques. However, the current thrust is more inclusive in that state space is no longer the dominant mathematical model, and ideas of graphs, game theory and distributed control have become more prevalent.

The object of this publication effort is to gather in a single volume a large spectrum of complex dynamic systems related papers written by experts in their fields, and strongly representative of current research trends. Complex systems present important challenges, in great part due to their sheer size which makes it difficult to grasp their dynamic behavior, optimize their operation, or study their reliability. Yet, we live in a world where due to increasing interdependencies, and networking of systems, complexity has become the norm.

A fundamental shift in engineering approaches has in all likelihood become essential, thereby propelling back to the forefront ideas of decomposition in space or along time scales, aggregation, distributed optimization (particularly as developed in the context of game theory), and decentralized control. The success of decomposition/aggregation approaches itself hinges on the ability to develop adequate building blocks for which a deep theoretical understanding can be achieved. Furthermore, analytical techniques, traditionally associated with fields of physics such as statistical mechanics, as well as the tools of probability theory, in particular the body of knowledge related to central limit theorems and large deviation theory, are becoming essential in the engineering of large systems, as they offer mechanisms for synthesizing order out of erratic individual behavior, and analyzing rare nevertheless important events in such systems. Finally, the traditional tools of dynamic optimization such as dynamic programming, and maximum principles, have to be revisited in conjunction with decomposition/aggregation ideas, or new numerical tools such as linear matrix inequalities, to extend their applicability in the large scale systems context.

This volume comprises two parts. The first part is dedicated to a spectrum of complex problems of decision and control encountered in the area of production and inventory systems. The second part is dedicated to large scale or multi agent system problems occurring in other areas of engineering such as telecommunication and electric power networks, as well more generic contexts. We now summarize contributions of individual papers.

The first paper by Beyer and Sethi corresponds to the development of a new building block in the optimization of inventory replenishment policies with fixed ordering costs, multi-state Markov chain demand, and lost sales when not satisfied immediately. The optimality of (s, S) replenishment policies is established. The second paper by Boukas and Rodrigues, introduces linear matrix inequality techniques in an inventory system, so as to develop robust inventory replenishment policies within the framework of the control theory of switched linear systems, in the face of an uncertain demand environment. The third paper, by Zhang and Yin, reviews how time scale separation techniques can be employed in decomposing and compressing large scale dynamic programs associated with the optimization of manufacturing systems operating in a random demand environment, so as to obtain near optimal performance at a much reduced computational cost. The fourth paper, by Li and Meerkov, considers the problem of evaluating the maximum throughput of an arbitrarily long serial production line, with unreliable individual machines that can have non exponential random failure and repair times. An analytical approach is proposed based on an understanding of the dynamics of two-machine building blocks, and a type of continuation approach whereby, non exponential line behavior is obtained as a weighted average of that of better understood extreme cases of lines with respectively two state deterministic, and two state Markovian machines. The last paper in that first part of the volume is contributed by Anli, Caramanis and Paschalidis, and is a telling example of how strong information sharing between the various parts of the supply chain of a manufacturing enterprise including in particular customers with their manufacturing facility state dependent lead times, can make for more efficient and better coordinated production and parts transportation schedules, as well as permit delivery of higher quality of service to customers. The computational complexity consequent to the reliance on increased information sharing is mitigated via a hierarchical and partially decentralized decision and optimization framework, which exploits spatial and time scale decompositions, as well as decision variables aggregation, and local optimization, at distinct levels of the hierarchy.

The second part of the book is concerned either with large scale system problems occurring in other engineering areas such as telecommunications or electric power networks, or large scale problems in a more generic context. The first paper here is contributed by Likhanov, Mazumdar and Théberge. It shows how the complex problem of admission control at an internet network router, under statistical quality of service constraints, for large volumes of heterogeneous traffic, can be approached based on notions of effective bandwidth, themselves deriving from probabilistic tools, in particular large deviation theory. This provides for surprisingly simple admission control laws, and is a first illustration in this volume, of what we shall call here the paradox of complexity: under the right conditions, system size and randomness, when looked at in the proper framework, can in fact become allies against complexity. The second telecommunications paper is contributed by El Azouzi, El Kamili, Abbad, Altman and. It is an illustration here of how complex optimization problems arising in internet and ATM networks can be simplified partially at least, by sacrificing global optimality, at the cost of a multi agent, non cooperative, game theoretic view of the optimization problem. Thus each agent has a limited view of the external world and aims at optimizing an individual criterion under constraints imposed by decisions made by other competing agents. In this particular problem, the objective is for individual sources to design their flow rates (rates as sources) on the various links (combined flow control and routing) so as to optimize individual cost functions (additive over the route links), in the presence of hard constraints dictated by other user choices, links total bandwidth, and the topology of the network. Existence and uniqueness of different types of Nash equilibria are studied for this class of multiagent problems.

Moving to electric power systems applications, Bouffard and Galiana study energy price setting problems in so-called pool-based electricity markets. The latter are power generation pools involving multiple self interested electricity producers agreeing to sell electricity at centrally coordinated prices and at levels consistent with their initially declared offerings over a fixed time horizon. The price time profile is initially set by the pool coordinator at a level dictated by a total pool maximizing profit criterion, under constraints dictated by available individual generation capacity, generator dynamic constraints, and network operating constraints. An important problem is that this social welfare dictated solution may be in conflict with the self interested solutions that would be hoped for by some of the producing agents in the pool. In the long run, instability of the pool may result. In game theoretic terms, this global optimum may not be a Nash equilibrium of the optimization problem formulated as a multi-agent non cooperative game, under global constraints. The authors propose uplifts (changes to the price and individual generation profiles) that would stir the solution towards a Nash equilibrium. In addition, the uplifts have to correspond to a minimum norm solution. In some sense, one is looking for a Nash (selfish) equilibrium closest to the social welfare dictated optimum. Thus the paper illustrates a coordination mechanism for best reconciling individual and collective interests in multi agent systems. The final paper in this volume is contributed by Huang, Malhamé and Caines. It considers a generic large scale control problem whereby a system of weakly coupled linear stochastic systems with random parameters, viewed as individual agents, seek to optimize individual quadratic costs over a finite or infinite time horizon. Individual costs are assumed to be dependent on a coupling term which is a function of the empirical average of the states of all agents in the system. This setup can be viewed as the game theoretic simplified, but suboptimal implementation of a corresponding complex centralized large scale optimal control problem involving the sum of all individual costs. It can be shown to be adequate for capturing power control dynamics in wireless communication systems from mobiles to their base stations. It is also able to capture individual economic agents reactions in a price driven market. It is established that, as the number of agents grows without limit, a unique dynamic Nash equilibrium exists, characterized by a set of coupled differential equations. Closed form solutions can be obtained when all coupling terms are linear. Furthermore, the large population limiting Nash equilibrium can be shown to remain a good approximation for phenomena occurring in the finite population setting.

In conclusion, this volume will have spanned but a limited number of application areas where large scale systems arise, as well as a limited number of mathematical techniques and concepts useful in grappling with complexity. It is by no means exhaustive and its limitations are those of its editors and time. We can only hope that the gathering of these techniques and applications in a single volume will stimulate further research in a systems area of growing importance, and help in providing it with a sense of unity.

References

[1] Jamshidi, M., *Large Scale Systems: Modelling and Control*, North Holland, New York, 1983.

[2] Singh, M.G. and Mamoud, M.S., *Large Scale Systems Modelling, International Series on Systems and Control*, Pergamemon Press, 1981.

EL KÉBIR BOUKAS
ROLAND MALHAMÉ
École Polytechnique de Montréal

I

PRODUCTION AND INVENTORY SYSTEMS

Chapter 1

AVERAGE COST OPTIMALITY IN INVENTORY MODELS WITH MARKOVIAN DEMANDS AND LOST SALES

Dirk Beyer
Suresh P. Sethi

Abstract This paper is concerned with long-run average cost minimization of a stochastic inventory problem with Markovian demand, fixed ordering cost, convex surplus cost, and lost sales. The states of the Markov chain represent different possible states of the environment. Using a vanishing discount approach, a dynamic programming equation and the corresponding verification theorem are established. Finally, the existence of an optimal state-dependent (s, S) policy is proved.

1. Introduction

In the literature of stochastic inventory models, there are two different assumptions about the excess demand unfilled from existing inventories: the backlog assumption and the lost sales assumption. The former is more popular in the literature partly because historically the inventory studies started with spare parts inventory management problems in military applications, where the backlog assumption is realistic. However, in many other business situations, it is quite often that demand that cannot be satisfied on time is lost. This is particularly true in a competitive business environment. For example in many retail establishments, such as a supermarket or a department store, a customer chooses a competitive brand or goes to another store if his/her preferred brand is out of stock.

In the presence of fixed ordering costs in inventory models under either assumption, an important issue has been to establish the optimality of (s, S)-type policies. There are many classical and recent papers that deal with this issue in the backlog case. These are cited and reviewed in Beyer and Sethi (1997, 1999); see also the forthcoming book of Beyer, Cheng and Sethi (2004). However, only Veinott (1966); Shreve (1976); Bensoussan et al. (1983); Cheng and Sethi (1999), to our knowledge, have considered the lost sales case explicitly. This is perhaps because the proofs of the results in the lost sales case are usually more complicated than those in the backlog case. Shreve (1976); Bensoussan et al. (1983) establish the optimality of an (s, S)-type policy by using the concept of K-convexity. Veinott (1966) provides a different proof for the optimality of (s, S)-type policies in the lost sales case. His proof is based on a different set of assumptions which neither implies nor is implied by those used in Shreve (1976); Bensoussan et al. (1983). It should be noted that all these results are obtained under the condition of zero lead time. Cheng and Sethi (1999) is discussed below.

Many efforts have been made to incorporate various realistic features in inventory models. However, most of them are carried out under the backlog assumption. One such feature is that of Markovian demand, which has been considered by Karlin and Fabens (1960); Song and Zipki (1993); Sethi and Cheng (1997); Ozekici and Parlar (1999); Beyer and Sethi (1997, 1999); Beyer, Sethi and Taksar (1998). Without exception, they all use the backlog assumption in their analysis. As a natural extension of and a flexible alternative to independent demands considered in the bulk of the classical inventory literature, Markovian demands can model demands that are dependent on randomly changing economic and market conditions. As the lost sales situation is often the case in competitive markets, it is interesting and worthwhile to extend the Markovian demand model to incorporate the lost sales case. Cheng and Sethi (1999) accomplish this in the discounted cost case. It is the purpose of this paper to treat that lost sales case with Markovian demands from the viewpoint of minimizing long-run average cost.

The plan of the paper is as follows. In the next section, we provide a precise formulation of the problem. Relevant results for the discounted cost problem obtained in Cheng and Sethi (1999) are summarized in Section 3. Also examined in this section is the asymptotic behavior of the differential discounted value function as the discount rate approaches zero. In Section 4, we develop the vanishing discount approach to establish the average cost optimality equation. The associated verification theorem is proved in Section 5, and the theorem is used to show that a state-dependent (s, S) policy, or simply an (s^i, S^i) policy, is optimal for the problem. Section 6 concludes the paper with suggestions for future research.

2. Formulation of the model

In order to specify the stationary, discrete time, infinite horizon inventory problem under consideration, we introduce the following notation and basic assumptions:

$(\Omega, \mathcal{F}, P) =$ the probability space;

$I = \{1, 2, \ldots, L\}$, finite collection of possible demand states;

$i_k =$ the demand state in period k, $k \in \mathbf{Z} = \{0, 1, 2, \ldots\}$;

$\{i_k\} =$ a Markov chain with the $(L \times L)$-transition matrix $P = \{p_{ij}\}$;

$\xi_k =$ the demand realized during the period k, ξ_k dependent on i_k, but not on k;

$\phi_i(\cdot) =$ the conditional density function of ξ_k when $i_k = i$;

$\Phi_i(\cdot) =$ the distribution function corresponding to $\phi_i(\cdot)$;

$u_k =$ the nonnegative order quantity in period k;

$x_k =$ the (non-negative) inventory level at the beginning of period k (or, at the end of period $k - 1$);

$c(i, u) =$ the cost of ordering $u \geq 0$ units in period k when $i_k = i$;

$f(i, x) =$ the inventory cost when $i_k = i$ and $x_k = x \geq 0$;

$q(i, x + u - \xi) =$ the shortage cost when $i_k = i$, $x_k = x$, $u_k = u$, and $\xi_k = \xi$;

$\delta(z) =$ 0, 1, respectively, when $z \leq 0$, $z > 0$, respectively.

We suppose that orders are placed at the beginning of a period, delivered instantaneously, and followed by the period's demand; see Figure 1.1. Unsatisfied demands are lost.

We make the following assumptions throughout the paper. While not all the results proved in this paper require all of the assumptions, we do use all of the assumptions to derive the main results of the paper in Sections 4 and 5. For specificity, we shall list the assumptions required in the statements of the results proved in this paper.

ASSUMPTION A1. The ordering cost is given by $c(i, u) = K\delta(u) + c_i u$, where the fixed ordering cost $K \geq 0$ and the variable cost $c_i \geq 0$.

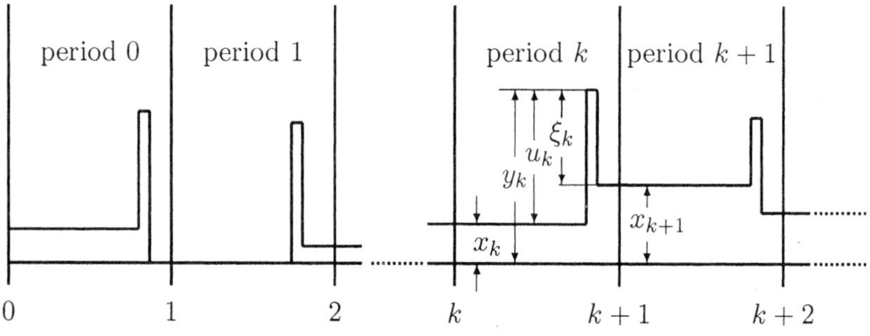

Figure 1.1. Temporal conventions used for the discrete-time inventory problem.

ASSUMPTION A2. For each i, the inventory cost function $f(i, \cdot)$ is convex, non-decreasing and of linear growth, i.e., $f(i, x) \leq C_f(1 + |x|)$ for some $C_f > 0$ and all x. Also $f(i, x) = 0$ for all $x \leq 0$.

ASSUMPTION A3. For each i, the shortage cost function $q(i, \cdot)$ is convex, non-increasing and of linear growth, i.e., $q(i, x) \leq C_q(1+|x|)$ for some $C_q > 0$ and all x. Also $q(i, x) = 0$ for all $x \geq 0$.

ASSUMPTION A4. There is a state $g \in I$ such that $f(g, x)$ is not identically zero.

ASSUMPTION A5. The production and inventory costs satisfy for all i,

$$c_i x + \sum_{j=1}^{L} p_{ij} \int_0^{\infty} f(j, (x - z)^+) \varphi_i(z)\, dz \to \infty \quad \text{as} \quad x \to \infty. \tag{1.1}$$

ASSUMPTION A6. For each i, $q'^-(i, 0) \leq \sum_{j=1}^{L} p_{ij}(f'^+(j, 0) - c^j)$, where the supersripts '+' and '−' denote the right-hand and left-hand derivatives.

ASSUMPTION A7. The Markov chain $(i_k)_{k=0}^{\infty}$ is irreducible.

ASSUMPTION A8. There is a state $h \in I$ such that $1 - \Phi_h(\varepsilon) = \rho > 0$ for some $\varepsilon > 0$.

ASSUMPTION A9. $E(\xi_i) < \infty$ for each i.

REMARK 1 Assumptions (A1)–(A3) reflect the usual structure of the production and inventory costs to prove the optimality of an (s^i, S^i) policy. Note that K is the same for all i. In the stationary case, this is equivalent to the condition $K_n^i \geq \sum_{j=1}^{L} p_{ij} K_{n+1}^j$ required in the non-stationary model for the existence of an optimal (s^i, S^i)-policy; see Cheng and Sethi (1999). Assumption (A4) rules out trivial cases where the optimal policy is never to order. Assumption (A5) will hold if either the unit ordering cost $c_i > 0$ or the second term in (1.1), which is the expected holding cost go to infinity as the surplus level x goes to infinity; obviously the *or* is inclusive here. While related, (A5) neither implies nor is implied by (A4). Condition (A5) is borne out of practical considerations and is not very restrictive. In addition, it rules out such unrealistic trivial cases as the one with $c_i = 0$ and $f(i, x) = 0, x \geq 0$, for each i, which implies ordering an infinite amount whenever an order is placed. Assumptions (A4) and (A5) *generalize* the usual assumption made by Scarf (1960) and others that the unit inventory holding cost $h > 0$.

REMARK 2 Assumption (A6) means that the marginal shortage cost in one period is larger than or equal to the expected unit ordering cost less the expected marginal inventory holding cost in any state of the next period. If this condition does not hold, that is, if $-q_n^{'-}(i, 0) < \sum_{j=1}^{L} p_{ij} [c_{n+1}^j - f_{n+1}^{'+}(j, 0)]$ for some i, a speculative retailer may find it attractive to meet a smaller part of the demand in period n than is possible from the available stock, carry the leftover inventories to period $n+1$, and order a little less as a result in period $n+1$ with the expectation that he will be better off. Thus, Assumption (A6) rules out this kind of speculation on the part of the retailer. But such a speculative behavior is not allowed in our formulation of the dynamics in any case, since the demand in any period must be satisfied to the extent of the availability of inventories. This suggests that it might be possible to prove our results without (A6).

REMARK 3 Assumptions (A7) and (A8) are needed to deplete any given initial inventory in a finite expected time. While (A8) says that in at least one state h, the expected demand is strictly larger than zero, (A7) implies that the state h would occur infinitely often with finite expected interval between successive occurrences.

Our objective is to minimize the expected long-run average cost

$$J(i,x;U) = \limsup_{N\to\infty} \frac{1}{N} E\left\{ \sum_{k=0}^{N-1} [c(i_k,u_k) + f(i_k,x_k) + q(i_k, x_k + u_k - \xi_k)] \right\}, \quad (1.2)$$

with $i_0 = i$ and $x_0 = x \geq 0$, where $U = (u_0, u_1, \ldots)$, $u_i \geq 0$, $i = 0, 1, \ldots,$ is a *history-dependent* or *non-anticipative* decision (order quantities) for the problem. Such a control U is termed admissible. Let $\mathcal{U}$ denote the class of all admissible controls. The surplus balance equations are given by

$$x_{k+1} = (x_k + u_k - \xi_{k+1})^+, \quad k = 0, 1, \ldots. \quad (1.3)$$

Our aim is to show that (i) there exists a constant λ^* termed the optimal average cost, which is independent of the initial i and x, (ii) a control $U^* \in \mathcal{U}$ such that

$$\lambda^* = J(i,x;U^*) \leq J(i,x;U), \quad \text{for all } U \in \mathcal{U}, \quad (1.4)$$

and (iii)

$$\lambda^* = \lim_{N\to\infty} \frac{1}{N} E\left\{ \sum_{k=0}^{N-1} [c(i_k,u_k^*) + f(i_k,x_k^*) + q(i_k, x_k + u_k - \xi_k)] \right\}, \quad (1.5)$$

where x_k^*, $k = 0, 1, \ldots,$ is the surplus process corresponding to U^* with $i_0 = i$ and $x_0 = x$.

To prove these results we will use the vanishing discount approach. That is, by letting the discount factor α in the discounted cost problem go to one, we will show that we can derive a dynamic programming equation whose solution provides an average optimal control and the associated minimum average cost λ^*.

For this purpose, we recapitulate relevant results for the discounted cost problem obtained in Cheng and Sethi (1999).

REMARK 4 It may be noted that the objective function (1.2) is slightly, but not essentially, different from that used in the classical literature. Whereas we base the surplus cost on the initial surplus in each period, the usual practice in the literature is to charge the cost on the ending surplus levels, which means to have $f(i_k, x_{k+1})$ instead $f(i_k, x_k)$ in (1.2). Note that the x_{k+1} is also the ending inventory in period k. It should be obvious that this difference in the objective functions does not change the long-run average cost for any admissible policy. By the same token we can justify our choice to charge shortage cost at the end of a given period.

3. Markovian demand model with discounted costs

Consider the model formulated above with the average cost objective (1.2) replaced by the following extended real-valued objective function:

$$J_\alpha(i, x; U) = \sum_{k=0}^{\infty} \alpha^k E[c(i_k, u_k) + f(i_k, x_k) + q(i_k, x_k + u_k - \xi_k)], \quad 0 \le \alpha < 1. \tag{1.6}$$

Define the value function with $i_0 = i$ and $x_0 = x$ as

$$v_\alpha(i, x) = \inf_{U \in \mathcal{U}} J_\alpha(i, x; U). \tag{1.7}$$

Let B_0 denote the class of all continuous functions from $I \times \mathbb{R}$ into $[0, \infty)$ and the point-wise limits of sequences of these functions; see Feller (1971). Note that it includes piecewise-continuous functions. Let B_1 denote the space of functions in B_0 that are of linear growth, i.e., for any $b \in B_1$, $0 \le b(i, x) \le C_b(1 + |x|)$ for some $C_b > 0$. Let B_2 denote the subspace of functions in B_1 that are uniformly continuous with respect to $x \in R$. For any $b \in B_1$, we define the notation

$$F(b)(i, y) = \sum_{j=1}^{L} p_{ij} \int_0^{\infty} b(j, (y - z)^+) \phi_i(z)\, dz\, dz. \tag{1.8}$$

THEOREM 1.1 *Let Assumptions* (A1)–(A3), (A5), (A6) *and* (A9) *hold. Then*

(i) *the value function* $v_\alpha(\cdot, \cdot)$ *is in* B_2 *and it solves the dynamic programming equation*

$$\begin{aligned} v_\alpha(i, x) &= f(i, x) + \inf_{u \ge 0} \left\{ c(i, u) + E\left[q(i, x + u - \xi^i) \right.\right. \\ &\qquad \left.\left. + \alpha \sum_{j=1}^{L} p_{ij} v(j, (x + u - \xi^i)^+) \right] \right\} \qquad (1.9) \\ &= f(i, x) + \inf_{u \ge 0} \{ c(i, u) + Eq(i, x + u - \xi^i) \\ &\qquad + \alpha F(v)(j, x + u) \}; \end{aligned}$$

(ii) $v_\alpha(i, \cdot)$ *is* K*-convex and there are real numbers* (s^i_α, S^i_α), $s^i_\alpha \le S^i_\alpha$, *such that the feedback policy* $\hat{u}_{k,\alpha}(i, x) = (S^i_\alpha - x)\delta(s^i_\alpha - x)$ *is optimal.*

Proof. Theorem 1.1 is stated but not proved in Cheng and Sethi (1999). The proof of part (i) follows the lines of the proof of Theorem 2.3 in Beyer, Sethi and Taksar (1998) by taking the limit of the n-period value function as n tends to infinity. Part (ii) follows immediately since the limit of a sequence of K-convex functions is K-convex. □

Hereafter, we shall omit the additional subscript α on the control policies for ease of notation. Thus, for example, $\hat{u}_{k,\alpha}(i,x)$ will be denoted simply as $\hat{u}_k(i,x)$. Since we do not consider the limits of the control variables as $\alpha \uparrow 1$, the practice of omitting the subscript α will not cause any confusion. In any case, the dependence of controls on α will always be clear from the context.

To insure a "smooth" limiting behavior for $\alpha \to 1$, we prove in Lemma 1.2 that $v_\alpha(i,\cdot)$ is locally equi-Lipschitzian, a term which is defined in the lemma itself.

For any given state $i_0 = l$ and $y > 0$, let

$$\tau_{l,y} := \inf\left\{ n\colon \sum_{k=1}^{n} \xi_k \geq y \right\}$$

be the first index for which the cumulative demand is not less than y. The following lemma is proved in Beyer and Sethi (1997, 1999):

LEMMA 1.1 *Let Assumptions* (A7) *and* (A8) *hold. Then* $E(\tau_{l,y}) < \infty$.

LEMMA 1.2 *Under Assumptions* (A1)–(A3), *and* (A7)–(A9), $v_\alpha(i,\cdot)$ *is locally equi-Lipschitzian, i.e., for* $X > 0$ *there is a positive constant* $C_1 < \infty$, *independent of* α, *such that*

$$|v_\alpha(i,x) - v_\alpha(i,\tilde{x})| \leq C_1 |x - \tilde{x}| \quad \textit{for all} \quad x,\ \tilde{x} \in [0, X]. \tag{1.10}$$

Proof. Consider the case $\tilde{x} \geq x$. Let us fix an $\alpha \in [0,1)$. It follows from Theorem 1.1 that there is an optimal feedback strategy U. Use the strategy U with initial surplus x, and the strategy $\widetilde{U}$ defined by

$$\tilde{u}_k = [u_k - (\tilde{x}_k - x_k)]^+ = \begin{cases} 0 & \text{if } u_k \leq \tilde{x}_k - x_k, \\ u_k + x_k - \tilde{x}_k & \text{if } u_k > \tilde{x}_k - x_k, \end{cases}$$

with initial $\tilde{x}$, where x_k and $\tilde{x}_k$ denote the inventory levels resulting from the respective strategies. It is easy to see that the following inequalities hold for all k:

$$0 \leq \tilde{x}_k - x_k \leq \tilde{x} - x \quad \text{and} \quad \tilde{u}_k \leq u_k.$$

Let $\tilde{\tau} := \inf\{n\colon \sum_{k=0}^{n} \xi_k \geq \tilde{x}\}$. If $\tilde{u}_k = 0$ for all $k \in [0, \tilde{\tau}]$, then $\tilde{x}_{\tilde{\tau}} = x_{\tilde{\tau}} = 0$ and the two trajectories are identical for all $k > \tilde{\tau}$. If $\tilde{u}_{k'} \neq 0$ for some $k' \in [0, \tilde{\tau}]$, then $\tilde{x}_{k'} = x_{k'}$ and the two trajectories are identical for all $k > k'$. In any case, the two trajectories are identical for all $k > \tilde{\tau}$. From Assumptions (A1)–(A3), we have

$$
\begin{gathered}
c(i_k, \tilde{u}_k) \leq c(i_k, u_k), \\
|f(i_k, \tilde{x}_k) - f(i_k, x_k)| \leq C_f |\tilde{x} - x|, \quad \text{and} \\
|q(i_k, \tilde{x}_k + \tilde{u}_k - \xi_k) - q(i_k, x_k + u_k - \xi_k)| \leq C_q |\tilde{x} - x|.
\end{gathered}
$$

Therefore,

$$
\begin{aligned}
v_\alpha(i, \tilde{x}) - v_\alpha(i, x) &\leq J_\alpha(i, \tilde{x}; \widetilde{U}) - J_\alpha(i, x; U) \\
&= E\Bigg(\sum_{k=0}^{\tilde{\tau}} \alpha^k \big(f(i_k, \tilde{x}_k) - f(i_k, x_k) \\
&\qquad + q(i_k, \tilde{x}_k + \tilde{u}_k - \xi_k) - q(i_k, x_k + u_k - \xi_k) \\
&\qquad\qquad + c(i_k, \tilde{u}_k) - c(i_k, u_k) \big) \Bigg) \\
&\leq E\left(\sum_{k=0}^{\tilde{\tau}} \alpha^k (C_f + C_q) |\tilde{x} - x| \right) \\
&\leq E(\tilde{\tau} + 1)(C_f + C_q)|\tilde{x} - x|.
\end{aligned}
$$

It follows immediately from Lemma 1.1 that $E(\tilde{\tau} + 1) = E(\tau_{i,\tilde{x}} + 1) \leq E(\tau_{i,X} + 1) < \infty$.

To complete the proof, it is sufficient to prove the above inequality for $\tilde{x} < x$. In this case, let us define the strategy $\widetilde{U}$ by

$$
\tilde{u}_k = \begin{cases} u_k + x - \tilde{x} & \text{if } u_k > 0, \\ 0 & \text{otherwise.} \end{cases}
$$

It is easy to see that the following inequalities hold for all k:

$$
0 \geq \tilde{x}_k - x_k \geq \tilde{x} - x \quad \text{and} \quad \tilde{u}_k - u_k \leq x - \tilde{x}.
$$

Let $\tau := \inf\{n\colon \sum_{k=0}^{n} \xi_k \geq x\}$. If $u_k = 0$ for all $k \in [0, \tau]$, then $\tilde{x}_\tau = x_\tau = 0$ and the two trajectories are identical for all $k > \tau$. If $u_{k'} \neq 0$ for some $k' \in [0, \tau]$, then $\tilde{x}_{k'} = x_{k'}$ and the two trajectories are identical for all $k > k'$. In any case, the two trajectories are identical for all $k > \tau$.

From Assumptions (A1)–(A3), we have

$$
c(i_k, \tilde{u}_k) - c(i_k, u_k) \leq \max\{c_i\} |x - \tilde{x}|,
$$

$$|f(i_k,\tilde{x}_k) - f(i_k,x_k)| \le C_f|\tilde{x} - x|, \quad \text{and}$$
$$|q(i_k,\tilde{x}_k + \tilde{u}_k - \xi_k) - q(i_k,x_k + u_k - \xi_k)| \le C_q|\tilde{x} - x|.$$

Therefore,

$$\begin{aligned}
v_\alpha(i,\tilde{x}) - v_\alpha(i,x) &\le J_\alpha(i,\tilde{x};\widetilde{U}) - J_\alpha(i,x;U) \\
&= E\Bigg(\sum_{k=0}^{\tau} \alpha^k \big(f(i_k,\tilde{x}_k) - f(i_k,x_k) \\
&\qquad\qquad + q(i_k,\tilde{x}_k + \tilde{u}_k - \xi_k) - q(i_k,x_k + u_k - \xi_k) \\
&\qquad\qquad\qquad + c(i_k,\tilde{u}_k) - c(i_k,u_k)\big)\Bigg) \\
&\le E\Bigg(\sum_{k=0}^{\tau} \alpha^k (C_f + C_q + \max\{c_i\})|\tilde{x} - x|\Bigg) \\
&\le E(\tau + 1)(C_f + C_q + \max\{c_i\})|\tilde{x} - x|.
\end{aligned}$$

Again, it follows immediately from Lemma 1.1 that $E(\tau+1) = E(\tau_{i,x} + 1) \le E(\tau_{i,X} + 1) < \infty$, and the proof is complete. □

LEMMA 1.3 *Under* (A1)–(A9), *there are constants* $\alpha_0 \in [0,1)$ *and* $C_2 > 0$ *such that for all* $\alpha \ge \alpha_0$, *we have* $S_\alpha^i \le C_2 < \infty$ *for any* i *for which* $s_\alpha^i > 0$.

Proof. Let us fix the initial state $i_0 = i$ for which $s_\alpha^i > 0$. Fix $\alpha_0 > 0$ and a discount factor $\alpha \ge \alpha_0$. Let $U = \big(u(i_0,x_0), u(i_1,x_1),\ldots\big)$ be an optimal strategy with parameters (s_α^j, S_α^j), $j \in I$. Let us fix a positive real number V and assume $S_\alpha^i > V$. In what follows, we specify a value of V, namely V^*, in terms of which we shall construct an alternative strategy $\widetilde{U}$ that is better than U.

For the demand state g specified in Assumption (A4), let

$$\tau^g := \inf\{n > 0 \colon i_n = g\}$$

be the first period (not counting the period 0) with the demand state g. Furthermore, let d be the state with the lowest per unit ordering cost, i.e., $c_d \le c_i$ for all $i \in I$. Then we define

$$\tau := \inf\{n \ge \tau^g \colon i_n = d\}.$$

Assume $x_0 = \tilde{x}_0 = \bar{x} := 0$ and consider the policy $\widetilde{U}$ defined as under:

$$\tilde{u}_k = 0, \quad k = 0,1,2,\ldots,\tau - 1,$$

$$\tilde{u}_T = x_T + u(i_T, x_T),$$
$$\tilde{u}_k = u(i_k, x_k), \quad k \geq \tau + 1.$$

The two policies and the resulting trajectories differ only in periods 0 through τ. Therefore, we have

$$\begin{aligned}
v_\alpha(i,\bar{x}) &- J_\alpha(i,\bar{x};\widetilde{U}) \\
&= J_\alpha(i,\bar{x};U) - J_\alpha(i,\bar{x};\widetilde{U}) \\
&= E\left(\sum_{k=0}^{\tau} \alpha^k \big(f(i_k,x_k) - f(i_k,\tilde{x}_k) + q(i_k, x_k+u_k-\xi_k)\right. \\
&\qquad\qquad \left. - q(i_k,\tilde{x}_k+\tilde{u}_k-\xi_k) + c(i_k,u_k) - c(i_k,\tilde{u}_k)\big)\right) \qquad (1.11)\\
&= E\left(\sum_{k=1}^{\tau} \alpha^k \big(f(i_k,x_k) + q(i_k,x_k+u_k-\xi_k) - q(i_k,-\xi_k)\big)\right) \\
&\qquad\qquad + E\left(\sum_{k=0}^{\tau}\alpha^k c(i_k,u_k)\right) - E\big(\alpha^\tau c(i_\tau,\tilde{u}_\tau)\big).
\end{aligned}$$

After ordering in period τ, the total accumulated ordered amount up to period τ is less for policy $\widetilde{U}$ than it is for U. Observe that the policy $\widetilde{U}$ orders only in period τ. The order of the policy $\widetilde{U}$ is executed at the lowest possible per unit cost c_d in period τ, which is not earlier than any of the ordering periods of policy U. Because $\widetilde{U}$ orders only once and U orders at least once in periods $0, 1, \ldots, \tau$, the total fixed ordering cost of $\widetilde{U}$ does not exceed the total fixed ordering cost of U. Thus,

$$E\left(\sum_{k=0}^{\tau}\alpha^k c(i_k,u_k)\right) \geq E\big(\alpha^\tau c(i_\tau,\tilde{u}_\tau)\big).$$

Furthermore, it follows from Assumptions (A3), (A7), and (A9) that

$$E\left(\sum_{k=1}^{\tau} q(i_k,-\xi_t)\right) < \infty.$$

Because $\tau \geq \tau^g$, we obtain

$$E\left(\sum_{k=1}^{\tau}\alpha^k\big(f(i_k,x_k) + q(i_k,x_k+u_k-\xi_k)\big)\right) \geq E\left(\sum_{k=1}^{\tau^g}\alpha^k f(i_k,x_k)\right),$$

and because $S_\alpha^i \geq V$, we obtain

$$x_k \geq V - \sum_{t=1}^{k} \xi_t.$$

Irreducibility of the the Markov chain $(i_n)_{n=0}^{\infty}$ implies existence of an integer m, $0 \leq m \leq L$, such that $P(i_m = g) > 0$. Let m_0 be the smallest such m. It follows that $\tau^g \geq m_0$ and, therefore,

$$\begin{aligned} E\left(\sum_{k=1}^{\tau^g} \alpha^k f(i_k, x_k)\right) &\geq \alpha^{m_0} E(f(i_{m_0}, x_{m_0})) \\ &\geq \alpha_0^{m_0} E\left(f\left(g, V - \sum_{t=}^{m_0} \xi_t\right) \middle| i_{m_0} = g\right) P(i_{m_0} = g), \\ &\qquad \text{for all } \alpha \geq \alpha_0. \end{aligned} \tag{1.12}$$

Using Assumptions (A2), (A4), and (A9), it is easy to show that the right-hand side of (1.12) tends to infinity as V goes to infinity. Therefore, we can choose V^*, $0 \leq V^* < \infty$, such that for all $\alpha \geq \alpha_0$,

$$\begin{aligned} v_\alpha(i, \bar{x}) - J_\alpha(i, \bar{x}; \widetilde{U}) \geq \alpha_0^{m_0} E\left(f\left(g, V^* - \sum_{t=1}^{m_0} \xi_t\right) \middle| i_{m_0} = g\right) P(i_{m_0} = g) \\ - E\left(\sum_{k=1}^{\tau} q(i_k, -\xi_t)\right) > 0. \end{aligned} \tag{1.13}$$

Note that the RHS of (1.13) is independent of α. Thus for $\alpha \geq \alpha_0$, a policy with $S_\alpha^i > C_2 := V^*$ cannot be optimal. □

4. Vanishing discount approach

LEMMA 1.4 *Under Assumptions* (A1)–(A9), *the differential discounted value function* $w_\alpha(i, x) := v_\alpha(i, x) - v_\alpha(1, 0)$ *is uniformly bounded with respect to* α *for all* x *and* i.

Proof. Since Lemma 1.2 implies

$$\begin{aligned} |w_\alpha(i, x)| = |v_\alpha(i, x) - v_\alpha(1, 0)| \leq |v_\alpha(i, x) - v_\alpha(i, 0)| + |v_\alpha(i, 0) \\ - v_\alpha(1, 0)| \leq C_3|x| + |w_\alpha(i, 0)|, \end{aligned} \tag{1.14}$$

it is sufficient to prove that $w_\alpha(i, 0)$ is uniformly bounded. Note that C_3 may depend on x, but it is independent of α.

First, we show that there is an $\underline{M} > -\infty$ with $w_\alpha(i,0) \geq \underline{M}$ for all α. Let α be fixed. From Theorem 1.1 we know that there is a stationary discount optimal feedback policy $U = \big(u(i,x), u(i,x), \ldots\big)$. With $k^* = \inf\{k\colon i_k = i\}$, we consider the cost for the initial state $(i_0, \tilde{x}_0) = (1,0)$ and the inventory policy $\widetilde{U}$, which does not order in periods $0, 1, \ldots, k^* - 1$ and follows U beginning with period k^*:

$$\begin{aligned} \tilde{u}_k &= 0 \quad \text{for} \quad k < k^*, \\ \tilde{u}_k &= u(i_k, x_k) \quad \text{for} \quad k \geq k^*. \end{aligned}$$

The cost corresponding to this policy is

$$J_\alpha(1,0;\widetilde{U}) = E\left(\sum_{k=0}^{k^*-1} \alpha^k q(i_k, -\xi_k) + \alpha^{k^*} v_\alpha(i,0) \right). \tag{1.15}$$

Because of Assumptions (A3), (A7) and (A9),

$$E\left(\sum_{k=0}^{k^*-1} q(i_k, -\xi_k) \right) \leq -\underline{M} < \infty.$$

Therefore, we have

$$\begin{aligned} w_\alpha(i,0) &= v_\alpha(i,0) - v_\alpha(1,0) \geq v_\alpha(i,0) - J_\alpha(1,0;\widetilde{U}) \\ &\geq v_\alpha(i,0) - E\left(\sum_{k=0}^{k^*-1} \alpha^k q(i_k, -\xi_k) + \alpha^{k^*} v_\alpha(i,0) \right) \\ &\geq v_\alpha(i,0)\big(1 - E(\alpha^{k^*})\big) + \underline{M} \geq \underline{M}. \end{aligned} \tag{1.16}$$

The opposite inequality $w_\alpha(i,0) \leq \bar{M}$ is shown analogously by changing the role of the states 1 and i. Thus, $|w_\alpha(i,x)| \leq C_3|x| + \max\{\underline{M}, \bar{M}\}$, and the proof is complete. □

Lemma 1.5 *Under Assumptions* (A3) *and* (A9), $(1-\alpha)v_\alpha(1,0)$ *is uniformly bounded on* $0 < \alpha < 1$.

Proof. Consider the strategy $\widetilde{U} = (0, 0, \ldots)$. Because $\widetilde{U}$ is not necessarily optimal,

$$0 \leq v_\alpha(1,0) \leq J_\alpha(1,0;\widetilde{U}) = E\left(\sum_{k=0}^{\infty} \alpha^k q(i_k, -\xi_k) \right).$$

Because of (A3) and (A9), $Eq(i, -\xi_k)$ is bounded for all i and there is a $C_4 < \infty$ such that $E\big(q(i_k, -\xi_k)\big) < C_4$. Therefore,

$$0 \leq (1-\alpha)v_\alpha(1,0) \leq (1-\alpha)\sum_{k=0}^{\infty} \alpha^k C_4 = C_4.$$

This completes the proof. □

THEOREM 1.2 *Let Assumptions* (A1)–(A9) *hold. There exist a sequence* $(\alpha_k)_{k=1}^{\infty}$ *converging to* 1*, a constant* λ^**, and a locally Lipschitz continuous function* $w^*(\cdot,\cdot)$ *such that*

$$(1-\alpha_k)v_{\alpha_k}(i,x) \to \lambda^* \quad \text{and} \quad w_{\alpha_k}(i,x) \to w^*(i,x),$$

locally uniformly in x *and* i *as* k *goes to infinity. Moreover,* (λ^*, w^*) *satisfies the average cost optimality equation*

$$w(i,x) + \lambda = f(i,x) + \inf_{u\geq 0}\{c(i,u) + Eq(i, x+u-\xi^i) + F(w)(i,x+u)\}. \quad (1.17)$$

Proof. From Lemma 1.2 and the definition of $w_\alpha(i,x)$, it is clear that $w_\alpha(i,\cdot)$ is locally equi-Lipschitzian for $\alpha \geq \alpha_0$, and therefore it is uniformly continuous in any finite interval. Additionally, according to Lemma 1.4, $w_\alpha(i,\cdot)$ is uniformly bounded, and by Lemma 1.5, $(1-\alpha)v_\alpha(1,0)$ is uniformly bounded. Therefore, from the Arzela–Ascoli Theorem (see, e.g., Yosida, 1980) and Lemma 1.2, there are a sequence $\alpha_k \to 1$, a locally Lipschitz continuous function $w^*(i,x)$, and a constant λ^* such that

$$(1-\alpha_k)v_{\alpha_k}(1,0) \to \lambda^*, \quad \text{and} \quad w_{\alpha_k}(i,x) \to w^*(i,x)$$

for each x locally uniformly in any given interval. By the diagonalization procedure , a subsequence can be found so that $w^*_{\alpha_k}(i,\cdot)$ converges to a locally Lipschitz continuous function $w^*(i,\cdot)$ on the entire real line.

Next, it is easy to see that

$$\lim_{k\to\infty}(1-\alpha_k)v_{\alpha_k}(i,x) = \lim_{k\to\infty}(1-\alpha_k)\big(w_{\alpha_k}(i,x) + v_{\alpha_k}(1,0)\big) = \lambda^*.$$

Substituting $v_{\alpha_k}(i,x) = w_{\alpha_k}(i,x) + v_{\alpha_k}(1,0)$ in (1.9) yields

$$\begin{aligned} w_{\alpha_k}(i,x) &+ (1-\alpha_k)v_{\alpha_k}(1,0) \\ &= f(i,x) + \inf_{u\geq 0}\{c(i,u) + Eq(i,x+u-\xi^i) \\ &\qquad + \alpha_k F(w_{\alpha_k})(i,x+u)\}. \quad (1.18) \end{aligned}$$

Since $w_{\alpha_k}(i,x)$ converges locally uniformly with respect to x and i and since for a given x, a minimizer u^* in (1.18) can be chosen such that $x+u^*-\xi \in [0, x+C_2]$ by Lemma 1.3, we can pass to the limit on both sides of (1.18), and obtain (1.17). This completes the proof. □

LEMMA 1.6 *Let λ^* be defined as in Theorem 1.2. Let Assumptions* (A1)–(A9) *hold. Then for any admissible strategy U, we have $\lambda^* \leq J(i, x; U)$.*

Proof. Let $U = (u_0, u_1, \dots)$ denote any admissible decision. Suppose

$$J(i, x; U) < \lambda^*. \tag{1.19}$$

Set

$$\tilde{f}(k) = E[f(i_k, x_k) + q(i_k, x_k + u_k - \xi_k) + c(i_k, u_k)].$$

From (1.19) it follows immediately that $\sum_{k=0}^{n-1} \tilde{f}(k) < \infty$ for each positive integer n, since otherwise we would have $J(i, x; U) = \infty$. Note that

$$J(i, x; U) = \limsup_{n \to \infty} \frac{1}{n} \sum_{k=0}^{n-1} \tilde{f}(k),$$

while

$$(1 - \alpha) J_\alpha(i, x; U) = (1 - \alpha) \sum_{k=0}^{\infty} \alpha^k \tilde{f}(k). \tag{1.20}$$

Since $\tilde{f}(k)$ is nonnegative for each k, the sum in (1.20) is well defined for $0 \leq \alpha < 1$, and we can use the Tauberian theorem (see, e.g., Sznadjer and Filar, 1992, Theorem 2.2), to obtain

$$\limsup_{\alpha \uparrow 1} (1 - \alpha) J_\alpha(i, x; U) \leq J(i, x; U) < \lambda^*.$$

On the other hand, we know from Theorem 1.2 that $(1 - \alpha_k) v_{\alpha_k}(i, x) \to \lambda^*$ on a subsequence $\{\alpha_k\}_{k=1}^{\infty}$ converging to one. Thus, there exists an $\alpha < 1$ such that

$$(1 - \alpha) J_\alpha(i, x; U) < (1 - \alpha) v_\alpha(i, x),$$

which contradicts the definition of the value function $v_\alpha(i, x)$. □

5. Verification theorem

DEFINITION 1.1 *Let (λ, w) be a solution of the average optimality equation* (1.17). *An admissible strategy $U = (u_0, u_1, \dots)$ is called stable with respect to w if for each initial inventory level $x \geq 0$ and for each initial demand state $i \in I$,*

$$\lim_{k \to \infty} \frac{1}{k} E\big(w(i_k, x_k)\big) = 0,$$

where x_k *is the inventory level in period* k *corresponding to the initial state* (i, x) *and the strategy* U.

LEMMA 1.7 *Assume* (A1)–(A9). *There are constants* $S^i < \infty$ *and* $0 \leq s_i \leq S^i$, $i \in I$, *such that*

$$u^*(i, x) = \begin{cases} S^i - x, & x < s^i, \\ 0, & x \geq s^i \end{cases}$$

attains the minimum on the RHS in (1.17) *for* $w = w^*$ *as defined in Theorem* 1.2. *Furthermore, the stationary feedback strategy* $U^* = (u^*, u^*, \dots)$ *is stable with respect to any continuous function* w.

Proof. Let $\{\alpha_k\}_{k=0}^{\infty}$ be the sequence defined in Theorem 1.2. Let

$$G_{\alpha_k}(i, y) = c_i y + Eq(i, y - \xi^i) + \alpha_k F(w_{\alpha_k})(i, y)$$

and

$$G(i, y) = c_i y + Eq(i, y - \xi^i) + F(w^*)(i, y). \tag{1.21}$$

Because $w^*(i, \cdot)$ is K-convex, we know that a minimizer in (1.17) is given by

$$u^*(i, x) = \begin{cases} S^i - x, & x < s^i, \\ 0, & x \geq s^i, \end{cases}$$

where $0 \leq S^i \leq \infty$ minimizes $G(i, \cdot)$, and s^i solves

$$G(s^i) = K + G(S^i),$$

if a solution exists or else $s^i = 0$. Note that if $s^i = 0$, it follows that $u^*(i, x) = 0$ for all nonnegative x. It remains to show that $S^i < \infty$.

We distinguish two cases.

Case 1. If there is a subsequence, still denoted by $\{\alpha_k\}_{k=0}^{\infty}$, such that $s^i_{\alpha_k} > 0$ for all $k = 0, 1, \dots$, then it follows from Lemma 1.3 that G_{α_k} attains its minimum in $[0, C_2]$ for all $\alpha_k > \alpha_0$. Thus G_{α_k}, $k = 0, 1, \dots$, are locally uniformly continuous and converge uniformly to G. Therefore, G attains its minimum also in $[0, C_2]$, which implies $S^i \leq C_2$.

Case 2. If there is no such sequence, then there is a sequence, still denoted by $\{\alpha_k\}_{k=0}^{\infty}$, such that $s^i_{\alpha_k} = 0$ for all $k = 0, 1, \dots$. It follows that for all $y > x$,

$$G_{\alpha_k}(x) < K + G_{\alpha_k}(y),$$

and therefore in the limit,

$$G(x) < K + G(y).$$

This implies that the infimum in (1.17) is attained for $u^*(i,x) \equiv 0$, which is equivalent to $s^i = 0$. But, if $s^i = 0$, we can choose S^i arbitrarily, say $S^i = C_2$.

It is immediate that the stationary policy U^* is stable with respect to any continuous function, since it implies $x_k \in [0, \max\{C_2, x_0\}]$ for all $k = 0, 1, \ldots$. □

THEOREM 1.3 (VERIFICATION THEOREM) *Let $(\lambda, w(\cdot,\cdot))$ be a solution of the average cost optimality equation* (1.17) *with w continuous on $[0,\infty)$.*

(i) *Then $\lambda \le J(i,x;U)$ for any admissible U.*

(ii) *Suppose there exists a $\hat{u}(i,x)$ for which the infimum in* (1.17) *is attained. Furthermore, let $\widehat{U} = (\hat{u}, \hat{u}, \ldots)$, the stationary feedback policy given by $\hat{u}$, be stable with respect to w. Then*

$$\begin{aligned}\lambda &= J(i,x;\hat{U}) = \lambda^* \\ &= \lim_{N\to\infty} \frac{1}{N} E\left(\sum_{k=0}^{N-1} f(i_k, \hat{x}_k) + q(i_k, x_k + u_k - \xi_k) + c(i_k, \hat{u}_k) \right),\end{aligned}$$

and $\widehat{U}$ is an average optimal strategy.

(iii) *Moreover, $\widehat{U}$ minimizes*

$$\liminf_{N\to\infty} \frac{1}{N} E\left(\sum_{k=0}^{N-1} f(i_k, x_k) + q(i_k, x_k + u_k - \xi_k) + c(i_k, u_k) \right)$$

over the class of admissible decisions which are stable with respect to w.

Proof. We start by showing that

$$\lambda \le J(i,x;U) \quad \text{for any } U \text{ stable w.r.t. } w. \tag{1.22}$$

We assume that U is stable with respect to w, and begin with a result on conditional expectations (see, e.g., Gikhman and Skorokhod, 1972), which gives

$$
\begin{aligned}
&E\{w(i_{n+1}, x_{n+1}) \mid i_1, \ldots, i_n, \xi_1, \ldots, \xi_n\} \\
&\quad = E\{w(i_{n+1}, (x_n + u_n - \xi_n)^+) \mid i_1, \ldots, i_n, \xi_1, \ldots, \xi_{n-1}\} \\
&\quad = E(w(i_{n+1}, (y - \xi_n)^+) \mid i_1, \ldots, i_n, \xi_1, \ldots, \xi_{n-1}\}_{y=x_n+u_n} \\
&\quad = E(w(i_{n+1}, (y - \xi_n)^+) \mid i_n\}_{y=x_n+u_n} \\
&\quad = F(w)(i_n, y)_{y=x_n+u_n} \\
&\quad = F(w)(i_n, x_n + u_n) \quad \text{a.s.}
\end{aligned}
\tag{1.23}
$$

Because u_k does not necessarily attain the infimum in (1.17), we have

$$
\begin{aligned}
w(i_k, x_k) + \lambda \le f(i_k, x_k) + c(i_k, u_k) + q(i_k, x_k + u_k - \xi_k) \\
+ F(w)(i_k, x_k + u_k) \quad \text{a.s.},
\end{aligned}
$$

and from (1.23) we derive

$$
\begin{aligned}
w(i_k, x_k) + \lambda \le f(i_k, x_k) + q(i_k, x_k + u_k - \xi_k) + c(i_k, u_k) \\
+ E(w(i_{k+1}, x_k + u_k - \xi_{k+1}) \mid i_k) \quad \text{a.s.}
\end{aligned}
$$

By taking the mathematical expectation on both sides, we obtain

$$
\begin{aligned}
E(w(i_k, x_k)) + \lambda \le E(f(i_k, x_k) + q(i_k, x_k + u_k - \xi_k) + c(i_k, u_k)) \\
+ E(w(i_{k+1}, x_{k+1})).
\end{aligned}
$$

Summing from 0 to $n-1$ yields

$$
\begin{aligned}
n\lambda \le E\left(\sum_{k=0}^{n-1} f(i_k, x_k) + q(i_k, x_k + u_k - \xi_k) + c(i_k, u_k)\right) \\
+ E(w(i_n, x_n)) - E(w(i_0, x_0)).
\end{aligned}
\tag{1.24}
$$

Divide by n, let n go to infinity, and use the fact that U is stable with respect to w, to obtain

$$
\lambda \le \liminf_{n\to\infty} \frac{1}{n} E\left(\sum_{k=0}^{n-1} f(i_k, x_k) + q(i_k, x_k + u_k - \xi_k) + c(i_k, u_k)\right). \tag{1.25}
$$

Note that if the above inequality holds for 'liminf', it certainly holds also for 'limsup'. This proves (1.22).

On the other hand, if there exists a $\hat{u}$ which attains the infimum in (1.17), we then have

$$w(i_k,\hat{x}_k)+\lambda = f(i_k,\hat{x}_k)+q(i_k,\hat{x}_k+\hat{u}_k-\xi_k)+c\big(i_k,\hat{u}(i_k,\hat{x}_k)\big) \\ + F(w)\big(i_k,\hat{x}_k+\hat{u}(i_k,\hat{x}_k)\big), \quad \text{a.s.},$$

and we obtain analogously

$$n\lambda = E\left(\sum_{k=0}^{n-1} f(i_k,\hat{x}_k)+q(i_k,\hat{x}_k+\hat{u}_k-\xi_k)+c\big(i_k,\hat{u}(i_k,\hat{x}_k)\big)\right) \\ + E\big(w(i_n,\hat{x}_n)\big)-E\big(w(i_0,\hat{x}_0)\big).$$

Because $\widehat{U}$ is assumed stable with respect to w, we get

$$\begin{aligned} \lambda &= \lim_{n\to\infty}\frac{1}{n}E\left(\sum_{k=0}^{n-1} f(i_k,\hat{x}_k)+q(i_k,\hat{x}_k+\hat{u}_k-\xi_k)+c\big(i_k,\hat{u}(i_k,\hat{x}_k)\right) \\ &= J(i,x;\widehat{U}). \end{aligned} \tag{1.26}$$

For the special solution (λ^*, w^*) defined in Theorem 1.2 and the strategy U^* defined in Lemma 1.7, we obtain

$$\lambda^* = J(i,x;U^*).$$

We know from Lemma 1.7 that U^* is stable with respect to any continuous function. It therefore follows that

$$\lambda \le J(i,x;U^*) = \lambda^*, \tag{1.27}$$

which, in view of Lemma 1.6, proves part (i) of the theorem.

Part (i) of the Theorem together with (1.26) proves the average-optimality of $\widehat{U}$ over all admissible strategies. Furthermore, since $\lambda = J(i,x;\widehat{U}) \ge \lambda^*$ by (1.26) and Lemma 1.6, it follows from (1.27) that $\lambda = \lambda^*$ and the proof of Part (ii) is completed.

Finally, Part (iii) immediately follows from Part (i) and (1.25). □

REMARK 5 It should be obvious that any solution (λ, w) of the average cost optimality equation and control u^* satisfying (i) and (ii) of Theorem 1.3 will have a unique λ, since it represents the minimum average cost. On the other hand, if (λ, w) is a solution, then $(\lambda, w + c)$, where c is any constant, is also a solution. For the purpose of this paper we do not require w to be unique up to a constant. If w is not unique up to a constant, then u^* may not be unique. We also do not need w^* in Theorem 1.2 to be unique either.

The final result of this section, namely, that there exists an average-optimal policy of (s, S)-type is an immediate consequence of Lemma 1.7 and Theorem 1.3.

THEOREM 1.4 *Assume* (A1)-(A9). *Let* s^i *and* S^i, $i \in I$, *be defined as in Lemma* 1.7. *The stationary feedback strategy* $U^* = (u^*, u^*, \dots)$ *defined by*

$$u^*(i,x) = \begin{cases} S^i - x, & x < s^i, \\ 0, & x \geq s^i, \end{cases}$$

is average optimal.

6. Concluding remarks

In this paper we have carried out a rigorous analysis of the average cost stochastic inventory problem with Markovian demand, fixed ordering cost and lost sales. We have proved a verification theorem for the average cost optimality equation, which we have used to establish the existence of an optimal state-dependent (s, S)-policy.

Throughout the analysis, we have assumed that the inventory and shortage cost functions are of linear growth. Further research should address extending this analysis to more general cost functions. An extension of the model for cost functions with polynomial growth in the backlog case was carried out in Beyer, Sethi and Taksar (1998). For other extensions, see the forthcoming book by Beyer, Cheng and Sethi (2004).

Acknowledgments This research was supported in part by NSERC grant A4619 and a grant from The University of Texas at Dallas.

References

Bensoussan, A., Crouhy, M., and Proth, J. (1983). *Mathematical Theory of Production Planning.* North-Holland.

Beyer, D., Cheng, F., and Sethi, S.P. (2004). *Inventory Models in Markovian Environments.* Kluwer. Forthcoming

Beyer, D. and Sethi, S.P. (1997). Average cost optimality in inventory models with Markovian demand. *Journal of Optimization Theory and Applications*, 92(3):497–526.

Beyer, D. and Sethi, S.P. (1999). The classical average-cost inventory models of Iglehart and Veinott–Wagner revisited. *Journal of Optimization Theory and Applications*, 101(3):523–555.

Beyer, D., Sethi, S.P., and Taksar, M. (1998). Inventory models with Markovian demands and cost functions of polynomial growth. *Journal of Optimization Theory and Application*, 98(2):281–323.

Cheng, F. and Sethi, S.P. (1999). Optimality of state-dependent (s, S) policies in inventory models with Markov-modulated demand and lost sales. *Production and Operations Management*, 8(2):183–192.

Federgruen, A. and Zipkin, P. (1984). An efficient algorithm for computing optimal (s, S) policies. *Operations Research*, 32:1268–1285.

Feller, W. (1971). *An Introduction to Probability Theory and its Application.* Vol. 2, 2nd Edition, Wiley, New York, NY.

Gikhman, I. and Skorokhod, A. (1972). *Stochastic Differential Equations.* Springer Verlag, Berlin, Germany.

Karlin, S. and Fabens, A. (1960). The (s, S) inventory model under Markovian demand process. In: *Mathematical Methods in the Social Sciences* (K. Arrow, S. Karlin, and P. Suppes, eds.), pp. 159–175, Stanford University Press, Stanford, CA.

Ozekici, S. and Parlar, M. (1999). Inventory models with unreliable suppliers in random environment. *Annals of Operations Research*, 91:123–136.

Scarf, H. (1960). The Optimality of (s, S) Policies in the Dynamic Inventory Problem. In: *Mathematical Methods in the Social Sciences* (K. Arrow, S. Karlin, and P. Suppes, eds.), pp. 196–202, Stanford University Press, Stanford, CA.

Sethi, S.P. and Cheng, F. (1997). Optimality of (s, S) policies in inventory models with Markovian demand processes. *Operations Research*, 45(6):931–939.

Shreve, S.E. (1976). Abbreviated proof in the lost sales case. In: *Dynamic Programming and Stochastic Control* (D.P. Bertsekas, ed.), pp. 105–106 Academic Press, New York, NY.

Song, J.S. and Zipkin, P. (1993). Inventory control in a fluctuating demand environment. *Operations Research*, 41:351–370.

Sznadjer, R. and Filar, J.A. (1992). Some comments on a theorem of Hardy and Littlewood. *Journal of Optimization Theory and Applications*, 75(1):201–208.

Veinott, A. (1966). On the optimality of (s, S) inventory policies: New conditions and a new proof. *SIAM Journal on Applied Mathematics*, 14(5):1067–1083.

Yosida, K. (1980). *Functional Analysis*, 6th Edition, Springer Verlag, New York, New York.

Chapter 2

INVENTORY CONTROL OF SWITCHED PRODUCTION SYSTEMS: LMI APPROACH

El Kébir Boukas
Luis Rodrigues

Abstract

This paper focuses on the problem of inventory control of production systems. The main contribution of the paper is that, for the first time, production systems are modeled as switched linear systems and the production problem is formulated as a switched $\mathscr{H}_\infty$ control problem with a piecewise-affine control law. The switching variable for the production systems modeled in this paper is the stock level. When the stock level is positive, some of the produced parts are being stored. The stocked parts may deteriorate with time at a given rate. When the stock level is negative it leads to backorders, which means that orders for production of parts are coming in and there is no stocked parts to immediately meet the demand. A switched linear model is used and it is shown that the inventory control problem can be solved using switched control theory. More specifically, a state feedback controller that forces the stock level to be kept close to zero, even when there are fluctuations in the demand, will be designed in this paper using $\mathscr{H}_\infty$ control theory. The synthesis of the gains of the state feedback controller that quadratically stabilizes the production dynamics and at the same time rejects the external demand fluctuation (treated as a disturbance) are determined by solving a given set of linear matrix inequalities (LMIs). A numerical example is provided to show the effectiveness of the developed method.

1. Introduction

Nowadays we are living in a world where the increasing competition between companies is dictating the business rules. Therefore, to survive, the companies are forced to focus seriously on how to produce high quality products at low cost and on how to respond quickly to rapid changes in the demand. The key competitive factors are the new technological advances and the ability to use them to quickly respond to rapid changes in the market. Production planning is one of the key ingredients that has a direct effect on the ability to quickly respond to rapid changes in the market. It is concerned with the optimal allocation of the system's production capacity to meet the demand efficiently. In general this problem is not easy and requires significant attention. This paper gives a first step in the direction of incorporating switching complexity in modeling production systems. The production dynamical model proposed in this paper captures the switching complexity originated from the fact that having a stock or running out of stock causes different dynamics of the system. In this paper it is proposed that the control policy (or decision making) should also include switching to cope with the switched nature of the system dynamics.

The problem of production planning has been tackled by many authors and many research results have been reported in the literature. Among them we quote the developments from (Grubbstrom and Wikner, 1996; Hennet, 2003; Towill, 1982; Towill, Evans, and Cheema, 1997; Wiendahl and Breithaupt, 2000; Axsater, 1985; Ridalls and Bennett, 2002; Simon, 1952; Gavish, and Graves, 1980; Disney, Naim, and Towill, 2000; Sethi and Zhang, 1994; Gershwin, 1994; Boukas and Liu, 2001) and the references therein. In these references, both stochastic and deterministic models have been proposed to handle the production planning and/or maintenance. Different approaches have been used to tackle production planning, such as, dynamic programming, linear programming, queuing theory, Petri nets, etc. In this paper we will deal only with deterministic production systems and we will depart considerably from previous research by showing how piecewise-affine control theory can be used to handle production planning of switched production systems.

To the best of our knowledge the methodology we are using in this paper to formulate and solve the problem has never been used in production planning before. The main contribution of the paper is therefore that, for the first time, production systems are modeled as switched linear systems and the production problem is formulated as a switched $\mathscr{H}_\infty$ control problem with a piecewise-affine control law. The switching variable for the production systems modeled in this paper is the stock level. When the stock level is positive, some of the produced parts are being stored. The stocked parts may deteriorate with time at a given rate. When the stock level is negative it means that orders for production of parts are coming in and there are no stocked parts to immediately meet the demand, which leads to backorders. A switched linear model is therefore used and it is shown that the inventory control problem can be solved using switched control theory. More specifically, a state feedback controller that forces the stock level to be kept close to zero, even when there are fluctuations in the demand, will be designed in this paper using $\mathscr{H}_\infty$ control theory. The gains of the state feedback controller that quadratically stabilizes the production dynamics and at the same time rejects the external demand fluctuation (treated as a disturbance) are determined by solving a given set of linear matrix inequalities (LMIs).

The rest of the paper is organized as follows. In Section 2, the production planning problem is stated for the case of production of a single part type. Then, Section 3 briefly reviews background material. Next, Section 4 presents the main contribution of the paper. First, the production planning problem is formulated as a switched $\mathscr{H}_\infty$ control problem and then solved using switched control theory. Finally, in Section 5, a numerical example is provided to show the effectiveness of the proposed methodology to handle the production planning problem. The paper then finishes by presenting possible extensions and the conclusions.

2. Problem statement

As a preliminary step, let us consider the case of a production system producing one part type and formulate the production control problem. Let $x_1(t) \in \mathbb{R}$, $u(t) \in \mathbb{R}, d(t) \in \mathbb{R}$ denote respectively the stock level, the production rate and the demand rate at time t. As mentioned in the introduction, the switching variable for the production system is the stock level $x_1(t)$. If $x_1(t) > 0$, some of the produced parts are being stored. It is assumed that the stocked parts may deteriorate with time with a rate factor ρ. The production model for $x_1(t) > 0$ is then,

$$\dot{x}_1(t) = -\rho x_1(t) + u(t) - d(t), \quad x_1(0) = x_1^0. \tag{2.1}$$

When $x_1(t) < 0$ it means that orders for production of parts are coming in and there are no stocked parts to immediately meet the demand, which leads to backorders. Therefore the production model for $x_1(t) < 0$ is simply

$$\dot{x}_1(t) = u(t) - d(t), \quad x_1(0) = x_1^0. \tag{2.2}$$

These two dynamical systems can be described together as a switched linear system if the rejection rate is made model dependent in the form

$$\rho = \rho(x_1) = \begin{cases} \rho_1, & x_1 \in \mathcal{R}_1 \\ \rho_2, & x_1 \in \mathcal{R}_2 \end{cases}, \tag{2.3}$$

where, $\mathcal{R}_1 = \{x_1 \in \mathbb{R} \mid x_1 > 0\}, \mathcal{R}_2 = \{x_1 \in \mathbb{R} \mid idx_1 < 0\}$, $\rho_1 = \rho$, $\rho_2 = 0$.

Using this definition, the switched system will have piecewise-affine dynamics described by

$$\dot{x}_1(t) = -\rho_i x_1(t) + u(t) - d(t), \quad i = 1, 2, \quad x_1(0) = x_1^0. \tag{2.4}$$

In order to capture the behavior of a real production system, the production rate cannot obviously be negative and must also be limited so the following constraints must be introduced

$$0 \leq u(t) \leq \bar{u} \tag{2.5}$$

where $\bar{u}$ is a known positive scalar corresponding to the maximum production rate. It is assumed that the demand rate $d(t)$ is composed of a known constant component $\hat{d}$ plus an unknown time-varying (possibly fluctuating) component $w(t)$ with finite energy (i.e, $\int_0^\infty w^2(\tau)\,d\tau < \infty$) so that a model for the demand rate is

$$d(t) = \hat{d} + w(t). \tag{2.6}$$

Inventory control is a complex hierarchical control problem with several levels of decision-making. In this paper it will be assumed that the managing top level decision-making has determined that the best policy to follow is one where the stock level is kept as close as possible to zero. Therefore, defining the $\mathcal{L}_2$-norm of a $\mathcal{L}_2$ integrable signal z to be

$$\|z\|_2 \triangleq \left[\int_0^\infty z^\top(t) z(t)\,dt\right]^{1/2} \tag{2.7}$$

the production control problem can now be stated as follows:

Switched production control problem. Given a production system with switched dynamics (2.4), input constraints (2.5) and demand rate given by (2.6), where $\hat{d}$ is a constant known factor, design the production rate (control input) $u(t)$ such that the stock level $x_1(t)$ converges to zero when there is no time-varying demand component $w(t)$ and such that it is kept as close as possible to zero when $w(t) \neq 0$. As close as possible here means that the $\mathcal{L}_2$-induced norm from $w(t)$ to $x_1(t)$ is minimized.

3. Review of background material

The problem stated in Section 2 is a piecewise-affine $\mathscr{H}_\infty$ control problem that will be formulated as a set of Linear Matrix Inequalities (LMIs). Therefore, before presenting the solution to the problem, a brief review of some definitions and results pertaining to $\mathscr{H}_\infty$ control theory, piecewise-affine control theory and LMIs are needed. Linear Matrix Inequalities are inequalities involving matrices that are linear in the decision variables according to the following definition.

DEFINITION 2.1 (BOYD, ET AL. (1994)) *A linear matrix inequality (LMI) has the form*

$$F(x) \triangleq F_0 + \sum_{i=1}^{m} x_i F_i > 0, \tag{2.8}$$

where $x \in \mathbb{R}^m$ is the variable and the symmetric matrices $F_i = F_i^\top \in \mathbb{R}^{n \times n}$, $i = 0, \ldots, m$ are given. The inequality (2.8) means that $F(x)$ is positive definite, i.e, $v^\top F(x) v > 0$ for all nonzero $v \in \mathbb{R}^n$.

Linear matrix inequalities appear naturally in the formulation of sufficient conditions for Lyapunov stability of open-loop linear systems as the next example illustrates.

EXAMPLE 2.1 Given a linear system with control input $u(t)$, disturbance input $w(t)$ and dynamics

$$\dot{x}(t) = Ax(t) + Bu(t) + B_w w(t), \quad z(t) = Cx(t) \tag{2.9}$$

with $x \in \mathbb{R}^n$, $u \in \mathbb{R}^u$, $w \in \mathbb{R}^w$, $z \in \mathbb{R}^z$, a necessary and sufficient condition to prove asymptotic stability of the trajectories of the homogeneous system (i.e, with $u \equiv 0$ and $w \equiv 0$) is to find a quadratic Lyapunov (energy-like) function of the form

$$V = x^\top P x \tag{2.10}$$

verifying the following positivity and negative time-rate conditions

$$V > 0, \tag{2.11}$$

$$\dot{V} < 0, \tag{2.12}$$

which can be rewritten as the set of LMIs

$$P > 0, \tag{2.13}$$

$$A^\top P + PA < 0. \tag{2.14}$$

Note that LMIs (2.13)–(2.14) can be put in the form (2.8) by representing the matrix P in a basis for symmetric matrices, the coefficients of the representation being the variables x_i in (2.8). This example motivates the following definition.

DEFINITION 2.2 (BOUKAS (2004)) *System (2.9) with $u(t) \equiv 0$, $w(t) \equiv 0$ is said to be internally quadratically stable if there exists a symmetric and positive definite matrix $P > 0$, satisfying*

$$A^\top P + PA < 0. \tag{2.15}$$

We now state a very important Lemma that will be repeatedly used in this paper.

LEMMA 2.1 (SCHUR COMPLEMENT BOUKAS (2004)) *The LMI*

$$\begin{bmatrix} H & S^\top \\ S & R \end{bmatrix} > 0$$

is equivalent to

$$R > 0, \quad H - S^\top R^{-1} S > 0$$

where $H = H^\top$, $R = R^\top$ and S is a matrix with appropriate dimension.

Having defined stability of the homogeneous system (without acting inputs or disturbances), stability with γ-disturbance attenuation is defined next.

DEFINITION 2.3 (BOUKAS (2004)) *Let $\gamma > 0$ be a given positive constant. System (2.9) with $u(t) \equiv 0$ is said to be stable with γ-disturbance attenuation if there exists a constant $M(x_0)$ with $M(0) = 0$, such that the following holds:*

$$\|z\|_2 \triangleq \left[\int_0^\infty z^\top(t) z(t)\, dt \right]^{1/2} \leq [\gamma^2 \|w\|_2^2 + M(x_0)]^{1/2}. \tag{2.16}$$

The following theorem gives a sufficient condition to guarantee stability of the system (2.9) and at the same time to guarantee disturbance rejection with a certain level $\gamma > 0$.

THEOREM 2.1 *Let γ be a given positive constant. If there exists a symmetric and positive definite matrix $P > 0$ such that*

$$\begin{bmatrix} A^\top P + PA + C^\top C & PB_w \\ B_w^\top P & -\gamma^2 \mathbb{I} \end{bmatrix} < 0, \tag{2.17}$$

then system (2.9) with $u(t) \equiv 0$, $w(t) \equiv 0$ is quadratically stable and for $u(t) \equiv 0$, $w(t) \neq 0$ it satisfies

$$\|z\|_2 \leq [\gamma^2 \|w\|_2^2 + x_0^\top P x_0]^{1/2}, \tag{2.18}$$

which means that the system with $u(t) = 0$ for all $t \geq 0$ is stable with γ-disturbance attenuation.

Proof. From (2.17) and using Schur complement we get

$$A^\top P + PA + C^\top C < 0.$$

Since $C^\top C \geq 0$ this inequality implies

$$A^\top P + PA < 0.$$

Based on Definition 2.2, this proves that the system under study is internally quadratically stable.

Let us now prove that (2.18) is satisfied. To this end, let us define the following performance function:

$$J_T = \int_0^T \left[z^\top(t)z(t) - \gamma^2 w^\top(t)w(t)\right] dt.$$

To prove (2.18), it suffices to establish that J_∞ is bounded. More specifically, we will show that $J_\infty \leq V(x_0) = x_0^\top P x_0$. First of all notice that for the Lyapunov candidate function $V(x(t)) = x^\top(t)Px(t)$

$$\dot{V}\big(x(t)\big) = x^\top(t)[A^\top P + PA]x(t) + x^\top(t)PB_w w(t) + w^\top(t)B_w^\top Px(t),$$

and

$$z^\top(t)z(t) - \gamma^2 w(t)w(t) = x^\top(t)C^\top Cx(t) - \gamma^2 w^\top(t)w(t)$$

which implies the following equality

$$z^\top(t)z(t) - \gamma^2 w^\top(t)w(t) + \dot{V}\big(x(t)\big) = \zeta^\top(t)\Theta\zeta(t),$$

with

$$\Theta = \begin{bmatrix} A^\top P + PA + C^\top C & PB_w \\ B_w^\top P & -\gamma^2\mathbb{I} \end{bmatrix}, \quad \zeta^\top(t) = \begin{bmatrix} x^\top(t) & w^\top(t) \end{bmatrix}.$$

Therefore, $J_T = \int_0^T [z^\top(t)z(t) - \gamma^2 w^\top(t)w(t) + \dot{V}(x(t))]\, dt - \int_0^T \dot{V}\big(x(t)\big)\, dt$. Using now Dynkin's formula, i.e: $\left[\int_0^T \dot{V}\big(x(t)\big)\, dt\right] = V\big(x(T)\big) - V(x_0)$, we get

$$J_T = \left[\int_0^T \zeta^\top(t)\Theta\zeta(t)\, dt\right] - V\big(x(T)\big) + V(x_0). \tag{2.19}$$

Since $\Theta < 0$ and $V\big(x(T)\big) \geq 0$, (2.19) implies $J_T \leq V(x_0)$, which taking the limit when $T \to \infty$ yields $J_\infty \leq V(x_0)$, i.e., $\|z\|_2^2 - \gamma^2\|w\|_2^2 \leq x_0^\top P x_0$.

This yields the desired result, i.e., $\|z\|_2 \leq [\gamma^2\|w\|_2^2 + x_0^\top P x_0]^{1/2}$, and ends the proof of the theorem. □

Finally, to end this section on review of background material, we will formally define what it is meant by a piecewise-affine (PWA) system, a special class of switched systems. A stability theorem for switched systems will also be presented.

DEFINITION 2.4 *A PWA system is a switched system that consists of a set of affine dynamic models together with a corresponding partition of the state space.*

- State Space Partition *is assumed to be composed of polytopic cells* $\mathcal{R}_i$, $i \in \mathcal{I} = \{1, \ldots, M\}$. *Each cell is constructed as the intersection of a finite number* (p_i) *of half spaces*

$$\mathcal{R}_i = \{x \mid H_i^\top x - g_i < 0\}, \tag{2.20}$$

where $H_i = [h_{i1} h_{i2} \ldots h_{ip_i}]$, $g_i = [g_{i1} g_{i2} \cdots g_{ip_i}]^\top$. *Moreover the sets* $\mathcal{R}_i$ *partition a subset of the state space* $\mathcal{X} \subset \mathbb{R}^n$ *such that* $\cup_{i=1}^M \bar{\mathcal{R}}_i = \mathcal{X}$, $\mathcal{R}_i \cap \mathcal{R}_j = \emptyset$, $i \neq j$, $\bar{\mathcal{R}}_i$ *denoting the closure of* $\mathcal{R}_i$.

- Affine Dynamical Models. *Within each cell the dynamics are affine and strictly proper of the form*

$$\dot{x}(t) = A_i x(t) + b_i + B_i u(t) + B_{w_i} w(t), \quad z(t) = C_i x(t) \tag{2.21}$$

where $x \in \mathbb{R}^n$, $u \in \mathbb{R}^u$, $w \in \mathbb{R}^w$, $z \in \mathbb{R}^z$.

For example, for the PWA system (2.4), $H_1 = -1$, $H_2 = 1$, $g_1 = g_2 = 0$. For system (2.21), the following adaptation of the definition of trajectories or solutions presented in Johansson (2003) is used.

DEFINITION 2.5 (JOHANSSON (2003)) *Let* $x(t) \in \mathcal{X}$ *be an absolutely continuous function. Then* $x(t)$ *is a trajectory of the system* (2.21) *on* $[t_0, t_f]$ *if, for almost all* $t \in [t_0, t_f]$ *and Lebesgue measurable* $u(t)$, $w(t)$, *the equation* $\dot{x}(t) = A_i x(t) + b_i + B_i u(t) + B_{w_i} w(t)$ *holds for* $x(t) \in \bar{\mathcal{R}}_i$.

Finally, the following result establishes global uniform (where uniform means for any switching) exponential stability of a homogeneous switched linear system (therefore, including piecewise-linear systems) when a globally quadratic Lyapunov function can be found.

THEOREM 2.2 (LIBERZON AND MORSE (1999)) *If there exists a symmetric matrix* $P > 0$ *such that*

$$A_i^\top P + P A_i < 0, \quad i = 1, \ldots, M \tag{2.22}$$

then the switched linear system $\dot{x} = A_i x$, $i = 1, \dots, M$ is globally uniformly (i.e, for any switching) exponentially stable.

Note that stability can be proven for any switching, which means that the dynamics at the boundaries of the polytopic regions in (2.21) can be arbitrary. For PWA systems, there are other results where piecewise-quadratic Lyapunov functions are searched for instead of the more conservative globally quadratic function. See Johansson (2003) for details.

4. Problem solution

To solve the control problem stated in Section 2, an $\mathscr{H}_\infty$ piecewise-affine state feedback controller with integral action is now designed. For that purpose, we need to have access to both the stock level and its integral (the cumulative stock level), so we define the state variables $x_1(t)$ and $x_2(t) = \int_0^t x_1(s)\,ds$ and the state vector $x(t) = [x_1(t)x_2(t)]^\top$. Using this state vector, the augmented dynamics with performance output $z(t) = x_1(t)$ can be described as

$$\begin{cases} \dot{x}(t) = A_i x(t) + b_i + B_i u(t) + B_{w_i} w(t), \\ z(t) = C_i x(t) \end{cases} \tag{2.23}$$

where

$$A_i = \begin{bmatrix} -\rho_i & 0 \\ 1 & 0 \end{bmatrix}, \quad b_i = \begin{bmatrix} -\hat{d} \\ 0 \end{bmatrix} B_i = \begin{bmatrix} 1 \\ 0 \end{bmatrix}, \quad B_{w_i} = -B_i, \quad C_i = \begin{bmatrix} 1 & 0 \end{bmatrix}$$

This is a PWA system. For this system, a PWA controller will be designed of the form

$$u(t) = K_i x(t) + \hat{d}. \tag{2.24}$$

The affine term $\hat{d}$ is included in the production rate (control signal) to compensate for the known constant demand rate $\hat{d}$. Combining (2.23) and (2.24) yields

$$\dot{x}(t) = [A_i + B_i K_i] x(t) + B_{w_i} w(t) = A_{\mathrm{cl}_i} x(t) + B_{w_i} w(t). \tag{2.25}$$

This is a piecewise-linear system. The result expressed in Theorem 2.2 and the reasoning followed in the proof of Theorem 2.1 will now be combined to prove stability of the closed-loop piecewise-linear system. More specifically, a globally quadratic candidate Lyapunov function $V(x) = x^\top P x$ will be used for the piecewise-linear system (2.25). Based on (2.17), using the Schur complement as before, the closed-loop dynamics will be stable with a guaranteed disturbance rejection of level

$\gamma > 0$ if there exists a symmetric and positive definite matrix $P > 0$ such that the following holds

$$A_{\text{cl}_i}^\top P + PA_{\text{cl}_i} + C_i^\top C_i + \gamma^{-2} PB_{w_i}B_{w_i}^\top P < 0, \quad i = 1, 2.$$

Using the expression of A_{cl_i}, we get for $i = 1, 2$

$$A_i^\top P + PA_i + K_i^\top B_i^\top P + PB_iK_i + C_i^\top C_i + \gamma^{-2} PB_{w_i}B_{w_i}^\top P < 0. \quad (2.26)$$

This matrix inequality is nonlinear in the design parameters P and K_i. To put it into LMI form, let $X = P^{-1}$ and make the change of variables $x = Xy$. This corresponds to pre- and post-multiplying inequality (2.26) by X, which yields

$$XA_i^\top + A_iX + XK_i^\top B_i^\top + B_iK_iX + XC_i^\top C_iX + \gamma^{-2} B_{w_i}B_{w_i}^\top < 0, \quad i = 1, 2.$$

Letting now $Y_i = K_iX$, $i = 1, 2$ and using Schur complement, yields

$$\begin{bmatrix} XA_i^\top + A_iX + Y_i^\top B_i^\top + B_iY_i & XC_i^\top & B_{w_i} \\ C_iX & -\mathbb{I} & 0 \\ B_{w_i}^\top & 0 & -\gamma^2\mathbb{I} \end{bmatrix} < 0, \quad i = 1, 2. \quad (2.27)$$

Based on this derivation, we can state the following theorem that gives a design method for the unconstrained state feedback controller that meets the control objectives in the absence of constraints on the production rate.

THEOREM 2.3 *Let γ be a positive constant. If there exist a symmetric matrix $X > 0$ and matrices Y_i, $i = 1, 2$ such that the LMIs (2.27) hold, then system (2.23) under the controller (2.24) with $K_i = Y_iX^{-1}$, $i = 1, 2$ is stable and moreover the closed-loop system satisfies the disturbance rejection level of $\gamma > 0$.*

The results of this theorem will allow us to determine the controller gain matrices K_i, $i = 1, 2$, but there is no guarantee that the control will meet the production rate bounds (2.5) that must always be satisfied. Extra constraints should then be added to the previous system of LMIs to force the control law to always satisfy the bounds. For this purpose, we note first that when the initial condition is known (which is the case in our problem), we can add the constraint $x_0^\top X^{-1} x_0 \leq 1$, which using Schur complement leads to the LMI

$$\begin{bmatrix} 1 & x_0^\top \\ x_0 & X \end{bmatrix} \geq 0. \quad (2.28)$$

Since the Lyapunov function is $V = x^\top X^{-1} x$, and the level sets of the Lyapunov function are invariant sets for the system trajectories, this LMI guarantees not only that the initial state is inside the set

$$\mathscr{D} = \{x \in \mathbb{R}^n \mid x^\top X^{-1} x \leq 1\} \tag{2.29}$$

but also that $x(t) \in \mathscr{D}$, $\forall t \geq 0$. In terms of the control input constraints, note that using (2.24), the constraints (2.5) are equivalent to

$$-\hat{d} \leq K_i x \leq \bar{u} - \hat{d}. \tag{2.30}$$

Since it is known that $x(t) \in \mathscr{D}$, $\forall t \geq 0$ if LMI (2.28) is verified, we will now develop a LMI to guarantee that the constraint $M_i x \leq m$ is met for $x(t) \in \mathscr{D}$ and arbitrary $m \in \mathbb{R}$. For the right inequality in (2.30) $M_i = K_i$, $m = \bar{u} - \hat{d}$ and for the left inequality $M_i = -K_i$, $m = \hat{d}$ after multiplication by -1. We start by changing variables again to $x = Xy$ so that in the new coordinates the constraint $Y_i y \leq m$ (where $Y_i = M_i X$) must be met for

$$y \in \mathscr{D} = \{y \in \mathbb{R}^n \mid y^\top X y \leq 1\}. \tag{2.31}$$

Notice now that for $X > 0$, the set (2.31) is an ellipsoid (or possibly a ball) and the condition $Y_i y \leq m$ represents a half space. Therefore, $Y_i y \leq m$ will be met for all $y \in \mathscr{D}_y$ if it is met for $y^\star$, where $y^\star$ is the solution to the following optimization problem:

$$\begin{array}{rl} \max & Y_i y \\ s.t. & y \in \mathscr{D}_y. \end{array} \tag{2.32}$$

The solution to problem (2.32) is the point at which the plane $Y_i y = m$ is a supporting hyperplane of the set $\mathscr{D}_y$ and is given by

$$y^\star = \lambda Y_i^\top, \quad 1 = {y^\star}^\top X y^\star$$

Combining these expressions yields

$$\lambda = \frac{1}{\sqrt{Y_i X Y_i^\top}}, \quad y^\star = \frac{1}{\sqrt{Y_i X Y_i^\top}} Y_i^\top.$$

Therefore, $Y_i y \leq m$ is met for $y \in \mathscr{D}_y$ and arbitrary $m \in \mathbb{R}$ if it is met for $y^\star$, i.e, if

$$Y_i Y_i^\top \leq m \sqrt{Y_i X Y_i^\top} \tag{2.33}$$

or, equivalently, if

$$Y_i [Y_i^\top Y_i] Y_i^\top \leq m^2 Y_i X Y_i^\top \iff Y_i [Y_i^\top Y_i - m^2 X] Y_i^\top \leq 0. \tag{2.34}$$

Condition (2.34) will be verified if

$$Y_i^\top Y_i - m^2 X \leq 0. \tag{2.35}$$

Using Schur complement, this condition is equivalent to the LMI

$$\begin{bmatrix} 1 & Y_i \\ Y_i^\top & m^2 X \end{bmatrix} \geq 0. \tag{2.36}$$

Finally, constraints (2.5) will be verified if the following two LMIs are verified:

- If LMI (2.36) is verified with m replaced by $\bar{u} - \hat{d}$ and
- If LMI (2.36) is verified with m replaced by $\hat{d}$

Recall that $Y_i = M_i X$ and either $M_i = K_i$ or $M_i = -K_i$. Note however that by using a Schur complement argument, Y_i can be replaced by $-Y_i$ in (2.36) without changing the inequality. Therefore, we can assume without loss of generality that $Y_i = M_i X = K_i X$. Incorporating these constraints, the following theorem can be stated, which will allow the design of an $\mathscr{H}_\infty$ piecewise-affine state feedback controller that stabilizes the system and guarantees the required disturbance rejection.

THEOREM 2.4 *Let γ be a positive constant. If there exist a symmetric matrix $X > 0$ and matrices Y_i, $i = 1, 2$ such that the LMIs (2.27), (2.28), (2.36) with $m = \bar{u} - \hat{d}$ and $m = -\hat{d}$ hold, then the system (2.23) under the controller (2.24) with $K_i = Y_i X^{-1}$, $i = 1, 2$ is stable, the closed-loop system satisfies the disturbance rejection of level $\gamma > 0$ and the control input satisfies constraints (2.5).*

From a practical point of view, the controller that quadratically stabilizes the system and at the same time guarantees the maximum disturbance rejection is of great interest. This controller can be obtained by solving the following optimization problem:

$$\mathcal{P}: \begin{cases} \min \nu \\ s.t \ (2.27), (2.28), (2.36), \\ \nu > 0, \quad X > 0, \\ \text{with } \nu = \gamma^2 \text{ in (2.27), } m = \bar{u} - \hat{d} \text{ and } m = \hat{d} \text{ in (2.36).} \end{cases}$$

Finally, the following corollary summarizes the results on the design of the controller that quadratically stabilizes the system (2.23) and simultaneously guarantees the smallest disturbance rejection level.

COROLLARY 2.1 *Let $\nu > 0$, $X > 0$, Y_1, and Y_2 be the solution of the optimization problem $\mathcal{P}$. Then, the controller* (2.24) *with $K_i = Y_i X^{-1}$, $i = 1, 2$ quadratically stabilizes the class of production systems we are considering and moreover the closed-loop system satisfies a disturbance rejection level of $\sqrt{\nu}$.*

5. Numerical example

To illustrate the effectiveness of the developed results, we consider in this section a manufacturing system producing one item.

EXAMPLE 2.2 The problem data for this example can be found in Table 2.1. For this problem, $x_1(t) \in \mathbb{R}$ and $d(t) \in \mathbb{R}$. The corresponding matrices are then given by

$$A_i = \begin{bmatrix} -\rho_i & 0 \\ 1 & 0 \end{bmatrix}, \quad b_i = \begin{bmatrix} -\hat{d} \\ 0 \end{bmatrix}, \quad B_i = \begin{bmatrix} 1 \\ 0 \end{bmatrix}, \qquad (2.37)$$
$$B_{w_i} = -B_i, \quad C_i = \begin{bmatrix} 1 \\ 0 \end{bmatrix}^\top .$$

Solving the optimization problem $\mathcal{P}$ yields

$$\nu = 0.84 K_1 = \begin{bmatrix} -1.257 & -3.26 \times 10^{-5} \end{bmatrix} K_2 = \begin{bmatrix} -1.260 & -3.26 \times 10^{-5}. \end{bmatrix}$$

The disturbance rejection is guaranteed to be at least $\gamma = \sqrt{\nu} = 0.9165$.

To simulate the performance of this controller we will consider 2 cases:

1. **Constant demand rate**: in this case $d(t) = \hat{d} = 1$. The simulation results are shown Fig. 2.1. As expected, the production rate is maintained at a constant value $u(t) = \hat{d}$ and all the produced parts are supplied to meet the demand. The stock level therefore remains at zero as desired. Note that the control remains always between the given bounds.

2. **Time varying demand rate**: in this case the demand rate is of the form $d(t) = \hat{d} + 0.01 \sin t$. The simulation results are shown in Fig. 2.2. As expected, the production rate is now oscillatory to

Table 2.1. Production system data.

rate ρ_1	rate ρ_2	control lower bound	control upper bound	demand rate ($\hat{d}$)
0.01	0	0	2	1

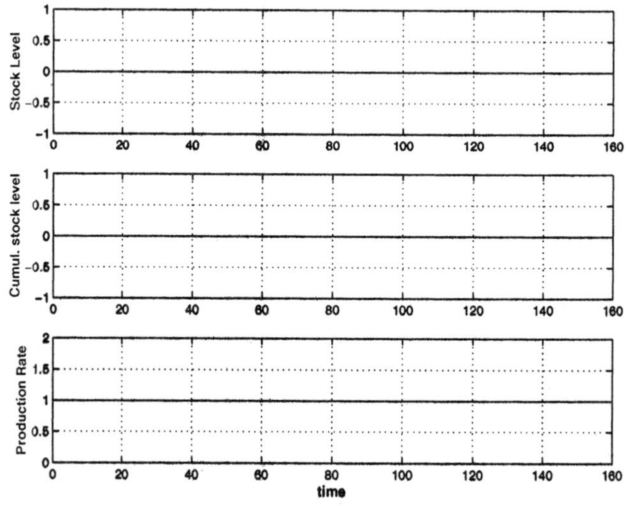

Figure 2.1. Stock level, cumulative stock level and production rate for constant unitary demand rate.

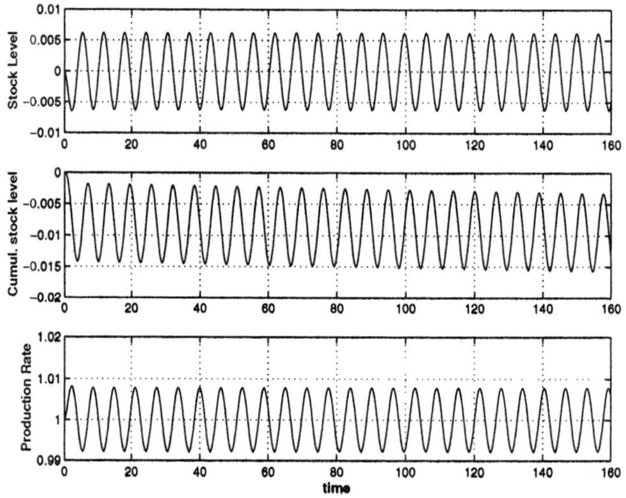

Figure 2.2. Stock level, cumulative stock level and production rate for oscillating demand rate.

compensate for the oscillating demand rate. Although the stock level cannot be kept at zero, there is an attenuation in the magnitude of the stock oscillations of 0.5 as compared to the magnitude of the demand rate oscillations. Note that again the control satisfies the imposed bounds.

6. Extensions of the proposed model

The model proposed earlier has shown that switched control theory can be applied to inventory control problems. This model can be extended to include several additional features such as, for example:

1. Production of multiple parts using multiple machines, each machine producing one part.

2. Limited storage capacity $x_1(t) < \bar{x}$.

3. Delay on the production rate $u(t-\tau)$, $\tau > 0$ to model the fact that the production policy is usually planned ahead of time.

In this section we will describe how these extensions could be incorporated. The first extension can be easily dealt with because a system with multiple (say N) machines, each machine producing one part, is a decoupled system of N subsystems and each of the N subsystems can be solved independently, as presented in this paper. Therefore, letting $\tilde{x}(t) = [x_1(t), \ldots, x_N(t)]^\top$, $\tilde{z}(t) = [z_1(t), \ldots, z_N(t)]^\top$, $\tilde{u}(t) = [u_1(t), \ldots, u_N(t)]^\top$, $\tilde{w}(t) = [w_1(t), \ldots, w_N(t)]^\top$ and $\mathrm{diag}(W)$ be the operator that constructs a matrix whose diagonal blocks correspond to W, the dynamics for N machines can be written as

$$\begin{cases} \dot{\tilde{x}}(t) = \tilde{A}_i \tilde{x}(t) + \widetilde{B}_i + \widetilde{B}_i \tilde{u}(t) + \widetilde{B}_{w_i} \tilde{w}(t), \\ \tilde{z}(t) = \widetilde{C}_i \tilde{x}(t) \end{cases}$$

where

$$\begin{aligned} \tilde{A}_i &= \mathrm{diag}[\rho_{1_i}, \ldots, \rho_{N_i}], \\ \widetilde{B}_i &= \mathrm{diag}[B_i, \ldots, B_i], \\ \widetilde{B}_i &= \mathrm{diag}[b_1, \ldots, b_N], \\ \widetilde{C}_i &= \mathrm{diag}[C_i, \ldots, C_i], \\ \widetilde{B}_{w_i} &= \mathrm{diag}[B_{w_i}, \ldots, B_{w_i}], \\ b_i &= [-\hat{d}_i 0]^\top \end{aligned}$$

and B_i, C_i, B_{w_i}, $i = 1, \ldots, N$ are as described in (2.37).

Extension number two can be dealt with by modifying region $\mathcal{R}_1$ to be of the form $\mathcal{R}_1 = \{x_1 \in \mathbb{R} \mid id0 < x_1 < \bar{x}\}$ and introducing a new region in the form $\mathcal{R}_3 = \{x_1 \in \mathbb{R} \mid idx_1 > \bar{x}\}$. Then the controllers for regions $\mathcal{R}_1$, $\mathcal{R}_2$ would be designed in the same way and the controller for region $\mathcal{R}_3$ would simply be of the form $u(t) = \hat{d}$. In other words, if the maximum storage capacity has been reached the system can only produce to meet the nominal (known) constant demand. Notice that because of the rejection rate, the dynamics in region $\mathcal{R}_3$ are stable and the system will eventually leave region $\mathcal{R}_3$ and come back to region $\mathcal{R}_1$.

Finally, to deal with extension number three, the system must be modeled as

$$\dot{x}_1(t) = -\rho_i x_1(t) + u(t-\tau) - d(t), \quad x_1(0) = x_1^0 \qquad (2.38)$$

where τ is the processing time that may be considered as time-varying with appropriate assumptions (see Boukas and Liu Boukas and Liu, 2002).

This is a PWA system with delay and is the subject of future research of the authors.

The models we presented in this paper deal with deterministic production systems. In practice, these models have some limitations since they do not include some important features such as, for example, breakdown of the machines. After breakdown, there is a no-activity time to get the machine repaired. To overcome this limitation in the model, we can extend it to include breakdowns in the machines and even the preventive maintenance or corrective maintenance as it was done in the literature. For more details refer to Sethi and Zhang (1994); Gershwin (1994); Boukas and Liu (2001). The used framework in the cited references is based on modeling of the breakdowns and the repairs as a continuous-time Markov process with finite state space.

7. Conclusions

This paper dealt with the production planning problem for a deterministic production system with given deterministic demand rate plus an unknown fluctuating demand rate. This problem has been modeled as a control problem of a switched (PWA) system and it has been solved using $\mathscr{H}_\infty$ control theory. Some important extensions of this problem have also been proposed.

References

Axsater, S. (1985). Control theory concepts in production and inventory control. *International Journal of Systems Science*, 16(2):161–169.

Boukas, E.K. (2004). *Stochastic Hybrid Systems: Analysis and Design*, Birkhäuser, Boston.

Boukas, E.K. and Liu, Z.K. (2002). *Deterministic and Stochastic Systems with Time-Delay*, Birkhäuser, Boston.

Boukas, E.K. and Liu, Z.K. (2001). Production and maintenance control for manufacturing systems. *IEEE Transactions on Automatic Control*, 46(9):1455–1460.

Boyd. S, Ghaoui, L.E., Feron, E., and Balakrishnan, V. (1994). *Linear Matrix Inequalities in System and Control Theory*, volume 15, Studies in Applied Mathematics, SIAM.

Disney, S.M., Naim, M.M., and Towill, D.R. (2000). Genetic algorithm optimization of a class of inventory control systems. *International Journal of Production Economics*, 68(3):259–278.

Gavish, B., and Graves, S. (1980). A one-product production/inventory problem under continuous review policy. *Operations Research*, 28:1228–1236.

Gershwin, S.B. (1994). *Manufacturing Systems Engineering*, Prentice Hall, New York.

Grubbstrom, R.W. and Wikner, J. (1996). Inventory trigger control policies developed in terms of control theory. *International Journal of Production Economics*, 154:397–406.

Hennet, J.C. (2003). A bimodal scheme for multi-stage production and inventory control. *Automatica*, 39:793–805.

Johansson, M. (1999). Piecewise linear control systems. *Ph.D thesis*, Lund Institute of Technology.

John, S., Naim, M.M., and Towill, D.R. (1994). Dynamic analysis of a WIP compensated decision support system. *International Journal of Manufacturing Systems Design*, 1(4).

Liberzon, D. and Morse, A.S. (1999). Basic problems in stability and design of switched systems. *Control Systems Magazine*, 19(5):59–70.

Ridalls, C.E. and Bennett, S. (2002). Production inventory system controller design and supply chain dynamics. *International Journal of Systems Science*, 33(3):181–195.

Sethi, S.P. and Zhang, Q. (1994). *Hierarchical Decision Making in Stochastic Manufacturing Systems*, Birkhäuser, Boston, Cambridge, MA.

Simon, H.A. (1952). On the application of servomechanism theory in the study of production control. *Economitra*, 20:247–268.

Towill, D.R. (1982). Dynamic analysis of an inventory and order based production control system. *International Journal of Production Research*, 63(4):671–687.

Towill, D.R., Evans, G.N., and Cheema, P. (1997). Analysis and design of an adaptive minimum reasonable inventory control system. *Production Planning and Control*, 8(6):545–557.

Wiendahl, H. and Breithaupt, J.W. (2000). Automatic production control applying control theory. *International Journal of Production Economics*, 63:33–46.

Chapter 3

A TWO-TIME-SCALE APPROACH FOR PRODUCTION PLANNING IN DISCRETE TIME

Qing Zhang
George Yin

Abstract This work develops asymptotically optimal production planning strategies for a class of discrete-time manufacturing systems. To reflect uncertainty, finite-state Markov chains are used in the formulation. The state space of the underlying Markov chain is decomposed into a number of recurrent classes and a group of transient states. Using a hierarchical control approach, by aggregating the states in each recurrent class into a single state, a continuous-time limit control problem in which the resulting limit Markov chain has much smaller state space is derived. Using the optimal control of the limit problem, control policies for the original problem are constructed. Moreover, it is shown that the strategies so designed are nearly optimal.

1. Introduction

This paper is concerned with nearly optimal production planning for a class of discrete-time manufacturing problems. We focus on a manufacturing system consisting of a number of machines and producing a number of parts. We assume that some of the machines are subject to breakdown and repair and the state of the machines is a finite-state Markov chain. The objective is to choose the production rates over time in order to minimize a cost function. To solve a control problem, a commonly used machinery is the dynamic programming (DP) principle. Using such a DP approach, one obtains a system of DP equations corresponding to various Markovian states. The optimal control can be obtained by solving the system of equations. Since the state space of the Markov chain is frequently very large in these manufacturing systems, a large number of DP equations have to be solved which is an insurmountable task. To overcome the difficulty, using time-scale separation, we introduce a small parameter in the underlying Markov chains to reflect the different rates of changes among different states resulting in a two-time scale (a fast-time scale and a slowly varying one) formulation. This leads to singularly perturbed Markovian models with weak and strong interactions; see for example, Abbad et al. (1992); Pervozvanskii and Gaitsgory (1988) among others. The asymptotic probability distributions of this class of Markov chains together with their structural properties have been studied comprehensively; see Yin and Zhang (2000); Yin et al. (2003) and references therein for related results and recent developments.

Research in continuous-time manufacturing systems and related two-time-scale problems are studied extensively; see Sethi and Zhang (1994) for the hierarchical control approach; Yin and Zhang (1998) for general dynamic setups in this connection; Akella and Kumar (1086); Bielecki and Kumar (1988); Boukas (1991); Boukas and Haurie (1990); Gershwin (1994) for related manufacturing systems. In many applications, one needs to deal with discrete-time systems since various system measurements are frequently taken in discrete time. It is equally important to develop optimal and near-optimal control policies for discrete-time hybrid linear systems that is our main focus of this paper.

In this paper, by exploring the interface between continuous-time and discrete-time problems, we first decompose the state space of the underlying Markov chain into a number of recurrent classes and a group of transient states according to the different jump rates. We then aggregate the states and replace the original system with its "average" and show that under suitable scaling, a limit control system that has fewer number of DP equations is obtained. One interesting aspect is that the corresponding limit problem is a continuous-time one. Using the optimal control law of the limit system, we construct controls for the original system, which leads to a feasible approximation scheme. We demonstrate that controls so constructed are asymptotically optimal. Since the construction of near-optimal control is completely determined by the solutions of limit problem, we can substantially reduce the complexity of the problem via reduction of dimensionality.

The rest of the paper is organized as follows. In the next section, we give the precise setup of the problem under consideration. In Section 3, we provide classification of state space of the underlying Markov chain and summarize related results. In Section 4, we derive basic properties of the value functions. In Section 5, we present the corresponding limit problem and demonstrate that the value function for the original problem converges to that of the limit problem. Construction of a near-optimal control policy using the solution of the limit problem is given in Section 6 together with the verification of the near optimality. A simple example is provided in Section 7. The paper is concluded in Section 8.

2. Problem formulation

Consider a manufacturing system consisting of a number of machines and producing several types of products. The machines are failure-prone, i.e., they are subject to breakdown and repair. For simplicity, we impose no conditions on internal buffers.

Let $x_n \in \mathbb{R}^{n_1}$ denote the surplus (inventory/shortage) and $u_n \in \Gamma \subset \mathbb{R}^{n_2}$ the rate of production. Let $\{\alpha_n^\varepsilon\colon\ n \geq 0\}$ be a finite-state Markov chain with state space $\mathcal{M} = \{1, 2, \ldots, m\}$ and transition matrix $P^\varepsilon = (p_{ij}^\varepsilon)_{m\times m}$, where ε is a small parameter to be specified later.

The discrete-time control system is governed by

$$x_{n+1} = x_n + \varepsilon\big(A(\alpha_n^\varepsilon)u_n + B(\alpha_n^\varepsilon)\big), \quad x_0 = x,\ n = 0, 1, \ldots, \tag{3.1}$$

where $A(\alpha) \in \mathbb{R}^{n_1\times n_2}$ and $B(\alpha) \in \mathbb{R}^{n_1\times 1}$, for each $\alpha \in \mathcal{M}$.

Let $u. = \{u_0, u_1, \ldots\}$ denote the control sequence. We consider the cost function

$$J^\varepsilon(x, \alpha, u.) = E\left[\sum_{n=0}^{\infty}(1-\gamma\varepsilon)^n \varepsilon G(x_n, \alpha_n^\varepsilon, u_n)\middle| x_0 = x, \alpha_0^\varepsilon = \alpha\right], \tag{3.2}$$

where $\gamma > 0$ is a constant and $G(x, \alpha, u)$ is bounded and Lipschitz in (x, u) for each $\alpha \in \mathcal{M}$.

The objective is to choose $u.$ to minimize J^ε. We use the DP approach to solve the problem. Let $v^\varepsilon(x, \alpha)$ denote the value function, i.e., $v(x, \alpha) = \inf_{u.} J^\varepsilon(x, \alpha, u.)$. The associated set of DP equations is given by

$$\begin{aligned} v^\varepsilon(x, \alpha) = \min_{u\in\Gamma}\Bigg\{ & \varepsilon G(x, \alpha, u) \\ & +(1-\gamma\varepsilon)\sum_{\beta\in\mathcal{M}} p_{\alpha\beta}^\varepsilon v^\varepsilon\big(x + \varepsilon\big(A(\alpha)u + B(\alpha)\big), \beta\big)\Bigg\}. \end{aligned} \tag{3.3}$$

For each x, let $u^*(x, \alpha)$ denote the minimizer of the right-hand side of (3.3). Then $u^*(x, \alpha)$ is optimal; see Bertsekas (1987).

For typimathcal Manufacturing systems, the number of states in $\mathcal{M}$ is large, so is the number of equations in (3.3). It is therefore difficult to solve these equations. In the rest of the paper, we study an approximate optimal scheme that requires solving simpler problems and yields near optimal controls. Let us give two examples as special cases to the general model.

EXAMPLE 3.1 (SINGLE-MACHINE SYSTEM) Consider a production system with one machine producing one part type. Let $c_n^\varepsilon \in \{0,1\}$ denote the machine state where 1 means the machine is up with maximum capacity 1 and 0 means that the machine is down. Let $z_n^\varepsilon \in \{z^1, z^2\}$ denote part demand rate. The system is given by

$$x_{n+1} = x_n + \varepsilon(c_n^\varepsilon u_n - z_n), \quad x_0 = x.$$

In this case, $n_1 = n_2 = 1$, $A(c, z) = c$, $B(c, z) = -z$, $\Gamma = [0, 1]$, and

$$\mathcal{M} = \{(1, z^1), (1, z^2), (0, z^1), (0, z^2)\}.$$

EXAMPLE 3.2 (TWO-MACHINE FLOWSHOP) Let $x_n = (x_n^1, x_n^2)'$ denote the surplus of the first machine and second machine and let $c^{\varepsilon,1} \in \{0,1\}$ and $c^{\varepsilon,2} \in \{0,1\}$ denote the capacity processes. Then $\alpha_n^\varepsilon = (c_n^{\varepsilon,1}, c_n^{\varepsilon,2}) \in \mathcal{M}$ with

$$\mathcal{M} = \{(0,0), (0,1), (1,0), (1,1)\}.$$

The system is given by

$$x_{n+1} = x_n + \varepsilon\big(A(c_n^{\varepsilon,1}, c_n^{\varepsilon,1})u_n + B(c_n^{\varepsilon,1}, c_n^{\varepsilon,1})\big),$$

with $u_n = (u_n^1, u_n^2) \in \Gamma = [0,1] \times [0,1]$ and

$$A(c^1, c^2) = \begin{pmatrix} c^1 & c^2 \\ c^2 & 0 \end{pmatrix} \quad \text{and} \quad B(c^1, c^2) = \begin{pmatrix} 0 \\ -z \end{pmatrix}, \quad \text{for all } (c^1, c^2) \in \mathcal{M}.$$

3. Singularly perturbed Markov chains

Suppose that the transition probability matrix P^ε of the Markov chain α_n^ε has the form

$$P^\varepsilon = P + \varepsilon Q, \tag{3.4}$$

where ε is a small parameter, $P = (p_{ij})_{m\times m}$ is a probability transition matrix (i.e., $p_{ij} \geq 0$ and, for each i, $\sum_j p_{ij} = 1$), and $Q = (q_{ij})_{m\times m}$ is a generator of a continuous-time Markov chain (i.e., $q_{ij} \geq 0$ for $i \neq j$ and, for each $i \in \mathcal{M}$, $\sum_j q_{ij} = 0$).

In view of (3.4), P is the dominating part so its structure is of crucial importance to the system behavior. Suppose that P has partitioned block form

$$P = \begin{pmatrix} P^1 & & & \\ & \ddots & & \\ & & P^l & \\ P^{*,1} & \cdots & P^{*,l} & P^* \end{pmatrix}, \tag{3.5}$$

where the matrices are such that $P^k = (p_{ij}^k) \in \mathbb{R}^{m_k \times m_k}$, $P^{*,k} = (p_{*ij}^k) \in \mathbb{R}^{m_* \times m_k}$, for $k = 1, \ldots, l$, $P^* = (P_{ij}^*) \in \mathbb{R}^{m_* \times m_*}$. As will be seen that P^k corresponds to recurrent states and P^* and $P^{*,k}$ correspond to transient states. Let $\mathcal{M}_k = \{s_{k1}, \ldots, s_{km_k}\}$ denote the recurrent sub-state space of α_n^ε corresponding to the block P^k, for $k = 1, \ldots, l$, and let $\mathcal{M}_* = \{s_{*1}, \ldots, s_{*m_*}\}$ denote the subspace of transient states. Then the state space of α_n^ε can be written as

$$\begin{aligned}\mathcal{M} &= \mathcal{M}_1 \cup \cdots \cup \mathcal{M}_l \cup \mathcal{M}_* \\ &= \{s_{11}, \ldots, s_{1m_1}\} \cup \cdots \cup \{s_{l1}, \ldots, s_{lm_l}\} \cup \{s_{*1}, \ldots, s_{*m_*}\}\end{aligned}$$

with $m_1 + \cdots + m_l + m_* = m$. For the Markov chain α_n^ε, assume the transition probability matrices satisfy

(a) P^k is a transition probability matrix and is irreducible and aperiodic, for each $k = 1, \ldots, l$;

(b) P^* has all of its eigenvalues inside the unit circle.

REMARK 6 The definition of irreducibility and aperiodicity can be found in many textbooks; see, for example, Iosifescu (1980). The assumption on P^ε is fairly general because any finite-state Markov chain has at least one recurrent state and by suitable rearrangement (p. 94 Iosifescu, 1980), the transition matrix can be either written as (3.5) (corresponding to the case of inclusion of l recurrent classes and a number of transient states), or similar to (3.5) with the last row deleted (corresponding to a chain with all recurrent states).

Note that for each $k = 1, \ldots, l$, $P^{*,k}$ represents the transition probabilities from the transient states to the kth recurrent class, and that P^* is the transition probabilities going into the transient states. Thus as $n \to \infty$, $(P^*)^n \to 0$. The natural condition for this is that all the eigenvalues are inside the unit circle.

For each k, there exists stationary distribution of the sub-Markov chain corresponding to P^k for each $k = 1, \ldots, l$. Denote such stationary distribution by $\nu^k = (\nu_1^k, \ldots, \nu_{m_k}^k)$. Then, ν^k is the unique solution of the following system of equations

$$\nu^k P^k = \nu^k, \quad \sum_{j=1}^{m_k} \nu_j^k = 1. \tag{3.6}$$

Note that the irreducibility can be replaced by the uniqueness of solutions to (3.6) for each k. Nevertheless, we still need the aperiodicity to guarantee the asymptotic expansion of probability vectors; see Yin and Zhang (2000) for details.

Moreover, since all the eigenvalues of P^* are inside the unit circle, $(P^* - I)$ is invertible. Define

$$a^k = (a_1^k, \dots, a_{m_*}^k)' = -(P^* - I)^{-1} P^{*,k}_{m_k}, \quad \text{for} \quad k = 1, \dots, l, \tag{3.7}$$

where $_k = (1, \dots, 1)' \in \mathbb{R}^{k\times 1}$. As shown in Liu et al. (2001), $a_j^k \geq 0$, and $\sum_{k=1}^l a_j^k = 1$ for each $j = 1, \dots, m_*$. This implies that for each j, $(a_j^1, \dots, a_j^l)$ is a probability row vector.

Partition Q as

$$Q = \begin{pmatrix} Q^{11} & Q^{12} \\ Q^{21} & Q^{22} \end{pmatrix},$$

where $Q^{11} \in \mathbb{R}^{(m-m_*)\times(m-m_*)}$, $Q^{12} \in \mathbb{R}^{(m-m_*)\times m_*}$, $Q^{21} \in \mathbb{R}^{m_*\times(m-m_*)}$, and $Q^{22} \in \mathbb{R}^{m_*\times m_*}$. Let

$$Q_* = (\nu^1, \dots, \nu^l)(Q^{11} + Q^{12}(a^1, \dots, a^l)), \tag{3.8}$$

where $(\Lambda^1, \dots, \Lambda^l)$ denotes a block diagonal matrix having matrix entries $\Lambda^1, \dots, \Lambda^l$, and

$$\tilde{} = ({}_{m_1}, \dots, {}_{m_l}) \in \mathbb{R}^{(m-m_*)\times l}.$$

It is easy to check that Q_* is a generator. Let $\alpha(\cdot) = \{\alpha(t) \colon t \geq 0\}$ be a continuous-time Markov chain generated by Q_*. The state space of $\alpha(\cdot)$ is $\mathcal{M} = \{1, \dots, l\}$.

Let ζ be a random variable uniformly distributed on $[0,1]$ that is independent of $\{\alpha_n^\varepsilon\}$. For each $j = 1, \dots, m_*$, define an integer-valued random variable ζ_j by

$$\zeta_j = I_{\{0\leq\zeta\leq a_j^1\}} + 2I_{\{a_j^1<\zeta\leq a_j^1+a_j^2\}} + \cdots + lI_{\{a_j^1+\cdots+a_j^{l-1}<\zeta\leq 1\}}.$$

Define an aggregated process α_n^ε by

$$\alpha_n^\varepsilon = \begin{cases} k & \text{if } \alpha_n^\varepsilon \in \mathcal{M}_k, \\ \zeta_j, & \text{if } \alpha_n^\varepsilon = s_{*j} \in \mathcal{M}_*. \end{cases}$$

Define also the continuous-time interpolations $\alpha^\varepsilon(\cdot)$ and $\alpha^\varepsilon(\cdot)$ as

$$\alpha^\varepsilon(t) = \alpha_n^\varepsilon \quad \text{and} \quad \alpha^\varepsilon(t) = \alpha_n^\varepsilon, \quad \text{for} \quad t \in [n\varepsilon, n\varepsilon + \varepsilon).$$

Note that the state space of $\alpha^\varepsilon(t)$ (resp. α_n^ε) is $\mathcal{M} = \{1, \dots, l\}$. In addition, $P(\alpha_n^\varepsilon = k \mid \alpha_n^\varepsilon = s_{*j}) = a_j^k$. As shown in Yin et al. (2003), the aggregated process $\alpha^\varepsilon(\cdot)$ converges weakly to $\alpha(\cdot)$ generated by Q_*. Moreover, as $\varepsilon \to 0$,

$$E\left(\varepsilon \sum_{n=1}^{\lfloor T/\varepsilon \rfloor -1} \left(I_{\{\alpha_n^\varepsilon = s_{kj}\}} - \nu_j^k I_{\{\alpha_n^\varepsilon \in \mathcal{M}_k\}}\right)\right)^2 \to 0,$$
$$\text{for} \quad k = 1, \ldots, l, \; j = 1, \ldots, m_k, \tag{3.9}$$
$$E\left(\varepsilon \sum_{n=1}^{\lfloor T/\varepsilon \rfloor -1} I_{\{\alpha_n^\varepsilon = s_{*j}\}}\right)^2 \to 0, \quad \text{for} \quad j = 1, \ldots, m_*,$$

where $\lfloor T/\varepsilon \rfloor$ denotes the integer part of T/ε.

In this paper, given a function $f(\cdot)$ and a generator $Q = (q_{ij})$, we use the notation $Qf(\cdot)(i)$ to denote

$$Qf(\cdot)(i) = \sum_{j \neq i} q_{ij}(f(j) - f(i)). \tag{3.10}$$

Note that $Qf(\cdot)(i)$ is the ith component of the vector $Q\big(f(1), \ldots, f(m)\big)'$.

REMARK 7 Note that the form of the decomposition, (3.4) is unique if the order of states in $\mathcal{M}$ is fixed and P^ε is given for sufficiently small $\varepsilon > 0$. However, to implement this approach in practice, one often needs to convert a given transition probability matrix P into the form of (3.4). In this connection, the demonstration of the conversion algorithm in Yin and Zhang (2003) can be used to obtain a "canonical form." Such a decomposition of P may not be unique. As a result, there will be more than one nearly optimal solutions. Nevertheless, in practice, one is often content with a feasible control policy which is close to optimal. Thus any of the nearly optimal controls will be good enough for practical purposes.

4. Properties of value functions

In this section, we derive basic properties of the value functions. These results are needed in the subsequent sections for convergence of value functions.

LEMMA 3.1 *$v^\varepsilon(x, \alpha)$ is uniformly Lipschitz in x, i.e., there exists a constant C, independent of ε and α such that*

$$|v^\varepsilon(x^1, \alpha) - v^\varepsilon(x^2, \alpha)| \le C|x^1 - x^2|, \quad \textit{for all } x^1, x^2 \in \mathbb{R}^{n_1}.$$

Proof. Let x^1 and x^2 in $\mathbb{R}^{n_1}$. Given u_n, let x_n^i be the corresponding states with $x_0^i = x^i$, for $i = 1, 2$, i.e.,

$$x_{n+1}^i = x_n^i + \varepsilon\big(A(\alpha_n)u_n + B(\alpha_n)\big), \quad x_0^i = x^i.$$

Then, we have

$$x^1_{n+1} - x^2_{n+1} = x^1_n - x^2_n, \quad n = 0, 1, \ldots$$

Hence, $x^1_n - x^2_n = x^1 - x^2$, for all n. It follows that

$$\begin{aligned}
&|J^\varepsilon(x^1, \alpha, u.) - J^\varepsilon(x^2, \alpha, u.)| \\
&\qquad \leq E \sum_{n=0}^{\infty} (1 - \beta\varepsilon)^n \varepsilon |G(x^1_n, \alpha_n, u_n) - G(x^2_n, \alpha_n, u_n)| \\
&\qquad \leq \sum_{n=0}^{\infty} (1 - \beta\varepsilon)^n \varepsilon K |x^1 - x^2| \\
&\qquad = \frac{K|x^1 - x^2|}{\beta},
\end{aligned}$$

where K is the Lipschitz constant of G. □

The next lemma is concerned with an probability inequality.

LEMMA 3.2 *If a probability transition matrix $P = (p^{ij})_{m\times m}$ is irreducible and aperiodic, and $f(i) \leq \sum_{j=1}^{m} p^{ij} f(j)$ for a function $f(i)$, $i = 1, \ldots, m$, then $f(1) = f(2) = \cdots = f(m)$.*

Proof. Note that $(P - I)$ is an irreducible generator. A direct application of Lemma A.39 in Yin and Zhang (1998) yields this result. □

LEMMA 3.3 *If there exists a subsequence of $\varepsilon \to 0$ (still denoted by ε for simplicity) such that $v^\varepsilon(x, \alpha) \to v^0(x, \alpha)$ for $\alpha \in \mathcal{M}$, the following assertions hold.*

(a) *For $\alpha \in \mathcal{M}_k$, the limit function $v^0(x, \alpha)$ depends only on k, i.e., $v^0(x, \alpha) = v(x, k)$ for some function $v(x, k)$.*

(b) *For $j = 1, \ldots, m_*$, denote the limit of $v^\varepsilon(x, s_{*j})$ by $v(x, *j) = v^0(x, s_{*j})$ and write $v(x, *) = \big(v(x, *1), \ldots, v(x, *m_*)\big)'$. Then*

$$v(x, *) = a^1 v(x, 1) + \cdots + a^l v(x, l), \tag{3.11}$$

where $v(x, i)$, for $i = 1, \ldots, l$, is given in Part (a).

Proof. Given $u \in \Gamma$, we have, for $\alpha \in \mathcal{M}$,

$$\begin{aligned}
\beta\varepsilon v^\varepsilon(x, \alpha) &\leq \varepsilon G(x, \alpha, u) \\
&\quad + (1 - \beta\varepsilon)\big(v^\varepsilon(x + \varepsilon(A(\alpha)u + B(\alpha)), \alpha) - v^\varepsilon(x, \alpha)\big) \\
&\quad + (1 - \beta\varepsilon) \sum_{\beta \in \mathcal{M}} (p^\varepsilon_{\alpha\beta} - \delta_{\alpha\beta}) v^\varepsilon\big(x + \varepsilon(A(\alpha)u + B(\alpha)), \beta\big),
\end{aligned} \tag{3.12}$$

where $\delta_{\alpha\beta} = 1$ if $\alpha = \beta$ and 0 otherwise. Using the hypothesis $v^\varepsilon(x, \alpha) \to v^0(x, \alpha)$ and sending $\varepsilon \to 0$ lead to

$$\sum_{\beta \in \mathcal{M}} (p^\varepsilon_{\alpha\beta} - \delta_{\alpha\beta}) v^0(x, \beta) \geq 0.$$

Therefore,

$$P^k v^k(x) \geq v^k(x), \quad \text{for} \quad k = 1, \dots, l.$$

where $v^k(x) = \big(v^0(x, s_{k1}), \dots, v^0(x, s_{km_k})\big)'$. Now, the irreducibility of P^k and Lemma 3.2 imply that

$$v(x, k) := v^0(x, s_{k1}) = v^0(x, s_{k2}) = \cdots = v^0(x, s_{km_k}).$$

This proves Part (a).

Next we establish Part (b). Let $u^\varepsilon \in \Gamma$ denote an optimal control. Then the equality in (3.12) holds under u^ε. Sending $\varepsilon \to 0$ in the last m_* equations leads to

$$P^{*,1} v^{0,1} + \cdots + P^{*,l} v^{0,l} + (P^* - I) v(x, *) = 0,$$

where $v^{0,k}(x) =_{m_k} v(x, k)$. This yields

$$v(x, *) = -(P^* - I)^{-1} \big(P^{*,1} v^{0,1}(x) + \cdots + P^{*,l} v^{0,l}(x)\big).$$

In view of the definition of a^i, we obtain

$$\begin{aligned} v(x, *) &= \sum_{i=1}^{l} -(P^* - I)^{-1} P^{*,i}_{m_i} v(x, i) \\ &= \sum_{i=1}^{l} a^i v(x, i), \end{aligned}$$

which proves Part (b). □

5. Limit problem

In this section, we show that there exists a limit problem as $\varepsilon \to 0$, which is a continuous-time control problem and is simple to solve. Then in next section, we use the optimal controls of the limit problem to construct controls for the original problem, which are asymptotically optimal.

Let $U^k = (u^{k1}, \dots, u^{km_k})$ with $u^{kj} \in \Gamma$ and let $U = (U^1, \dots, U^l)$ denote control for the limit problem. Define the running cost function

$$G(x, k, U^k) = \sum_{j=1}^{m_k} \nu^k_j G(x, s_{kj}, u^{kj}).$$

Moreover, define

$$F(x,k,U) = \sum_{j=1}^{m_k} \nu_j^k (A(s_{kj})u^{kj} + B(s_{kj})).$$

Note that $F(x,k,U)$ depends only on U^k, i.e., $F(x,k,U) = F(x,k,U^k)$.

The DP equation for the limit problem has the following form:

$$\gamma v(x,k) = \min_{U^k} \left\{ G(x,k,U^k) + \left(\frac{\partial v(x,k)}{\partial x} \right) \cdot F(\bar{x},k,U^k) \right\} + Q_* v(x,\cdot)(k). \quad (3.13)$$

Let $\bar{\alpha}(t)$ be the Markov chain generated by Q_*. Let $\mathcal{A}^0$ denote a class of admissible controls; see Sethi and Zhang (1994) for related definition and properties. The corresponding limit control problem can be written as

$$\mathcal{P}^0: \begin{cases} \text{minimize: } J(x,k,U(\cdot)) = E \int_0^\infty e^{-\gamma t} G(\bar{x}(t), \bar{\alpha}(t), U(t))\, dt, \\ \text{subject to: } \frac{d\bar{x}(t)}{dt} = F(\bar{x}(t), \bar{\alpha}(t), U(t)), \\ t \geq 0,\ \bar{x}_0 = i,\ \bar{\alpha}(0) = k,\ U(\cdot) \subset \mathcal{A}^0, \\ \text{value function: } v(x,k) = \inf_{U(\cdot)\in\mathcal{A}^0} J(x,k,U(\cdot)). \end{cases}$$

THEOREM 3.1 *For each $\alpha \in \mathcal{M}_k$ and $k = 1, \ldots, l$,*

$$\lim_{\varepsilon \to 0} v^\varepsilon(x,\alpha) = v(x,k), \quad (3.14)$$

and for $\alpha = s_{*j}$,

$$\lim_{\varepsilon \to 0} v^\varepsilon(x,*) = a^1 v(x,1) + \cdots + a^l v(x,l), \quad (3.15)$$

where $v^\varepsilon(x,*) = (v^\varepsilon(x,s_{1*}), \ldots, v^\varepsilon(x,s_{m_* *}))$,

Proof. It suffices to verify (3.14). By Lemma 3.1, for each sequence of $\{\varepsilon \to 0\}$, there exists a further subsequence (still indexed by ε for notational simplicity) such that $v^\varepsilon(x,\alpha)$ converges. Denote the limit by $v^0(x,\alpha)$. Then by Lemma 3.3, $v^0(x,\alpha) = v(x,k)$. That is, the exact value of α is unimportant and only k counts.

Fix $k = 1, \ldots, l$. For any $\alpha = s_{kj} \in \mathcal{M}_k$, let $v(x,k)$ be a limit of $v^\varepsilon(x,s_{kj})$ for some subsequence of ε. Given x_0, let a function $\phi(x,k) \in C^1(\mathbb{R}^n)$ such that $v(x,k) - \phi(x,k)$ has a strictly lomathcal Maximum at x_0 in a neighborhood $N(x_0)$. Choose $x_{kj}^\varepsilon \in N(x_0)$ such that for each $\alpha = s_{kj} \in \mathcal{M}_k$,

$$v^\varepsilon(x_{kj}^\varepsilon, s_{kj}) - \phi(x_{kj}^\varepsilon, k) = \max_{x\in N(x_0)} \{v^\varepsilon(x,s_{kj}) - \phi(x,k)\}.$$

Then it follows that $x^\varepsilon_{kj} \to x_0$ as $\varepsilon \to 0$. Given $u^{kj} \in \Gamma$, let

$$X^\varepsilon_{kj} = x^\varepsilon_{kj} + \varepsilon\big(A(s_{kj})u^{kj} + B(s_{kj})\big).$$

Then $X^\varepsilon_{kj} \in N(x_0)$ for ε small enough. We have

$$\gamma v^\varepsilon(x^\varepsilon_{kj}, s_{kj}) \le G(x^\varepsilon_{kj}, s_{kj}, u) + \left(\frac{1-\gamma\varepsilon}{\varepsilon}\right)\big(v(X^\varepsilon_{kj}, s_{kj}) - v^\varepsilon(x^\varepsilon_{kj}, s_{kj})\big)$$
$$\times \left(\frac{1-\gamma\varepsilon}{\varepsilon}\right) \sum_{\beta\in\mathcal{M}} (p^\varepsilon_{s_{kj}\beta} - \delta_{s_{kj}\beta}) v^\varepsilon(X^\varepsilon_{kj}, \beta). \quad (3.16)$$

Recall the definition of x^ε_{kj}. We have

$$v^\varepsilon(X^\varepsilon_{kj}, s_{kj}) - \phi(X^\varepsilon_{kj}, k) \le v^\varepsilon(x^\varepsilon_{kj}, s_{kj}) - \phi(x^\varepsilon_{kj}, k).$$

In addition, recall that $v^\varepsilon \to v$ and $x^\varepsilon_{kj} \to x_0$. It follows that as $\varepsilon \to 0$,

$$\left(\frac{1-\beta\varepsilon}{\varepsilon}\right) \sum_{j=1}^{m_k} \nu^k_j \big(v^\varepsilon(X^\varepsilon_{kj}, s_{kj}) - v^\varepsilon(x^\varepsilon_{kj}, s_{kj})\big)$$
$$\le \left(\frac{1-\beta\varepsilon}{\varepsilon}\right) \sum_{j=1}^{m_k} \nu^k_j \left(\phi(X^\varepsilon_{kj}, k) - \phi(x^\varepsilon_{kj}, k)\right)$$
$$\to \frac{\partial\phi(x_0,k)}{\partial x} \cdot \sum_{j=1}^{m_k} \nu^k_j \big(A(s_{kj})u^{kj} + B(s_{kj})\big).$$

Note also that

$$v^\varepsilon(X^\varepsilon_{kj}, s_{kj'}) - \phi(X^\varepsilon_{kj}, k) \le v^\varepsilon(x^\varepsilon_{kj'}, s_{kj'}) - \phi(x^\varepsilon_{kj'}, k).$$

We have

$$\sum_{j=1}^{m_k} \nu^k_j \sum_{j'=1}^{m_k} (p^k_{s_{kj}s_{kj'}} - \delta_{s_{kj}s_{kj'}}) v^\varepsilon(X^\varepsilon_{kj}, s_{kj'})$$
$$= \sum_{j=1}^{m_k} \nu^k_j \left(\sum_{j'=1}^{m_k} p^k_{s_{kj}s_{kj'}} v^\varepsilon(X^\varepsilon_{kj}, s_{kj'}) - v^\varepsilon(X^\varepsilon_{kj}, s_{kj}) \right)$$
$$\le \sum_{j=1}^{m_k} \nu^k_j \sum_{j'=1}^{m_k} p^k_{s_{kj}s_{kj'}} \Big(\big(v^\varepsilon(x^\varepsilon_{kj'}, s_{kj'}) - \phi(x^\varepsilon_{kj'}, k)\big)$$
$$- \big(v^\varepsilon(X^\varepsilon_{kj}, s_{kj}) - \phi(X^\varepsilon_{kj}, k)\big)\Big) = 0.$$

Write (3.16) in vector form, multiply both sides by

$$\nu = \begin{pmatrix} \nu^1 & & & \\ & \ddots & & \\ & & \nu^l & \\ 0_{m_* \times m_1} & \cdots & 0_{m_* \times m_l} & 0_{m_* \times m_*} \end{pmatrix},$$

and use Lemma 3.3 to obtain that $v(x, k)$ is a viscosity subsolution to (3.13).

Similarly, v is also a viscosity supersolution to (3.13). Moreover, the uniqueness of solution of (3.13) (see Yin and Zhang, 1998, p. 311) implies $v(x, k)$ the value for $\mathcal{P}^0$. Thus, for any subsequence of ε (indexed also by ε), $v^\varepsilon(x, \alpha) \to v^0(x, k)$. The desired result thus follows. □

6. Asymptotic optimality

In this section, we consider the asymptotic optimality of our approximation scheme.

Let

$$U^o(x) = \left(U^{o,1}(x), \ldots, U^{o,l}(x)\right)$$

with $U^{o,k}(x) = \left(u^{o,k1}(x), \ldots, u^{o,km_k}(x)\right)$ denote an optimal control for the limit problem $\mathcal{P}^0$.

Pick $u^* \in \Gamma$. Construct a control for the original discrete-time control problem

$$u_n^\varepsilon = \sum_{i=1}^{l} \sum_{j=1}^{m_k} I_{\{\alpha_n^\varepsilon = s_{ij}\}} u^{o,ij}(x_n) + \sum_{j=1}^{m_*} I_{\{\alpha_n^\varepsilon = s_{*j}\}} u^*. \tag{3.17}$$

Let $J^\varepsilon(x, \alpha)$ denote the cost under this control, i.e.,

$$J^\varepsilon(x, \alpha) = J^\varepsilon(x, \alpha, u^\varepsilon).$$

LEMMA 3.4 *Assume that $U^o(x)$ is Lipschitz. Then $J^\varepsilon(x, \alpha)$ is uniformly continuous.*

Proof. Given x^1 and x^2 in $\mathbb{R}^{n_1}$, let x_n^i denote the corresponding states with $x_0^i = x^i$, for $i = 1, 2$, i.e.,

$$\begin{aligned} x_{n+1}^i &= x_n^i \\ &+ \varepsilon \left\{ A(\alpha_n) \left(\sum_{i=1}^{l} \sum_{j=1}^{m_k} I_{\{\alpha_n^\varepsilon = s_{ij}\}} u^{o,ij}(x_n^i) + \sum_{j=1}^{m_*} I_{\{\alpha_n^\varepsilon = s_{*j}\}} u^* \right) + B(\alpha_n) \right\}, \end{aligned}$$

with $x_0^i = x^i$. It follows that

$$\begin{aligned} x_{n+1}^1 - x_{n+1}^2 &= x_n^1 - x_n^2 \\ &\quad + \varepsilon A(\alpha_n)\left(\sum_{i=1}^{l}\sum_{j=1}^{m_k} I_{\{\alpha_n^\varepsilon = s_{ij}\}}\big(u^{o,ij}(x_n^1) - u^{o,ij}(x_n^2)\big) \right). \end{aligned}$$

Let $\phi_n(x) = |x_n^1 - x_n^2|$. Then,

$$\phi_{n+1} \le \phi_n + K\varepsilon\phi_n = (1 + K\varepsilon)\phi_n,$$

with $\phi_0 = |x^1 - x^2|$. This implies that

$$\phi_n \le (1 + K\varepsilon)^n |x^1 - x^2|.$$

Recall that G is bounded. We have

$$\begin{aligned} &|J^\varepsilon(x^1,\alpha) - J^\varepsilon(x^2,\alpha)| \\ &\quad \le E \sum_{n=0}^{\lfloor T/\varepsilon \rfloor - 1} (1-\beta\varepsilon)^n \varepsilon (1+K\varepsilon)^n |x^1 - x^2| + \sum_{n=\lfloor T/\varepsilon \rfloor}^{\infty} (1-\beta\varepsilon)^n \varepsilon K \\ &\qquad\qquad = |x^1 - x^2| O(e^{(K-\beta)T}) + O(e^{-\gamma T}). \end{aligned}$$

For any $\eta > 0$, choose T large enough such that $O(e^{-\gamma T}) < \eta/2$ and $\delta = \eta O(e^{(K-\beta)T})/2$. Then whenever $|x^1 - x^2| < \delta$, we have

$$|J^\varepsilon(x^1,\alpha) - J^\varepsilon(x^2,\alpha)| < \eta. \qquad \square$$

REMARK 8 The boundedness assumption of G can be relaxed to the condition that G has at most polynomial growth in x.

REMARK 9 Typically, the Lipschitz condition on $U^o(x)$ can be obtained when G is smooth and strictly convex; see (Yin and Zhang, 1998, p. 254).

THEOREM 3.2 *The control u^ε is asymptotically optimal for the original control system* (3.1) *and* (3.2) *in the sense that*

$$\lim_{\varepsilon \to 0} |J^\varepsilon(x,\alpha,u^\varepsilon) - v^\varepsilon(x,\alpha)| = 0, \quad \textit{for all } \alpha \in \mathcal{M}. \tag{3.18}$$

Proof. Similarly as in (3.3), we can show (see Bertsekas, 1987) that $J^\varepsilon(x,\alpha)$ satisfies

$$\begin{aligned} J^\varepsilon(x, s_{ij}) &= \varepsilon G\big(x, s_{ij}, u^{o,ij}(x)\big) \\ &\quad + (1-\gamma\varepsilon) \sum_{\beta\in\mathcal{M}} p^\varepsilon_{s_{ij}\beta} J^\varepsilon\big(x + \varepsilon\big(A(s_{ij})u^{o,ij}(x) + B(s_{ij})\big), \beta\big). \end{aligned}$$

$$J^\varepsilon(x, s_{*j}) = \varepsilon G(x, s_{*j}, u^*) \\ + (1-\gamma\varepsilon) \sum_{\beta \in \mathcal{M}} p^\varepsilon_{s_{*j}\beta} J^\varepsilon\big(x + \varepsilon\big(A(s_{*j})u^* + B(s_{*j})\big), \beta\big).$$

As in Lemma 3.3, we can show that if $v^\varepsilon(x, \alpha) \to v^0(x, \alpha)$ for some sequence of ε, then the limit is only k dependent for $\alpha \in \mathcal{M}_k$. It can be shown similarly as in the proof of Theorem 3.1, together with Lemma 3.4, that $J^\varepsilon(x, \alpha)$ convergence to $v^0(x, k)$, for $\alpha \in \mathcal{M}_k$. Therefore,

$$|J^\varepsilon(x,\alpha) - v^\varepsilon(x,\alpha)| \le |J^\varepsilon(x,\alpha) - v^0(x,k)| + |v^0(x,k) - v^\varepsilon(x,\alpha)| \to 0. \quad \square$$

Finally in this section, we give a technical lemma needed in the example section.

LEMMA 3.5 *Let $\{f_n(x)\}$ denote a sequence of functions on a compact subset of $\mathbb{R}^{n_1}$. Assume $\{f_n(x)\}$ to be uniformly bounded and there exists a constant K such that*

$$|f_n(x) - f_n(y)| \le K\left(\frac{1}{n} + |x - y|\right), \quad \text{for all } x, y.$$

Then there exists a uniformly convergent subsequence.

Proof. The proof is a slight variation of that given in (Strichartz, 1995, p. 312,) with the equicontinuous condition replaced by the near Lipschitz condition. $\square$

7. Examples

We continue our study of Example 3.1. Our objective is to choose a control u to minimize the surplus costs

$$J^\varepsilon(x, \alpha, u) = E \sum_{n=0}^{\infty} (1 - \beta\varepsilon)^n \varepsilon \big(c^+(x_n)^+ + c^-(x_n)^-\big),$$

where c^+ and c^- are positive constants, $x^+ = \max\{0, x\}$, and $x^- = \max\{0, -x\}$.

Consider the case in which the demand fluctuates more rapidly than the capacity process. In this case, z^ε_n is the fast changing process, and $c^\varepsilon_n = c_n$ is the slowly varying capacity process being independent of ε. The idea is to derive a limit problem in which the fast fluctuating demand is replaced by its average. Thus one may ignore the detailed changes in the demand when making an average production planning decision.

Let

$$\mathcal{M} = \{s_{11}, s_{12}, s_{21}, s_{22}\} = \{(1, z_1), (1, z_2), (0, z_1), (0, z_2)\}.$$

Consider the transition matrix P^ε given by

$$P^\varepsilon = \begin{pmatrix} 1-\lambda_z & \lambda_z & 0 & 0 \\ \mu_z & 1-\mu_z & 0 & 0 \\ 0 & 0 & 1-\lambda_z & \lambda_z \\ 0 & 0 & \mu_z & 1-\mu_z \end{pmatrix} + \begin{pmatrix} -\lambda_c & 0 & \lambda_c & 0 \\ 0 & -\lambda_c & 0 & \lambda_c \\ \mu_c & 0 & -\mu_c & 0 \\ 0 & \mu_c & 0 & -\mu_c \end{pmatrix},$$

where $0 < \lambda_z < 1$ is the jump rate of the demand from z^1 to z^2 and $0 < \mu_z < 1$ is the rate from z^2 to z^1; λ_c and μ_c are the breakdown and repair rates, respectively.

In this example,

$$Q_* = \begin{pmatrix} -\lambda_c & \lambda_c \\ \mu_c & -\mu_c \end{pmatrix}.$$

Moreover, the control set for the limit problem

$$\{(u^{11}, u^{12}, 0, 0) : 0 \le u^{11}, u^{12} \le 1\},$$

since when $c_n = 0$ the system is independent of the values of u^{21} and u^{22}. Furthermore, since G is independent of u, we have $G = G$. Therefore, the system of equations in the limit problem $\mathcal{P}^0$ is given by

$$\frac{dx(t)}{dt} = c(t)u(t) - \bar{z}, \quad x(0) = x,$$

where $z = \nu_1^1 z_1 + \nu_2^1 z_2$ with

$$(\nu_1^1, \nu_2^1) = \left(\frac{\mu_z}{\lambda_z + \mu_z}, \frac{\lambda_z}{\lambda_z + \mu_z} \right),$$

and $\bar{\alpha}(t) = c(t)$ is a Markov chain generated by Q_*. Theorem 3.1 implies that $v^\varepsilon(x, \alpha) \to v^0(x, k)$, for $\alpha \in \mathcal{M}_k$, $k = 1, 2$.

Let

$$A_1 = \begin{pmatrix} -\frac{\rho+\mu_c}{\bar{z}} & \frac{\mu_c}{\bar{z}} \\ -\frac{\lambda_c}{1-\bar{z}} & \frac{\rho+\lambda_c}{1-\bar{z}} \end{pmatrix}.$$

It is easy to see that A_1 has two real eigenvalues, one greater than 0 and the other less than 0. Let $a_- < 0$ denote the negative eigenvalue of the matrix A_1 and define

$$\tilde{x} = \max\left(0, \frac{1}{a_-} \log\left[\frac{c^+}{c^+ + c^-} \left(1 + \frac{\rho \bar{z}}{\lambda_c \bar{z} - (\rho + \mu_c + \bar{z} a_-)(1 - \bar{z})} \right) \right] \right).$$

The optimal control for $\mathcal{P}^0$ is given by

$$\text{If } c(t)=0, \quad u^{\text{o}}(x)=0, \quad \textit{and} \quad \text{if } c(t)=1,$$

$$u^{\text{o}}(x)=\begin{cases} 0, & \text{if } x>\tilde{x}, \\ \bar{z}, & \text{if } x=\tilde{x}, \\ 1, & \text{if } x<\tilde{x}. \end{cases}$$

Let

$$U^{\text{o}}(x)=\big(u^{\text{o},11}(x), u^{\text{o},12}(x), u^{\text{o},21}(x), u^{\text{o},22}(x)\big)$$

denote the optimal control for $\mathcal{P}^0$. Note that $\big(u^{\text{o},11}(x),\ u^{\text{o},12}(x)\big)$ corresponds to $c(t)=1$ and $\big(u^{\text{o},21}(x), u^{\text{o},22}(x)\big)$ corresponds to $c(t)=0$. Naturally, $\big(u^{\text{o},21}(x), u^{\text{o},22}(x)\big)=0$, since, when $c(t)=0$, there should be no production. When $c(t)=1$, let $\nu_1^1 u^{\text{o},11}(x)+\nu_2^1 u^{\text{o},12}(x)=u^{\text{o}}(x)$. It should be pointed out that in this case the solution $\big(u^{\text{o},11}(x), u^{\text{o},12}(x)\big)$ is not unique.

Using $u^{\text{o},11}(x)$ and $u^{\text{o},12}(x)$, we construct a control for $\mathcal{P}^0$ as

$$u_n^\varepsilon = u^\varepsilon(x_n, \alpha_n^\varepsilon) = u^\varepsilon(x_n, c_n^\varepsilon, z_n^\varepsilon),$$

where

$$\begin{aligned} u^\varepsilon(x,c,z) &= I_{\{c=1\}}\big(I_{\{z=z_1\}}u^{\text{o},11}(x)+I_{\{z=z_2\}}u^{\text{o},12}(x)\big) \\ &\qquad + I_{\{c=0\}}\big(I_{\{z=z_1\}}u^{\text{o},21}(x)+I_{\{z=z_2\}}u^{\text{o},22}(x)\big) \\ &= I_{\{c=1\}}\big(I_{\{z=z_1\}}u^{\text{o},11}(x)+I_{\{z=z_2\}}u^{\text{o},12}(x)\big). \end{aligned}$$

Note that in this example, the optimal control $U^{\text{o}}(x)$ is not Lipschitz. Therefore the conditions in Theorem 3.2 are not satisfied. However, noting that

$$|x_n^1 - x_n^2| = O\big(\sqrt{n}\varepsilon + |x^1 - x^2|\big),$$

which implies

$$|J^\varepsilon(x^1,\alpha) - J^\varepsilon(x^2,\alpha)| = O(\varepsilon^{1/4} + |x^1 - x^2|),$$

we can still show, as in Theorem 3.2 using Lemma 3.5 that the constructed control u_n^ε in (3.17) is asymptotically optimal.

One may also consider the case in which the capacity process changes rapidly, whereas the random demand is relatively slowly varying. Similar to the previous case, assume c_n^ε is the capacity process and $z_n^\varepsilon = z_n$ is the demand. Using exactly the same approach, one may resolve this problem. The discussion is analogous to the previous case; the details are omitted.

8. Conclusions

This paper focuses on approximation schemes for a class of discrete-time production planning systems. It provides a systematic approach to reduce the complexity of the underlying systems. The computation load is reduced considerably compared with the optimal solution to the original problem. This is the most attractive feature of our approach. Furthermore, the asymptotic optimality ensures that such an approximation is almost as good as the optimal one for sufficiently small ε.

Acknowledgments

This research of was supported in part by the National Science Foundation.

References

Abbad, M., Filar, J.A., and Bielecki, T.R. (1992). Algorithms for singularly perturbed limiting average Markov control problems. *IEEE Transactions on Automatic Control*, AC-37:1421–1425.

Akella, R. and Kumar, P.R. (1986). Optimal control of production rate in a failure-prone manufacturing system. *IEEE Transactions on Automatic Control*, AC-31:116–126.

Bertsekas, D. (1987). *Dynamic Programming: Deterministic and Stochastic Models*. Prentice-Hall, Englewood Cliffs, NJ.

Bielecki, T.R. and Kumar, P.R. (1988). Optimality of zero-inventory policies for unreliable manufacturing systems. *Operations Research*, 36:532–541.

Boukas, E.K. (1991). Techniques for flow control and preventive maintenance in manufacturing systems. *Control and Dynamics*, 36:327–365.

Boukas, E.K. and Haurie, A. (1990). Manufacturing flow control and preventive maintenance: A stochastic control approach. *IEEE Transactions on Automatic Control*, AC-35:1024–1031.

Gershwin, S.B. (1994). *Manufacturing Systems Engineering*. Prentice Hall, Englewood Cliffs, NJ.

Iosifescu, M. (1980). *Finite Markov Processes and Their Applications*. Wiley, Chichester.

Liu, R.H., Zhang, Q., and Yin, G. (2001). Nearly optimal control of singularly perturbed Markov decision processes in discrete time. *Applied Mathematics and Optimization*, 44:105–129.

Pervozvanskii, A.A. and Gaitsgory, V.G. (1988). *Theory of Suboptimal Decisions: Decomposition and Aggregation.* Kluwer, Dordrecht.

Sethi, S.P. and Zhang, Q. (1994). *Hierarchical Decision Making in Stochastic Manufacturing Systems.* Birkhäuser, Boston.

Strichartz, R.S. (1995). *The Way of Analysis.* Jones and Bartlett Publishers, Boston.

Yin, G. and Zhang, Q. (1998). *Continuous-time Markov Chains and Applications: A Singular Perturbation Approach.* Springer-Verlag, New York.

Yin, G. and Zhang, Q. (2000). Singularly perturbed discrete-time markov chains, *SIAM Journal on Applied Mathematics*, 61:834–854.

Yin, G. and Zhang, Q. (2003). Discrete-time singularly perturbed Markov chains. In: D. Yao, H.Q. Zhang, and X.Y. Zhou (eds.), *Stochastic Models and Optimization*, pages 1–42, Springer-Verlag.

Yin, G., Zhang, Q., and Badowski, G. (2003). Discrete-time singularly perturbed Markov chains: Aggregation, occupation measures, and switching diffusion limit. *Advances in Applied Probability*, 35:449–476.

Chapter 4

EVALUATION OF THROUGHPUT IN SERIAL PRODUCTION LINES WITH NON-EXPONENTIAL MACHINES

Jingshan Li
Semyon M. Meerkov

Abstract This paper provides an analytical method for evaluating production rates in serial lines having finite buffers and unreliable machines with arbitrary unimodal distributions of up- and downtime. Provided that each buffer is capable of accommodating at least one downtime of all machines in the system, we show that the production rate (a) is relatively insensitive to the type of up- and downtime distributions and (b) can be approximated by a linear function of their coefficients of variation. The results obtained are verified using Weibull, gamma, and log-normal probability distributions of up- and downtime.

1. Introduction

Analytical methods for evaluating throughput in serial production lines are available only if the up- and downtime of machines obey either exponential (in discrete time, geometric) or coaxial (phase type) probability distributions (see reviews by Koenigsberg, 1959; Buxey et al., 1973; Buzacott and Hanifin, 1978; Dallery and Gershwin, 1992; Papadopoulos and Heavey, 1996, monographs by Viswanadham and Narahari (1992); Buzacott and Shanthikumar (1993); Gershwin (1994); Altiok (1997) and representative papers by Sevast'yanov (1962); Buzacott (1967); Sheskin (1976); Soytser et al. (1979); Wijingaard (1979); Gershwin and Berman (1981); Altiok (1985, 1989); Buzacott and Kotelski (1987); Choong and Gershwin (1987); Gershwin (1987); Jafari and Shanthikumar (1987); De Koster (1987, 1988); Terracol and David (1987); Dallery et al. (1988, 1989); Altiok and Ranjan (1989); Liu and Buzacott (1989); Lim et al. (1990); Hiller and So (1991a,b); Glassey and Hong (1993); Powell (1994);

Jacobs and Meerkov (1995a,b); Tan and Yeralan (1997); Chiang et al. (1998, 2000, 2001); Yamshita and Altiok (1998); Dallery and Le Bihan (1999); Vidalis and Papdopoulos (1999); Tempelmeier and Burger (2001); Enginarlar et al. (2002); Sadr and Malhame (2003); Tempelmeier (2003)). The present paper is intended to offer an analytical method for calculating throughput in serial lines with machines having arbitrary unimodal distributions of up- and downtime, provided that each buffer is capable of accommodating at least one downtime of all machines in the system. Specifically, we show that the production rate, PR, of a serial line (i.e., the average number of parts produced by the last machine per unit of time) can be evaluated as follows:

$$PR = e_{\min} - (e_{\min} - PR^{\exp}) \sum_{i=1}^{M} \frac{CV_{up,i} + CV_{down,i}}{2M},$$
$$CV_{up,i} \in [0,1], \quad CV_{down,i} \in [0,1], \tag{4.1}$$

where $PR^{\exp}$ is the production rate of the line if all machines were exponential, $CV_{up,i}$ and $CV_{down,i}$ are the coefficients of variation of up- and downtime of the ith machine, $i = 1, \ldots, M$, and $e_{\min}$ is the smallest efficiency in isolation among all the machines in the system, i.e.,

$e_{\min} = \min_{i=1,\ldots,M} e_i = \min_{i=1,\ldots,M} \frac{T_{up,i}}{T_{up,i}+T_{down,i}}$,

$T_{up,i} =$ average uptime of machine i,

$T_{down,i} =$ average downtime of machine i,

$CV_{up,i} = \frac{\sigma_{up,i}}{T_{up,i}}$, $CV_{down,i} = \frac{\sigma_{down,i}}{T_{down,i}}$,

$\sigma_{up,i} =$ standard deviation of uptime of machine i,

$\sigma_{down,i} =$ standard deviation of downtime of machine i,

$M =$ number of machines in the system.

Using Weibull, gamma, and log-normal probability distributions, we show that the accuracy of this method is within 6%. Along with providing a quantitative result, expression (4.1) indicates that the production rate depends mostly on the first two moments of up- and downtime, rather than on complete distributions of these random variables.

The CVs considered in this paper are less than 1 because, according to the empirical evidence of Inman (1999), the equipment on the factory floor often satisfies this condition. In addition, it has been shown in Li and Meerkov (2003) that CVs are less than 1 if the breakdown and

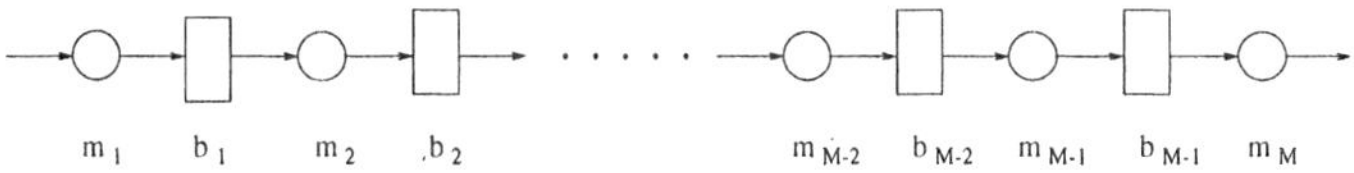

Figure 4.1. Serial production line.

repair rates of the machines are increasing functions of time, which often take place in reality.

The outline of this paper is as follows: In Section 2, the model of the production system under consideration is introduced and the problems addressed are formulated. Sections 3 and 4 introduce some analytical expressions and describe the approach of the study, respectively. Sections 5–7 present the main results, and in Section 8 the conclusions are formulated. The proofs are given in the Appendix.

2. Model and problem formulation

2.1 Model

The block diagram of the production system considered in this work is shown in Figure 4.1, where the circles represent the machines and the rectangles are the buffers. Assumptions on the machines and buffers are as follows:

(i) Each machine m_i, $i = 1, \dots, M$, has two states: up and down. When up, the machine is capable of processing one part per cycle time; when down, no production takes place. The cycle times of all machines are the same.

(ii) The up- and downtime of each machine are continuous random variables, $t_{\text{up},i}$ and $t_{\text{down},i}$, $i = 1, \dots, M$, with arbitrary unimodal probability density functions, $f_{t_{\text{up},i}}(t)$ and $f_{t_{\text{down},i}}(t)$, $t \geq 0$, $i = 1, \dots, M$, respectively. It is assumed that these random variables are mutually independent. For convenience, it is also assumed that the up- and downtime are measured in units of the cycle time. In other words, uptime (respectively, downtime) of length $t \geq 0$ implies that the machine is up (respectively, down) during t cycle times.

(iii) Buffer b_i, $i = 1, \dots, M-1$, is of capacity N_i such that $\max_{i=1,\dots,M} T_{\text{down},i} \leq N_i < \infty$, where $T_{\text{down},i}$ is the average downtime of machine m_i. In other words, it is assumed that

$$k_i = \frac{N_i}{\max_{i=1,\dots,M} T_{\text{down},i}} \geq 1, \qquad (4.2)$$

where parameter k_i is referred to as the level of buffering.

(iv) Machine m_i, $i = 2, \ldots, M$, is starved at time t if it is up at time t, buffer b_{i-1} is empty at time t and m_{i-1} does not place any work in this buffer at time t. Machine m_1 is never starved.

(v) Machine m_i, $i = 1, \ldots, M-1$, is blocked at time t if it is up at time t, buffer b_i is full at time t and m_{i+1} fails to take any work from this buffer at time t. Machine m_M is never blocked.

The production rate, PR, of the serial line (i)–(v) is the average number of parts produced by the last machine, m_M, per cycle time. As it was pointed out above, no analytical method for its evaluation are available in the literature, except for exponential and coaxial distributions of up- and downtime of the machines.

2.2 Notations

Each machine considered in this paper is denoted by a pair

$$[f_{t_{\text{up},i}}, f_{t_{\text{down},i}}], \quad i = 1, \ldots, M, \tag{4.3}$$

where, as before, $f_{t_{\text{up},i}}$ and $f_{t_{\text{down},i}}$ are the probability density functions of up- and downtime of machine i, respectively. The serial line with M machines is denoted as

$$\{[f_{t_{\text{up},1}}, f_{t_{\text{down},1}}], \ldots, [f_{t_{\text{up},M}}, f_{t_{\text{down},M}}]\}. \tag{4.4}$$

If all machines have identical distributions of up- and downtime, the notation for the line is:

$$\{[f_{t_{\text{up}}}, f_{t_{\text{down}}}]_i, i = 1, \ldots, M\}. \tag{4.5}$$

2.3 Problems addressed

Using the model (i)–(v) and notations (4.4), (4.5), this paper is intended to:

- Develop an analytical method for calculating the production rate in serial lines (4.4) and (4.5) under the assumption that the average uptime and downtime of all machines are identical and, in addition, the coefficients of variation of uptime and downtime of all machines are the same and, moreover, equal to each other, i.e.,

$$\begin{aligned} T_{\text{up},i} = T_{\text{up}}, \quad T_{\text{down},i} = T_{\text{down}}, \qquad & i = 1, \ldots, M, \\ CV_{\text{up},i} = CV_{\text{down},i} = CV, \quad & i = 1, \ldots, M. \end{aligned} \tag{4.6}$$

This is referred to as the case of *identical machines.* Note that the machines may have different distributions of up- and downtime but are identical in the sense (2.3).

- Extend this method to the case where $T_{\text{up},i}$ and $T_{\text{down},i}$ are arbitrary and the coefficients of variation of uptime and downtime of all machines are the same but may be nonequal to each other, i.e.,

$$\text{CV}_{\text{up},i} = \text{CV}_{\text{up}}, \quad \text{CV}_{\text{down},i} = \text{CV}_{\text{down}}, \qquad i = 1, \dots, M, \quad (4.7)$$

  and, in general,

$$\text{CV}_{\text{up}} \neq \text{CV}_{\text{down}}.$$

  This is referred to as the case of *identical coefficients of variation.*

- Finally, extend this method to the case where all $T_{\text{up},i}$, $T_{\text{down},i}$, $\text{CV}_{\text{up},i}$ and $\text{CV}_{\text{down},i}$, $i = 1, \dots, M$, are arbitrary. This is referred to as the *general case.*

3. Analytical expressions

3.1 Production rate for $\text{CV}_{\text{up}} = \text{CV}_{\text{down}} = 0$

In the case of $\text{CV}_{\text{up}} = \text{CV}_{\text{down}} = 0$, the production rate of the line (i)–(v) can be evaluated as follows:

THEOREM 4.1 *Consider a serial production line defined by assumption* (i)–(v) *and assume that* $\text{CV}_{\text{up}} = \text{CV}_{\text{down}} = 0$. *Then its production rate is given by*

$$\text{PR} = \min_{i=1,\dots,M} \frac{T_{\text{up},i}}{T_{\text{up},i} + T_{\text{down},i}},$$

i.e., the PR *of the line is equal to the smallest efficiency in isolation among all machines in the system.*

Proof. See the Appendix. □

It should be pointed out that the main reason why Theorem 4.1 holds is that the level of buffering $k_i \geq 1$, $\forall i = 1, \dots, M-1$.

3.2 Production rate for $\text{CV}_{\text{up}} = \text{CV}_{\text{down}} = 1$

Assume that all machines have up- and downtime distributed exponentially and, therefore, $\text{CV}_{\text{up}} = \text{CV}_{\text{down}} = 1$. As it was pointed above, PR in serial line (i)–(v) with exponential machines can be evaluated using a number of analytical techniques instance, (see, for Gershwin, 1987; De Koster, 1987, 1988; Dallery et al., 1988, 1989; Chiang et al., 2000, 2001; Sadr and Malhame, 2003).

Although all of them are relatively precise, each has a certain error in comparison with the real production rate (which can be obtained, for example, by numerical simulations). Because of this error, to determine the accuracy of the method developed in this paper, we evaluate PR of serial lines with exponential machines using simulations, rather than analytical calculations. We denote this production rate as $\mathrm{PR}^{\mathrm{exp}}$.

3.3 Production rate for $0 < \mathrm{CV}_{\mathrm{up}}, \mathrm{CV}_{\mathrm{down}} < 1$

Based on the above, PR of serial lines with CV = 0 and CV = 1 can be easily evaluated. For all other CVs, we formulate

HYPOTHESIS 4.1 *In the case of identical machines* (2.3), *the production rate of serial lines* (i)–(v) *can be evaluated as follows*:

$$\mathrm{PR} = e - (e - \mathrm{PR}^{\mathrm{exp}})\mathrm{CV}, \tag{4.8}$$

where e is the machine efficiency in isolation, i.e., $e = \frac{T_{\mathrm{up}}}{T_{\mathrm{up}}+T_{\mathrm{down}}}$.

HYPOTHESIS 4.2 *In the case of identical coefficients of variation* (4.7), *the production rate of serial lines* (i)–(v) *can be evaluated as follows:*

$$\mathrm{PR} = e_{\mathrm{min}} - (e_{\mathrm{min}} - \mathrm{PR}^{\mathrm{exp}})\frac{\mathrm{CV}_{\mathrm{up}} + \mathrm{CV}_{\mathrm{down}}}{2}, \tag{4.9}$$

where $e_{\mathrm{min}} = \min_{i=1,\dots,M} e_i = \min_{i=1,\dots,M} \frac{T_{\mathrm{up},i}}{T_{\mathrm{up},i}+T_{\mathrm{down},i}}$.

HYPOTHESIS 4.3 *In the general case, the production rate of serial lines* (i)–(v) *can be evaluated as follows*:

$$\mathrm{PR} = e_{\mathrm{min}} - (e_{\mathrm{min}} - \mathrm{PR}^{\mathrm{exp}})\sum_{i=1}^{M}\frac{\mathrm{CV}_{\mathrm{up},i} + \mathrm{CV}_{\mathrm{down},i}}{2M}. \tag{4.10}$$

Verifications of these Hypotheses are given in Sections 5–7, while the approach to the verification is described in Section 4.

4. Approach

4.1 Distributions considered

For the verification of (4.8)–(4.10) we consider the following distributions:

(a) Weibull, i.e.,

$$\begin{aligned} f_{t_{\mathrm{up},i}}(t) &= p^P e^{-(pt)^P} P t^{P-1}, \\ f_{t_{\mathrm{down},i}}(t) &= r^R e^{-(rt)^R} R t^{R-1}. \end{aligned} \tag{4.11}$$

Here and in all subsequent distributions, (p, P) and (r, R) are positive real numbers. These distributions are denoted as $W(p, P)$ and $W(r, R)$, respectively.

(b) Gamma, i.e.,

$$f_{t_{\text{up},i}}(t) = pe^{-pt}\frac{(pt)^{P-1}}{\Gamma(P)}, \quad f_{t_{\text{down},i}}(t) = re^{-rt}\frac{(rt)^{R-1}}{\Gamma(R)}, \tag{4.12}$$

where $\Gamma(x)$ is the gamma function, $\Gamma(x) = \int_0^\infty s^{x-1}e^{-s}ds$. These distributions are denoted as $g(p, P)$ and $g(r, R)$, respectively.

(c) Log-normal, i.e.,

$$\begin{aligned} f_{t_{\text{up},i}}(t) &= \frac{1}{\sqrt{2\pi}Pt}e^{-\frac{(ln(t)-p)^2}{2P^2}}, \\ f_{t_{\text{down},i}}(t) &= \frac{1}{\sqrt{2\pi}Rt}e^{-\frac{(ln(t)-r)^2}{2R^2}}. \end{aligned} \tag{4.13}$$

We denote these distributions as $\text{LN}(p, P)$ and $\text{LN}(r, R)$, respectively.

Specific realizations of downtime distributions analyzed in this work are given in Table 4.1. They are classified according to their coefficients of variation, CV_{down}, which take values from the set {0.1,0.25,0.5,0.75,1}, and according to their average values, which are 10 and 20.

Table 4.1. Downtime distributions considered

CV_{down}	$T_{\text{down}} = 10$
0.1	$W(0.0959, 12.15)$, $g(10, 100)$, $\text{LN}(2.30, 0.1)$
0.25	$W(0.0913, 4.542)$, $g(1.6, 16)$, $\text{LN}(2.27, 0.25)$
0.5	$W(0.0886, 2.1013)$, $g(0.4, 4)$, $\text{LN}(2.19, 0.47)$
0.75	$W(0.0917, 1.3475)$, $g(0.18, 1.78)$, $\text{LN}(2.08, 0.67)$
1.00	$\text{LN}(1.96, 0.83)$

CV_{down}	$T_{\text{down}} = 20$
0.1	$W(0.0479, 12.15)$, $g(5, 100)$, $\text{LN}(2.99, 0.1)$
0.25	$W(0.0457, 4.542)$, $g(0.8, 16)$, $\text{LN}(2.97, 0.25)$
0.5	$W(0.0443, 2.1013)$, $g(0.2, 4)$, $\text{LN}(2.88, 0.47)$
0.75	$W(0.0459, 1.3475)$, $g(0.09, 1.78)$, $\text{LN}(2.77, 0.67)$
1.00	$\text{LN}(2.65, 0.83)$

The uptime distributions, corresponding to the downtime distributions of Table 4.1, have been selected as follows:

For the case of identical machines, given machine efficiency, e, the average uptime was chosen as

$$T_{\text{up}} = \frac{e}{1-e}T_{\text{down}}.$$

Next, $\mathrm{CV_{up}}$ was selected as $\mathrm{CV_{up}} = \mathrm{CV_{down}}$ and, using these T_{up} and $\mathrm{CV_{up}}$, the distribution of uptime was selected to be the same as that of the downtime, if the case of identical distributions was analyzed; otherwise it was selected randomly and equiprobably from the set $\{W, g, \mathrm{LN}\}$.

For the case of non-identical machines (4.7), the values of e_i, $T_{\mathrm{down},i}$ and the distributions of up- and downtime were selected randomly and equiprobably from the sets $\{0.55, 0.65, 0.75, 0.85, 0.9, 0.95\}$, $\{10, 20\}$ and $\{w, g, \mathrm{LN}\}$, respectively.

4.2 Evaluation of the production rate

To evaluate the production rate of serial lines (i)–(v) with up- and downtime distributed according to the distributions described above, a `MATLAB` code was constructed, which simulated the operation of the production line (i)–(v). In all simulation runs, zero initial conditions of all buffers have been assumed and the states of all machines at the initial time moment have been selected "up". The first 10,000 cycle times were considered the warm-up period. The subsequent 100,000 cycle times were used for statistical evaluation of PR. Each simulation was repeated 20 times, which resulted in 95% confidence intervals of less than 0.003.

4.3 Parameters selected

In all systems analyzed, particular values of M, e, and N have been selected as follows:

(a) The number of machines in the system, M: The number of machines in the system was selected to be 3, 5 and 10.

(b) Machine efficiency, e: Although in practice e may have widely different values (e.g., smaller in machining operations and much larger in assembly), to obtain a manageable set of systems, e was selected from the set $\{0.55, 0.65, 0.75, 0.85, 0.9, 0.95\}$.

(c) Level of buffering, k_i: the value of k_i was selected to be 1 ("small" buffer capacity) or 3 ("large" buffer capacity).

4.4 Systems analyzed

We consider two groups of systems. The first one consists of machines with identical types of up- and downtime distributions. For the case of identical machines, this group is given by

$$\begin{gathered}\{[W(p,P), W(r,R)]_i, i = 1, \ldots, 10\},\\ \{[g(p,P), g(r,R)]_i, i = 1, \ldots, 10\},\\ \{[\mathrm{LN}(p,P), \mathrm{LN}(r,R)]_i, i = 1, \ldots, 10\}.\end{gathered} \tag{4.14}$$

We use systems (4.14) in order to evaluate the sensitivity of PR to different distributions of up- and downtime.

For the case of identical coefficients of variation and for the general case, this group is denoted as

$$\begin{gathered}\{[W(p_1,P_1),W(r_1,R_1)],\ldots,[W(p_{10},P_{10}),W(r_{10},R_{10})]\},\\ \{[g(p_1,P_1),g(r_1,R_1)],\ldots,[g(p_{10},P_{10}),g(r_{10},R_{10})]\},\\ \{[\mathrm{LN}(p_1,P_1),\mathrm{LN}(r_1,R_1)],\ldots,[\mathrm{LN}(p_{10},P_{10}),\mathrm{LN}(r_{10},R_{10})]\}.\end{gathered} \tag{4.15}$$

The second group consists of machines with different up- and downtime distributions. These lines have been formed as follows: For each machine m_i, $i = 1,\ldots,M$, the up- and downtime distributions were chosen from the set $\{W, g, \mathrm{LN}\}$ equiprobably and independently of each other and all other machines in the system. As a result, the following lines have been selected:

$$\begin{aligned}\text{Line 1}: \{&[g,W],[\mathrm{LN},\mathrm{LN}],[W,g],[g,\mathrm{LN}],[g,W],[\mathrm{LN},g],\\ &[W,W],[g,g],[\mathrm{LN},W],[g,\mathrm{LN}]\}\\ \text{Line 2}: \{&[W,\mathrm{LN}],[g,W],[\mathrm{LN},W],[W,g],[g,\mathrm{LN}],[g,W],\\ &[W,W],[\mathrm{LN},g],[g,W],[\mathrm{LN},\mathrm{LN}]\}\end{aligned} \tag{4.16}$$

For $M = 3$ (respectively, $M = 5$), the first 3 (respectively, first 5) machines of lines (4.14)–(4.16) have been used.

We will employ the notations $A \in \{(4.14)\}$ or $A \in \{(4.14),(4.16)\}$ or $A \in \{(4.15),(4.16)\}$ to indicate, respectively, that line A is one of (4.14) or one of (4.14), (4.16) or one of (4.15), (4.16).

Specific parameters of the distributions involved in (4.14)–(4.16) are selected in a manner consistent with the problem analyzed; they are described in Subsections 5.1, 6.1 and 7.1.

4.5 Metrics for sensitivity and accuracy analysis

The analysis of the sensitivity of PR to the type of up- and downtime distribution is carried out using the following metric:

$$\varepsilon = \max_{A,B\in\{(4.14)\}} \frac{|\mathrm{PR}^A - \mathrm{PR}^B|}{\mathrm{PR}^A}\cdot 100\%, \tag{4.17}$$

where PR^A and PR^B are the production rates of systems from set (4.14) evaluated by simulations.

The accuracy of Hypothesis i, $i = 1,2,3$, is estimated using the following metrics:

$$\Delta_1 = \max_{A\in\{(4.14),(4.16)\}} \frac{|\mathrm{PR}^A - \mathrm{PR}_1|}{\mathrm{PR}^A}\cdot 100\%, \tag{4.18}$$

$$\Delta_i = \max_{A \in \{(4.15),(4.16)\}} \frac{|\mathrm{PR}^A - \mathrm{PR}_i|}{\mathrm{PR}^A} \cdot 100\%, \quad i = 2, 3, \tag{4.19}$$

where PR^A is, as before, the production rates of line from (4.14), (4.16) or (4.15), (4.16) evaluated by simulation and PR_i, $i = 1, 2, 3$, is the production rate calculated using Hypothesis i.

5. Production rate evaluation for the case of identical machines

5.1 Parameters of systems analyzed

Since in this case all machines have identical T_{up}, T_{down}, $\mathrm{CV}_{\mathrm{up}}$ and $\mathrm{CV}_{\mathrm{down}}$ and, moreover, $\mathrm{CV}_{\mathrm{up}} = \mathrm{CV}_{\mathrm{down}} = \mathrm{CV}$, the parameters of the systems analyzed coincide with those introduced in Section 4, i.e.,

$$\begin{aligned} \mathrm{CV} &\in \{0.1, 0.25, 0.5, 0.75, 1\}, \\ T_{\mathrm{down}} &\in \{10, 20\}, \\ e &\in \{0.55, 0.65, 0.75, 0.85, 0.9, 0.95\}, \\ M &\in \{3, 5, 10\}, \\ k &\in \{1, 3\}. \end{aligned}$$

Taking into account that these parameters have been used for all five systems (4.14), (4.16), this implies that the total of 1800 different production lines have been analyzed.

5.2 Results

Tables 4.2 and 4.3 present the production rates of serial lines (4.14) and (4.16), evaluated by simulations and by Hypothesis 4.1 (broken lines). In these Tables, the rows and columns correspond to $e \in \{0.55, 0.65, 0.75, 0.85, 0.9, 0.95\}$ and $M \in \{3, 5, 10\}$, respectively. Each entry of the Tables contains the data for $k = 1$ and $k = 3$. Based on these data, we conclude the following:

(a) The type of up- and downtime distributions does not affect PR in any significant manner. This phenomenon is quantified by the values of metric ε (calculated according to (4.17)) given in Tables 4.4 and 4.5. As one can see, they take values within 6%. In addition, Tables 4.4 and 4.5 exhibits qualitative effects of system parameters on the sensitivity of PR to distributions of up- and downtime. These effects can be summarized as follows:

$$\begin{aligned} \mathrm{CV} \uparrow &\Rightarrow \text{Sensitivity } \uparrow, \\ M \uparrow &\Rightarrow \text{Sensitivity } \uparrow, \end{aligned}$$

Table 4.2. Production rates evaluated by simulations and by Hypothesis 4.1: $T_{\text{down}} = 10$.

	M=3	M=5	M=10
e=0.55	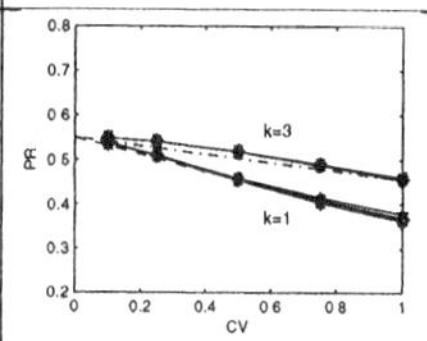	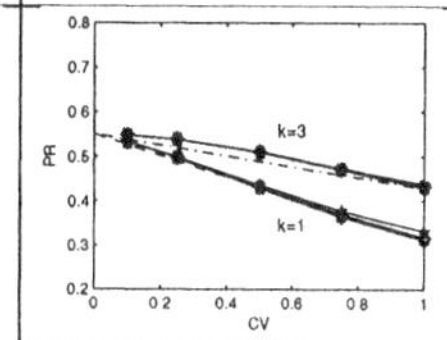	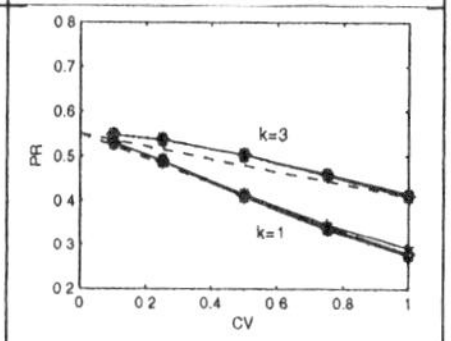
e=0.65	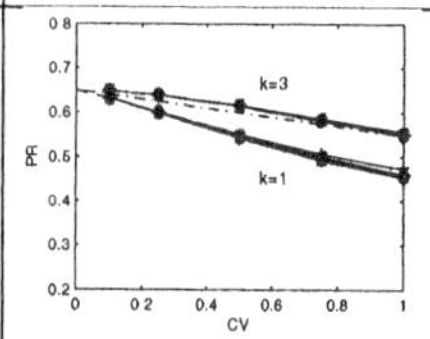	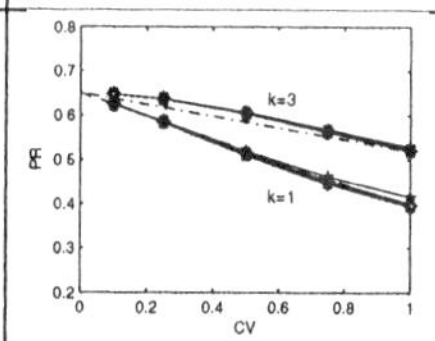	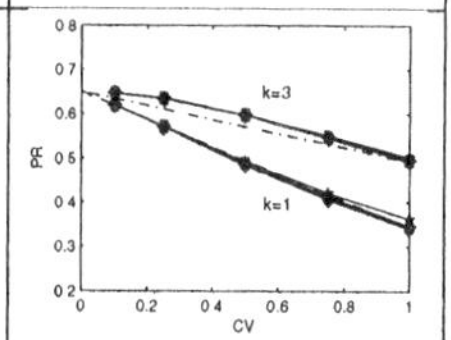
e=0.75	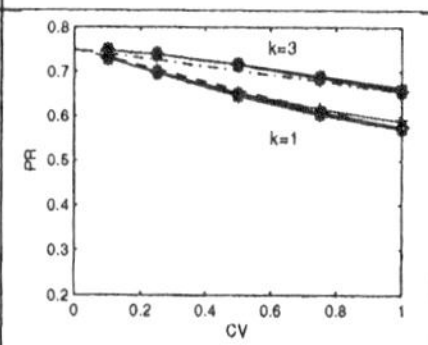	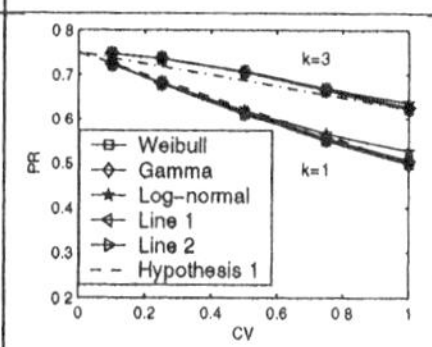	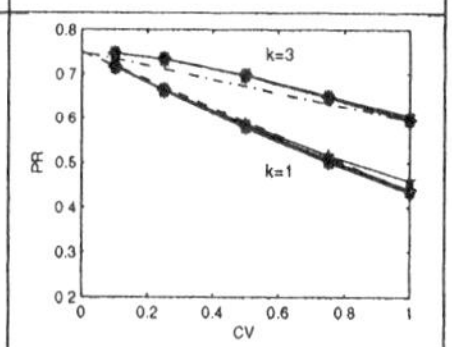
e=0.85	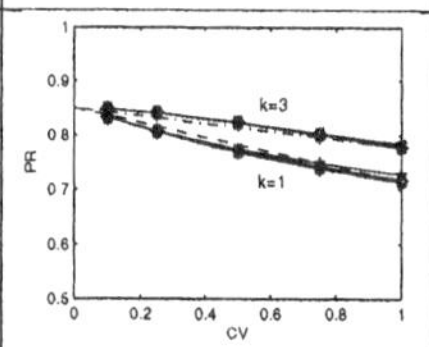	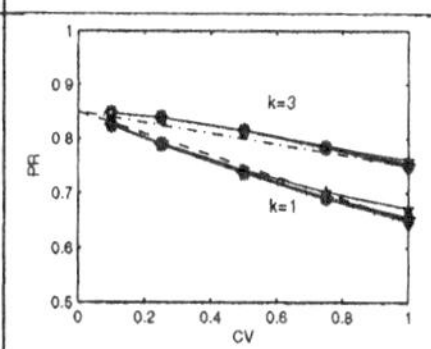	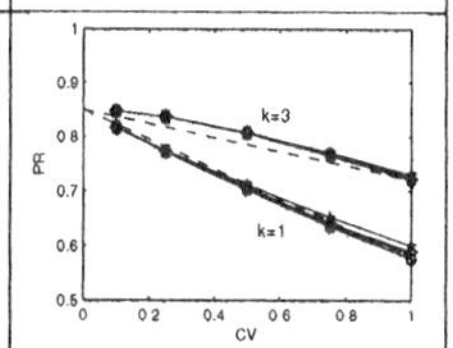
e=0.9	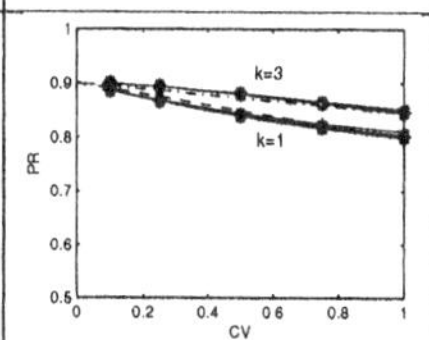	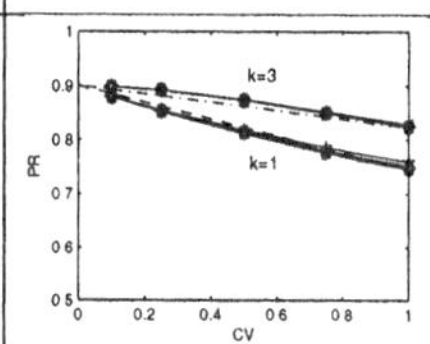	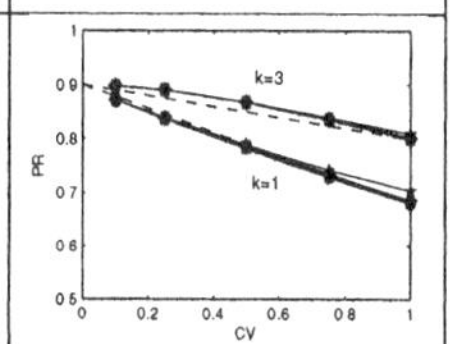
e=0.95	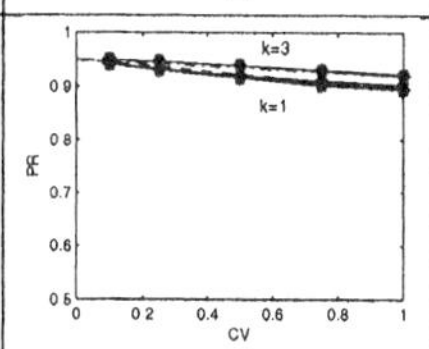	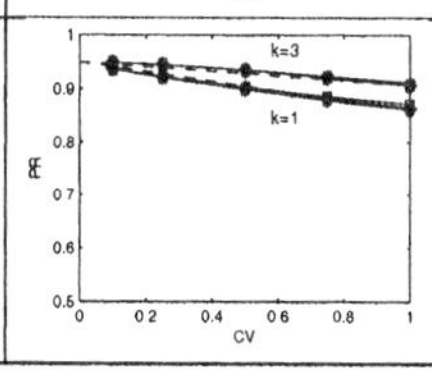	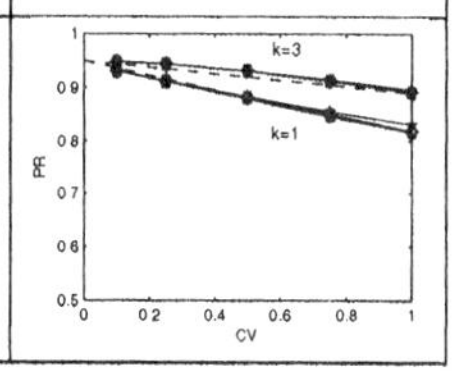

Table 4.3. Production rates evaluated by simulations and by Hypothesis 4.1: $T_{\text{down}} = 20$.

	M=3	M=5	M=10
e=0.55	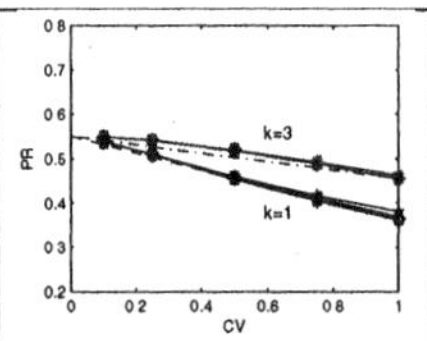	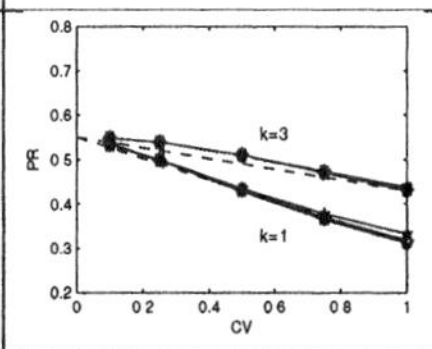	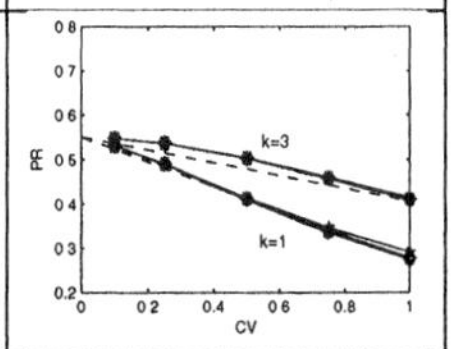
e=0.65	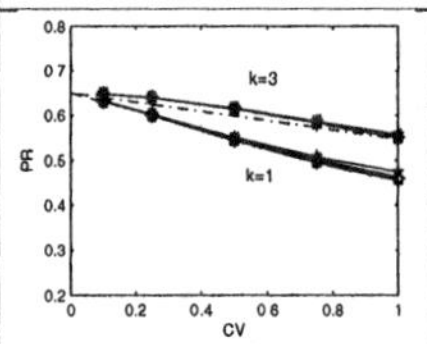	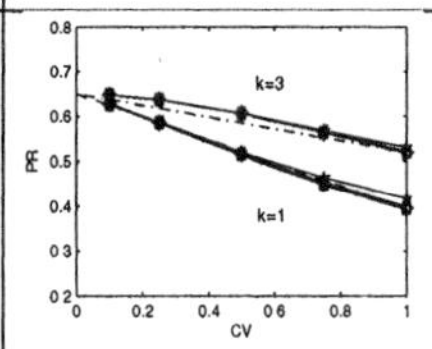	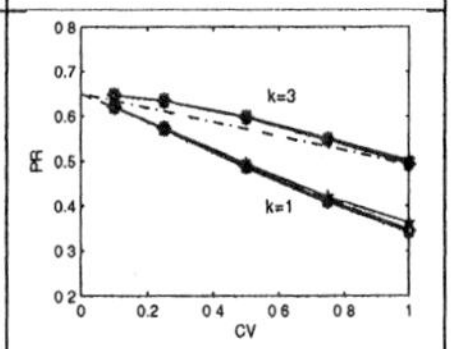
e=0.75	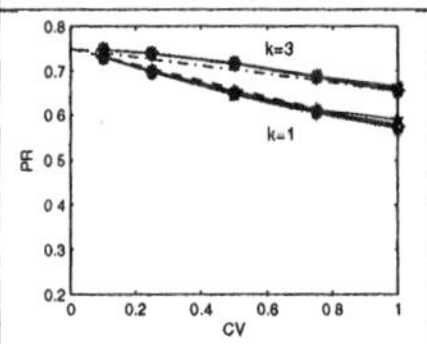	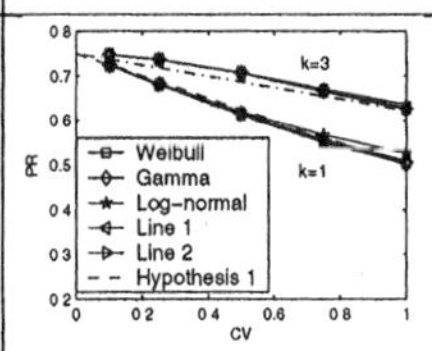	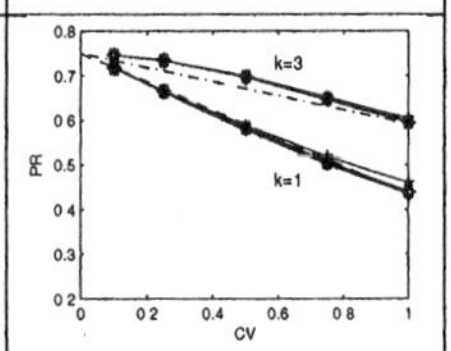
e=0.85	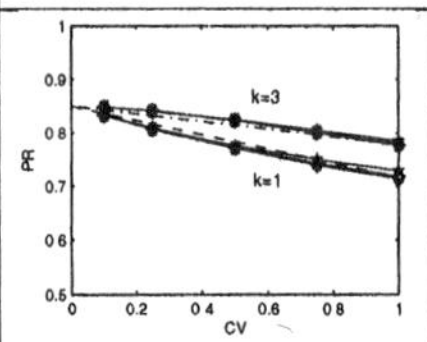	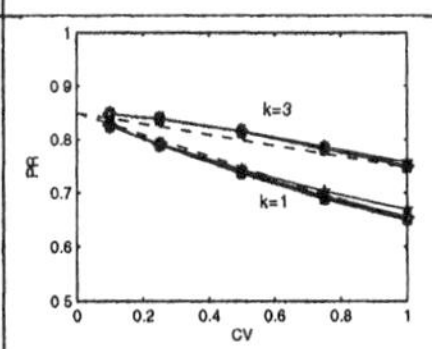	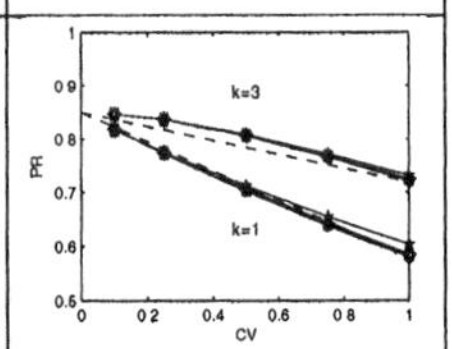
e=0.9	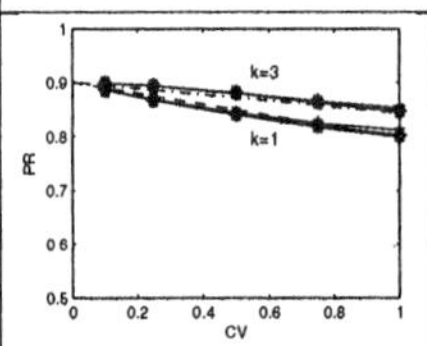	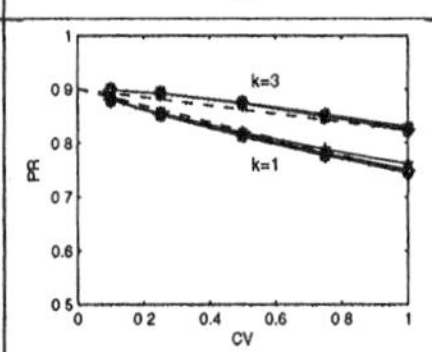	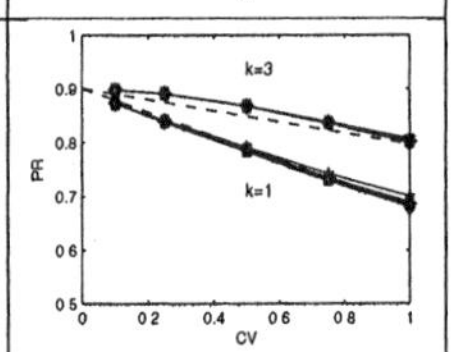
e=0.95	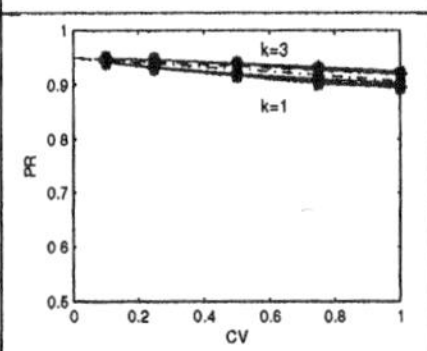	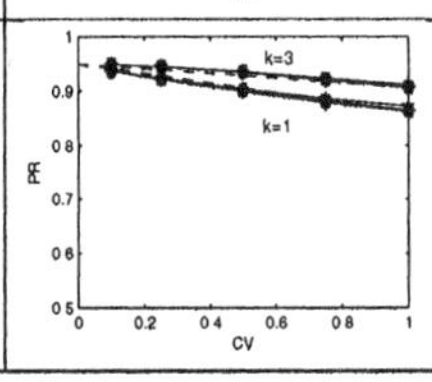	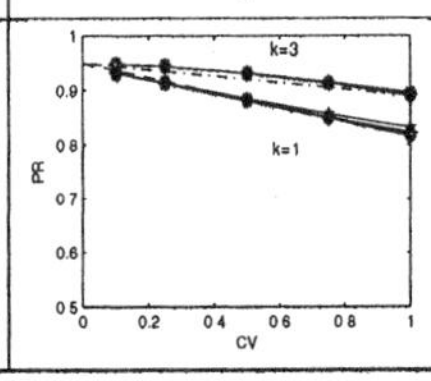

Table 4.4. Sensitivity of production rate to the types of up- and downtime distributions (the values of ε calculated according to (4.17)) (a). $T_{\text{down}} = 10$.

		CV = 0.1			CV = 0.25			CV = 0.5		
	M	3	5	10	3	5	10	3	5	10
$e = 0.55$	$k = 1$	0.06	0.06	0.02	0.22	0.30	0.33	1.10	1.73	1.40
	$k = 3$	0.05	0.02	0.02	0.04	0.02	0.04	0.17	0.22	0
$e = 0.65$	$k = 1$	0.05	0.03	0.03	0.35	0.36	0.33	1.59	1.86	1.65
	$k = 3$	0.02	0.03	0.02	0.05	0.05	0	0.31	0.38	0.15
$e = 0.75$	$k = 1$	0.07	0.01	0.06	0.33	0.28	0.33	1.13	1.47	1.68
	$k = 3$	0.04	0.03	0	0.03	0.01	0.01	0.20	0.18	0.24
$e = 0.85$	$k = 1$	0.02	0.05	0.09	0.17	0.34	0.39	0.78	0.77	1.28
	$k = 3$	0.01	0.01	0.01	0.06	0.05	0.05	0.15	0.05	0.24
$e = 0.9$	$k = 1$	0.03	0.02	0.01	0.15	0.28	0.33	0.56	0.81	0.87
	$k = 3$	0.02	0.02	0	0.03	0.09	0.03	0.16	0.08	0.08
$e = 0.95$	$k = 1$	0.02	0.01	0.01	0.06	0.18	0.13	0.21	0.41	0.47
	$k = 3$	0.01	0	0.01	0.02	0.01	0.03	0.04	0.07	0.06

		CV = 0.75			CV = 1		
	M	3	5	10	3	5	10
$e = 0.55$	$k = 1$	2.93	3.54	3.68	4.50	6.25	6.08
	$k = 3$	0.47	0.74	0.68	1.21	1.77	1.64
$e = 0.65$	$k = 1$	3.11	3.87	3.75	4.44	5.54	6.37
	$k = 3$	0.84	1.05	1.01	1.66	1.56	2.09
$e = 0.75$	$k = 1$	2.20	2.91	3.02	3.19	5.63	5.86
	$k = 3$	0.63	0.57	0.06	1.44	2.25	1.97
$e = 0.85$	$k = 1$	1.29	2.00	2.69	2.00	3.65	4.29
	$k = 3$	0.35	0.47	0.69	1.03	1.62	1.43
$e = 0.9$	$k = 1$	0.97	1.30	1.87	1.38	1.84	3.62
	$k = 3$	0.36	0.27	0.50	0.75	0.64	1.26
$e = 0.95$	$k = 1$	0.40	0.87	1.17	0.54	1.23	1.79
	$k = 3$	0.14	0.26	0.26	0.22	0.60	0.59

Table 4.5. Sensitivity of production rate to the types of up- and downtime distributions (the values of ε calculated according to (4.17)) (b). $T_{\text{down}} = 20$.

		CV = 0.1			CV = 0.25			CV = 0.5		
	M	3	5	10	3	5	10	3	5	10
$e = 0.55$	$k = 1$	0.06	0.11	0.06	0.12	0.34	0.37	1.46	1.56	1.25
	$k = 3$	0.09	0.04	0.05	0.24	0.06	0.09	0.31	0.12	0.16
$e = 0.65$	$k = 1$	0.08	0.05	0.03	0.33	0.48	0.47	1.27	1.39	1.92
	$k = 3$	0.05	0.03	0.03	0	0.13	0.02	0.23	0.08	0.37
$e = 0.75$	$k = 1$	0.04	0	0.06	0.34	0.31	0.30	1.16	1.13	1.88
	$k = 3$	0.01	0.04	0.01	0.11	0.07	0.05	0.18	0.07	0.36
$e = 0.85$	$k = 1$	0.01	0.05	0.04	0.17	0.29	0.32	0.61	1.00	1.45
	$k = 3$	0.01	0.02	0.01	0.05	0.01	0.06	0.17	0.14	0.33
$e = 0.9$	$k = 1$	0.03	0.03	0.02	0.20	0.34	0.39	0.46	0.94	1.06
	$k = 3$	0.02	0.01	0.01	0.03	0.06	0.09	0.16	0.10	0.13
$e = 0.95$	$k = 1$	0.02	0.06	0.05	0.14	0.16	0.14	0.31	0.47	0.52
	$k = 3$	0.01	0.01	0.01	0.06	0.03	0.02	0.12	0.01	0.06

		CV = 0.75			CV = 1		
	M	3	5	10	3	5	10
$e = 0.55$	$k = 1$	3.12	3.68	3.74	5.34	6.70	6.05
	$k = 3$	0.86	0.62	0.94	1.91	4.02	1.66
$e = 0.65$	$k = 1$	2.69	3.65	3.38	4.22	6.27	5.94
	$k = 3$	0.55	0.80	0.75	1.64	2.18	1.62
$e = 0.75$	$k = 1$	1.47	2.77	3.45	3.44	4.39	5.43
	$k = 3$	0.32	0.81	0.90	1.67	1.61	1.49
$e = 0.85$	$k = 1$	1.42	2.27	2.67	2.10	3.04	4.00
	$k = 3$	0.54	0.58	0.80	0.09	1.04	1.43
$e = 0.9$	$k = 1$	0.93	1.52	1.89	1.57	2.23	3.06
	$k = 3$	0.36	0.45	0.30	0.75	1.00	1.04
$e = 0.95$	$k = 1$	0.64	0.88	0.98	1.01	1.00	2.07
	$k = 3$	0.30	0.39	0.24	0.65	0.47	0.81

$$k \uparrow \Rightarrow \text{ Sensitivity } \downarrow,$$
$$e \uparrow \Rightarrow \text{ Sensitivity } \downarrow.$$

(b) Hypothesis 4.1 (i.e., expression (4.8)) approximates well the production rate of all systems considered. Indeed, the values of metric Δ_1 (calculated according to (4.18) shown in Tables 4.6 and 4.7 are within 6%. Since this precision is commensurable with the accuracy of the data available on the factory floor with regard to machine and buffer parameters, we conclude that Hypothesis 4.1 can be used as a tool for evaluating the production rate in serial lines with identical machines obeying Weibull, gamma, and log-normal reliability models. We have also investigated the applicability of (4.8) for systems with Rayleigh and Erlang reliability models and obtained similar results. Based on this, we conjecture that expression (4.8) can be used for evaluating production rates in serial lines with identical machines obeying any reliability model, provided that buffer capacity is at least one downtime and probability density functions of up- and downtime are unimodal.

6. Production rate for the case of identical coefficients of variation

6.1 Parameters of systems analyzed

In the case of machines with arbitrary e_i and $T_{\text{up},i}$ but with identical coefficients of variation, i.e.,

$$CV_{\text{up},i} = CV_{\text{up}}, \quad CV_{\text{down},i} = CV_{\text{down}}, \qquad i = 1, \ldots, M,$$

we consider lines (4.15) and (4.16) with parameters selected randomly and equiprobably from the sets:

$$T_{\text{down},i} \in \{10, 20\}, \tag{4.20}$$
$$e_i \in \{0.85, 0.9, 0.95\} \tag{4.21}$$

or

$$e_i \in \{0.55, 0.65, 0.75, 0.85, 0.9, 0.95\}. \tag{4.22}$$

We use the sets (4.21) and (4.22) in order to investigate production lines with relatively efficient and relatively inefficient machines, respectively. As a results, the following parameter sets (PS) have been selected:

$$\begin{aligned} \text{PS1}: T_{\text{down}} &= [20, 10, 10, 20, 20, 10, 20, 20, 10, 10], \\ e &= [0.95, 0.85, 0.9, 0.95, 0.85, 0.9, 0.95, 0.9, 0.9, 0.85], \qquad (4.23) \\ \text{PS2}: T_{\text{down}} &= [10, 20, 20, 20, 10, 10, 20, 20, 10, 10], \end{aligned}$$

Table 4.6. Accuracy of Hypothesis 4.1 (the values of Δ_1 calculated according to (4.18)) (a). $T_{\text{down}} = 10$.

		CV = 0.1			CV = 0.25			CV = 0.5		
	M	3	5	10	3	5	10	3	5	10
$e = 0.55$	$k = 1$	1.14	1.33	1.34	1.36	1.76	1.72	0.61	1.10	1.00
	$k = 3$	1.40	1.82	2.09	2.64	3.51	4.09	3.04	4.22	4.67
$e = 0.65$	$k = 1$	0.17	0.16	0.08	0.40	0.38	0.42	1.79	2.23	2.28
	$k = 3$	1.22	1.58	1.90	2.27	3.06	4.03	2.70	3.56	4.40
$e = 0.75$	$k = 1$	0.33	0.35	0.43	1.18	1.24	1.27	2.18	2.36	2.61
	$k = 3$	0.96	1.31	1.62	1.78	2.48	3.15	2.01	2.91	3.74
$e = 0.85$	$k = 1$	0.53	0.63	0.79	1.30	1.43	1.40	1.91	1.78	1.64
	$k = 3$	0.62	0.90	1.20	1.14	1.72	2.33	1.18	1.91	2.81
$e = 0.9$	$k = 1$	0.48	0.65	0.81	0.98	1.17	1.06	1.25	1.46	0.99
	$k = 3$	0.45	0.66	0.87	0.86	1.29	1.76	0.97	1.41	2.10
$e = 0.95$	$k = 1$	0.36	0.49	0.62	0.63	0.75	0.55	0.76	0.71	0.35
	$k = 3$	0.22	0.34	0.49	0.40	0.64	1.01	0.40	0.74	1.27

		CV = .75			CV = 1		
	M	3	5	10	3	5	10
$e = 0.55$	$k = 1$	1.52	1.86	2.99	4.23	5.85	5.59
	$k = 3$	2.20	3.21	3.49	0.91	2.01	1.76
$e = 0.65$	$k = 1$	2.36	2.79	3.19	4.04	5.28	5.88
	$k = 3$	2.05	2.85	3.45	1.46	1.65	1.95
$e = 0.75$	$k = 1$	1.92	2.36	2.17	3.06	5.03	5.16
	$k = 3$	1.64	2.14	2.72	1.35	2.04	1.72
$e = 0.85$	$k = 1$	1.40	1.81	1.71	1.68	3.13	3.76
	$k = 3$	0.95	1.54	2.24	0.83	1.29	1.30
$e = 0.9$	$k = 1$	0.91	1.23	0.99	1.52	1.79	3.51
	$k = 3$	0.80	1.14	1.74	0.82	0.65	1.33
$e = 0.95$	$k = 1$	0.55	0.52	0.80	0.46	1.27	1.89
	$k = 3$	0.30	0.61	1.09	0.25	0.49	0.75

Table 4.7. Accuracy of Hypothesis 4.1 (the values of Δ_1 calculated according to (4.18)) (b). $T_{\text{down}} = 20$.

		CV = 0.1			CV = 0.25			CV = 0.5		
	M	3	5	10	3	5	10	3	5	10
$e = 0.55$	$k = 1$	1.23	1.44	1.49	1.33	1.85	1.86	0.74	0.99	0.84
	$k = 3$	1.47	1.78	2.09	2.79	3.47	4.11	3.13	3.91	5.06
$e = 0.65$	$k = 1$	0.35	0.32	0.27	0.17	0.31	0.14	1.45	1.83	2.14
	$k = 3$	1.34	1.64	1.90	2.42	3.12	3.78	2.76	3.58	4.59
$e = 0.75$	$k = 1$	0.23	0.14	0.24	1.10	1.06	1.07	2.07	1.99	2.46
	$k = 3$	1.00	1.35	1.62	1.84	2.56	3.28	2.09	2.93	3.98
$e = 0.85$	$k = 1$	0.45	0.44	0.61	1.21	1.26	1.15	1.62	1.66	1.57
	$k = 3$	0.65	0.94	1.19	1.16	1.82	2.42	1.25	2.07	3.03
$e = 0.9$	$k = 1$	0.42	0.58	0.69	0.93	1.14	0.98	1.12	1.43	0.88
	$k = 3$	0.49	0.66	0.89	0.93	1.29	1.87	1.02	1.48	2.28
$e = 0.95$	$k = 1$	0.31	0.48	0.58	0.58	0.72	0.48	0.59	0.76	0.34
	$k = 3$	0.25	0.35	0.45	0.46	0.66	1.00	0.58	0.82	1.25

		CV = .75			CV = 1		
	M	3	5	10	3	5	10
$e = 0.55$	$k = 1$	1.25	1.99	2.64	4.65	6.25	5.94
	$k = 3$	2.76	2.77	3.94	1.64	1.91	2.09
$e = 0.65$	$k = 1$	1.58	2.46	2.55	4.47	5.52	5.88
	$k = 3$	2.30	2.89	3.50	1.90	2.07	1.81
$e = 0.75$	$k = 1$	1.82	1.97	2.30	3.12	5.04	5.72
	$k = 3$	1.38	2.37	3.13	1.52	2.12	1.98
$e = 0.85$	$k = 1$	1.36	1.51	1.36	1.67	3.15	4.28
	$k = 3$	1.07	1.73	2.66	0.82	1.27	1.73
$e = 0.9$	$k = 1$	0.83	1.06	1.01	1.64	1.93	3.00
	$k = 3$	0.90	1.29	1.85	1.06	0.82	1.30
$e = 0.95$	$k = 1$	0.67	0.58	0.84	1.00	1.19	1.84
	$k = 3$	0.66	0.67	1.10	0.67	0.56	0.69

Table 4.8. Production rates evaluated by simulation and Hypothesis 4.2.

	$M = 3$	$M = 5$	$M = 10$
PS1	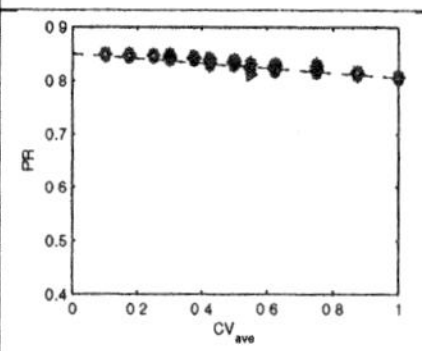	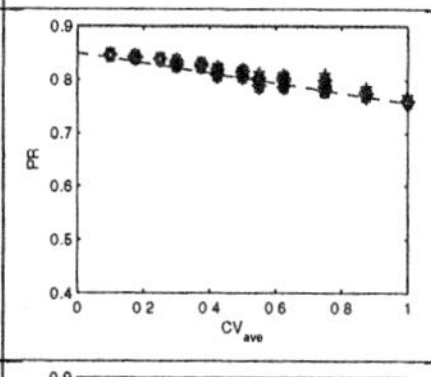	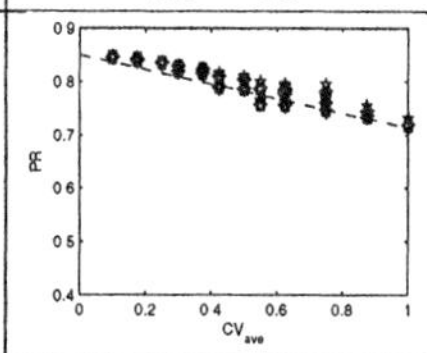
PS2	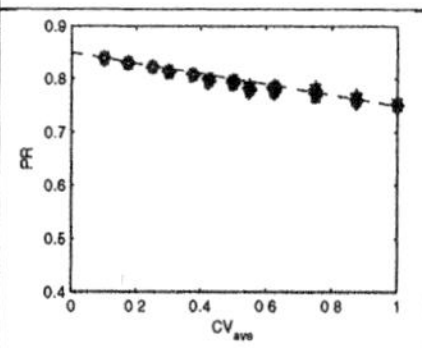	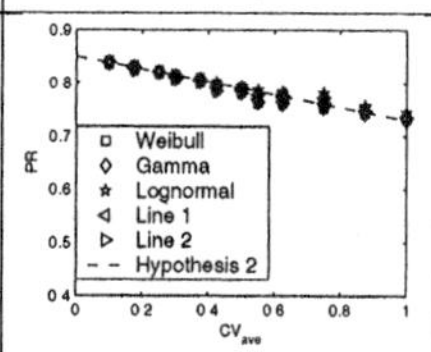	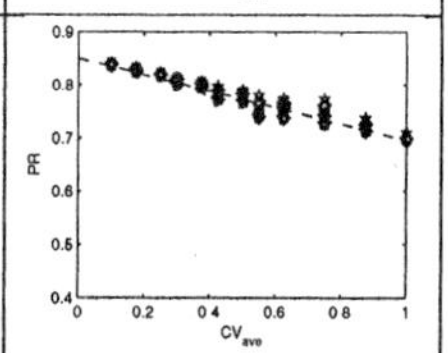
PS3	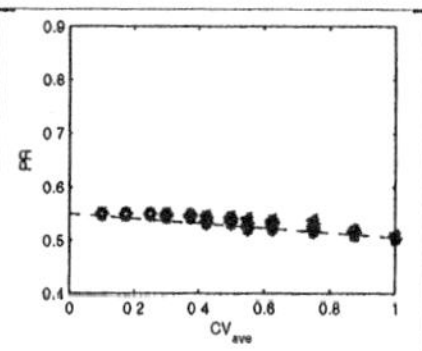	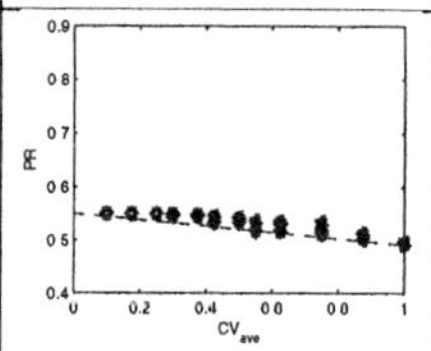	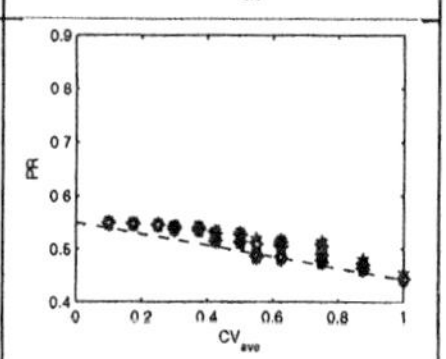

$$e = [0.9, 0.85, 0.85, 0.95, 0.9, 0.95, 0.9, 0.85, 0.9, 0.95], \quad (4.24)$$

$$\text{PS3}: T_{\text{down}} = [10, 20, 10, 20, 10, 20, 20, 10, 10, 20],$$

$$e = [0.55, 0.75, 0.9, 0.85, 0.65, 0.9, 0.55, 0.95, 0.65, 0.75], \quad (4.25)$$

where T_{down} and e are vectors with elements $T_{\text{down},i}$ and e_i, $i = 1, \ldots, 10$, respectively.

Each of these parameter sets is used to calculate PR in all lines (4.15), (4.16) with $M \in \{3, 5, 10\}$, $k = 1$, CV_{up} and $\text{CV}_{\text{down}} \in \{0.1, 0.25, 0.5, 0.75, 1\}$. Thus, the total of 1125 production lines have been analyzed.

6.2 Results

Table 4.8 presents the simulation results for all systems analyzed along with the dashed line corresponding to Hypothesis 4.2, i.e.,

$$\text{PR} = e_{\min} - (e_{\min} - \text{PR}^{\text{exp}})\text{CV}_{\text{ave}},$$

where

$$\text{CV}_{\text{ave}} = \frac{\text{CV}_{\text{up}} + \text{CV}_{\text{down}}}{2}.$$

Table 4.9 characterizes the accuracy of Hypothesis 4.2 (i.e., metric Δ_2 defined in (4.19)). As one can see, Δ_2 is within 6%. Thus, we

Table 4.9. Accuracy of Hypothesis 4.2 (the values of Δ_2 calculated according to (4.19)).

	CV_{ave}	0.1	0.175	0.25	0.3	0.375	0.425
PS1	$M=3$	0.34	0.73	0.87	1.21	1.25	1.12
	$M=5$	0.64	1.16	1.44	1.61	1.90	1.73
	$M=10$	1.12	1.91	2.37	2.57	3.02	2.91
PS2	$M=3$	0.13	0.44	0.45	1.06	0.93	1.76
	$M=5$	0.06	0.32	0.17	0.99	0.53	1.91
	$M=10$	0.47	0.81	1.00	1.22	1.59	1.70
PS3	$M=3$	0.82	1.43	1.97	2.43	2.80	2.76
	$M=5$	1.07	1.88	2.65	3.13	3.67	3.90
	$M=10$	1.82	3.05	4.17	4.68	5.66	5.43

	CV_{ave}	0.5	0.55	0.625	0.75	0.875	1
PS1	$M=3$	1.37	1.67	1.23	1.02	0.65	0.81
	$M=5$	2.06	1.62	2.21	2.20	2.07	1.64
	$M=10$	3.50	3.17	3.86	3.81	3.27	2.55
PS2	$M=3$	1.28	2.35	1.69	1.20	1.04	1.27
	$M=5$	1.21	2.73	1.93	1.13	1.55	2.07
	$M=10$	2.15	3.51	2.56	3.05	3.06	2.15
PS3	$M=3$	3.10	2.79	2.99	2.56	1.86	1.35
	$M=5$	4.21	3.96	4.25	4.30	3.28	2.04
	$M=10$	6.16	5.52	6.58	6.35	5.11	2.72

conjecture that the production rate for systems with identical coefficients of variation and any distribution of up- and downtime can be evaluated using Hypothesis 4.2 (i.e., expression (4.9)).

REMARK 10 In Table 4.8, some of the CV_{ave} correspond to multiple values of PR. This is because different selection of CV_{up} and CV_{down} may result in same CV_{ave}. For instance, $\{CV_{up} = 0.25, CV_{down} = 0.75\}$, $\{CV_{up} = CV_{down} = 0.5\}$, and $\{CV_{up} = 0.75, CV_{down} = 0.25\}$ all lead to $CV_{ave} = 0.5$. As one can see, the differences among all values of PR corresponding to the same CV_{ave} are not significant, and Hypothesis 4.2 approximates well to all values of PR. This again verifies that the PR depends mostly on the first two moments of up- and downtime distribution.

7. Production rate evaluation for the general case

7.1 Parameters of systems analyzed

In the general case, we again consider production lines (4.15) and (4.16) and parameter sets (4.23)–(4.25) but with $CV_{up,i}$ and $CV_{down,i}$ selected randomly. Specifically, we select $CV_{up,i}$ and $CV_{down,i}$ using the following probability mass functions defined on $\{0.1, 0.25, 0.5, 0.75, 1\}$:

- uniform distribution, i.e., both CV_{up} and CV_{down} are selected equiprobably;
- increasing triangular distribution, i.e., both CV_{up} and CV_{down} are selected according to triangular distributions with higher values of CVs being selected with larger probabilities than lower values;
- decreasing triangular distribution, i.e., both CV_{up} and CV_{down} are selected according to triangular distributions with lower values of CVs being selected with larger probabilities than higher values;
- increasing/decreasing triangular distributions, i.e., CV_{up} (respectively, CV_{down}) is selected according to triangular distributions where higher values (respectively, lower values) are more probable than lower (respectively, higher) values.

As a result, we obtained the following sequences (S) of CV_{up} and CV_{down}:

$$\begin{aligned}
S1: \quad & CV_{up} = [1, 0.25, 0.75, 0.5, 1, 0.75, 0.5, 0.1, 1, 0.5], \\
& CV_{down} = [0.25, 0.5, 0.75, 0.75, 0.25, 0.5, 1, 0.1, 0.75, 0.1], \quad (4.26) \\
S2: \quad & CV_{up} = [1, 0.5, 1, 0.75, 0.25, 1, 0.75, 1, 1, 1], \\
& CV_{down} = [1, 1, 0.25, 1, 0.75, 1, 1, 0.75, 1, 0.5], \quad (4.27) \\
S3: \quad & CV_{up} = [0.1, 0.25, 0.1, 0.5, 0.1, 0.1, 0.1, 0.75, 0.1, 0.1], \\
& CV_{down} = [0.75, 0.1, 0.25, 0.1, 0.5, 0.1, 0.1, 0.25, 0.1, 0.1], \quad (4.28) \\
S4: \quad & CV_{up} = [0.25, 0.1, 0.1, 0.1, 0.25, 0.1, 0.1, 0.5, 0.1, 0.1], \\
& CV_{down} = [0.1, 0.25, 0.1, 0.1, 0.1, 0.1, 0.5, 0.1, 0.1, 0.1], \quad (4.29) \\
S5: \quad & CV_{up} = [0.75, 1, 1, 0.75, 1, 1, 1, 0.5, 1, 1], \\
& CV_{down} = [1, 1, 0.75, 1, 1, 1, 0.5, 1, 1, 1], \quad (4.30) \\
S6: \quad & CV_{up} = [1, 0.75, 0.5, 1, 1, 0.25, 1, 0.75, 1, 0.1], \\
& CV_{down} = [0.1, 0.25, 0.5, 0.1, 0.1, 0.75, 0.1, 1, 0.1, 0.1], \quad (4.31) \\
S7: \quad & CV_{up} = [0.1, 0.25, 0.1, 0.1, 0.5, 0.1, 1, 0.1, 0.1, 0.75], \\
& CV_{down} = [1, 0.5, 0.75, 1, 0.5, 1, 0.75, 0.25, 0.1, 1], \quad (4.32)
\end{aligned}$$

where CV_{up} and CV_{down} are vectors with components $CV_{up,i}$ and $CV_{down,i}$, $i = 1, \ldots, 10$, respectively.

Each of these sequences is used for all systems (4.15), (4.16) with parameter sets (4.23)–(4.25), $M \in \{3, 5, 10\}$ and $k = 1$. Thus, the total of 315 systems have been analyzed.

7.2 Results

Table 4.10 presents the results for all systems analyzed by simulations and by Hypothesis 4.3, i.e.,

$$PR = e_{min} - (e_{min} - PR^{exp})CV_{ave},$$

where

$$CV_{ave} = \sum_{i=1}^{M} \frac{CV_{up,i} + CV_{down,i}}{2M},$$

represented by the broken line. Table 4.11 characterizes the accuracy of Hypothesis 4.3 by showing the values of Δ_3. Clearly, this accuracy is again within 4%. Thus, we conjecture that expression (4.10) can be used for evaluating the throughput in serial production lines with arbitrary unimodal distributions of up- and downtime.

8. Conclusions

The method developed in this paper can be used as a tool for a quick analytic evaluation of the performance of serial production lines with arbitrary models of machines reliability. Its application does not require more information than that necessary for such an evaluation under the exponential assumption. Indeed, since under the exponential assumption one needs the data to evaluate the average up- and downtime of the machines, the same data can be used to evaluate their coefficients of variation. Thus, all the information necessary for (4.8)–(4.10) is available, and the production rate may be evaluated without the knowledge of the types of up- and downtime distributions. This is the main advantage of the method developed.

The drawbacks of this method are as follows:

1. All buffers must be large enough to accommodate at least one largest downtime of all machines in the system.

2. No analytical proofs of relationships (4.8)–(4.10) are available.

These drawbacks can be commented upon as follows:

Table 4.10. Production rates evaluated by simulation and Hypothesis 4.3.

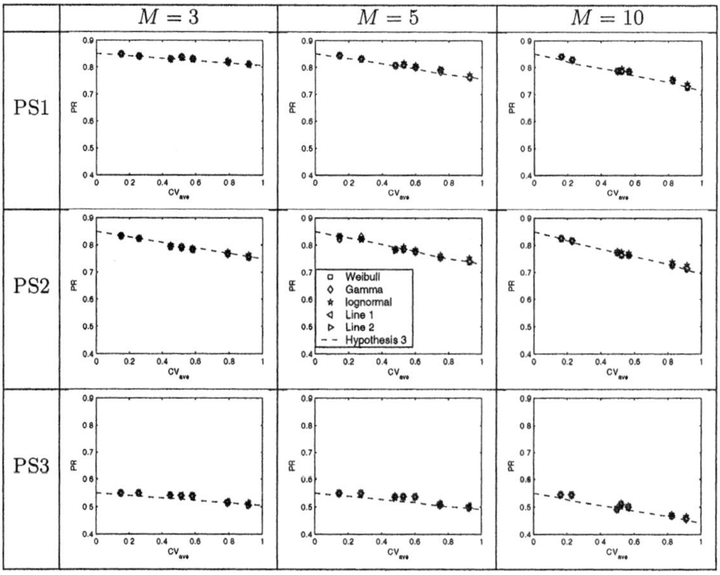

Table 4.11. Accuracy of Hypothesis 4.3 (the values of Δ_3 calculated according to (4.19)).

$M = 3$	CV_{ave}	0.15	0.2583	0.45	0.5167	0.5833	0.7917	0.9167
	PS1	0.60	0.26	0.22	1.51	1.22	1.18	0.47
	PS2	0.18	0.16	1.62	1.12	1.03	0.61	0.93
	PS3	1.22	2.00	1.96	2.61	3.00	0.85	1.22
$M = 5$	CV_{ave}	0.145	0.275	0.48	0.53	0.6	0.75	0.925
	PS1	0.82	0.67	0.45	1.98	1.59	2.65	1.23
	PS2	0.05	0.47	1.40	0.98	0.60	1.28	1.95
	PS3	1.55	2.73	3.35	3.93	3.81	2.12	1.94
$M = 10$	CV_{ave}	0.1625	0.2275	0.4975	0.5225	0.565	0.825	0.9225
	PS1	2.45	1.77	0.70	2.20	2.10	2.19	1.81
	PS2	0.06	0.93	0.09	0.97	1.24	1.53	2.64
	PS3	2.28	4.09	1.12	3.89	3.17	1.98	3.34

We believe that the method developed in this paper can be extended to production lines with buffers smaller than those necessary for accommodating the largest downtime of all the machines. Since in this case Theorem 4.1 does not hold, to evaluate the production rate for $CV_{up} = CV_{down} = 0$ simulations would be necessary. An investigation of such an extension is a topic for future work.

As for the lack of analytical proofs of (4.8)–(4.10), we believe that, at present, they are all but impossible.

Appendix: Proof of Theorem 4.1

The proof of Theorem 4.1 is based on the following two lemmas:

LEMMA 4.1 *Consider a two-machine line defined by assumptions* (i)–(v) *with* $CV_{up,i} = CV_{down,i} = 0$, $i = 1, 2$, *and assume that* $T_{up,1} = T_{up,2} = T_{up}$, $T_{down,1} = T_{down,2} = T_{down}$. *Then its production rate is given by*

$$PR = e = \frac{T_{up}}{T_{up} + T_{down}}.$$

Proof. Let $s_i(t)$, $i = 1, 2$ denote the states of machine m_i, $i = 1, 2$, at time t,

$$s_i(t) = \begin{cases} 0, & m_i \text{ is down,} \\ 1, & m_i \text{ is up,} \end{cases}$$

and $h(t)$ be the buffer occupancy at time t. Without loss of generality, $h(0) = 0$.

Assume that $s_i(0) = 0$, $i = 1, 2$, and m_i changes its state at time $t_i \leq T_{down}$, $i = 1, 2$, i.e.,

$$s_i(t) = \begin{cases} 0, & t \in [0, t_i), \\ 1, & t \in [t_i + lT, t_i + lT + T_{up}), \\ 0, & t \in [t_i + lT + T_{up}, t_i + (l+1)T), \\ & i = 1, 2, l = 0, 1, 2, \ldots, \end{cases}$$

where $T = T_{up} + T_{down}$. Then, the following three cases are possible:

Case 1. $t_1 = t_2$. Obviously, in this case machine m_1 is never blocked and m_2 is never starved, and therefore, $PR = e$.

Case 2. $t_1 > t_2$. In this case,

(a) If $t_2 + T_{\rm up} \geq t_1$, the following hold:

$$\begin{array}{ll}
s_1(t)=0,\, s_2(t)=0,\, h(t)=0, & t \in [0, t_2), \\
s_1(t)=0,\, s_2(t)=1,\, h(t)=0, & t \in [t_2, t_1), \\
s_1(t)=1,\, s_2(t)=1,\, h(t)=0, & t \in [t_1, t_2 + T_{\rm up}), \\
s_1(t)=1,\, s_2(t)=0,\, h(t)=t - T_{\rm up} - t_2, & t \in [t_2 + T_{\rm up}, t_1 + T_{\rm up}), \\
s_1(t)=0,\, s_2(t)=0,\, h(t)=t_1 - t_2, & t \in [t_1 + T_{\rm up}, t_2 + T), \\
s_1(t)=0,\, s_2(t)=1,\, h(t)=t_1 + T - t, & t \in [t_2 + T, t_1 + T), \\
s_1(t)=1,\, s_2(t)=1,\, h(t)=0, & t \in [t_1 + T, t_2 + T + T_{\rm up}), \\
s_1(t)=1,\, s_2(t)=0,\, h(t)=t - t_2 - T - T_{\rm up}, & t \in [t_2 + T + T_{\rm up}, t_1 + T + T_{\rm up}), \\
s_1(t)=0,\, s_2(t)=0,\, h(t)=t_1 - t_2, & t \in [t_1 + T + T_{\rm up}, t_2 + 2T), \\
\ldots,\quad \ldots,\quad \ldots, & \ldots.
\end{array}$$

By induction, we obtain

$$\begin{array}{ll}
s_1(t)=0,\, s_2(t)=0,\, h(t)=t_1 - t_2, & t \in [t_1 + (l-1)T + T_{\rm up}, t_2 + lT), \\
s_1(t)=0,\, s_2(t)=1,\, h(t)=t_1 - t + lT, & t \in [t_2 + lT, t_1 + lT), \\
s_1(t)=1,\, s_2(t)=1,\, h(t)=0, & t \in [t_1 + lT, t_2 + lT + T_{\rm up}), \\
s_1(t)=1,\, s_2(t)=0,\, h(t)=t - t_2 - lT - T_{\rm up}, & t \in [t_2 + lT + T_{\rm up}, t_1 + lT + T_{\rm up}), \\
\qquad l=1, 2, \ldots. &
\end{array}$$

Since $k \geq 1$, i.e., $N \geq T_{\rm down}$, machine m_1 is never blocked and machine m_2 is not starved (except for the initial period $t \in [t_2, t_1)$). Thus, after the period $[0, t_2 + T_{\rm up})$, the line is producing during the interval of length $T_{\rm up}$ and is not producing during the interval of length $T_{\rm down}$.

(b) If $t_2 + T_{\rm up} < t_1$, which implies that $T_{\rm up} < T_{\rm down}$, the following hold:

$$\begin{array}{ll}
s_1(t)=0,\, s_2(t)=0,\, h(t)=0, & t \in [0, t_2), \\
s_1(t)=0,\, s_2(t)=1,\, h(t)=0, & t \in [t_2, t_2 + T_{\rm up}), \\
s_1(t)=0,\, s_2(t)=0,\, h(t)=0, & t \in [t_2 + T_{\rm up}, t_1), \\
s_1(t)=1,\, s_2(t)=0,\, h(t)=t - t_1, & t \in [t_1, t_1 + T_{\rm up}), \\
s_1(t)=0,\, s_2(t)=0,\, h(t)=T_{\rm up}, & t \in [t_1 + T_{\rm up}, t_2 + T), \\
s_1(t)=0,\, s_2(t)=1,\, h(t)=T_{\rm up} - t + t_2 + T, & t \in [t_2 + T, t_2 + T + T_{\rm up}), \\
s_1(t)=0,\, s_2(t)=0,\, h(t)=0, & t \in [t_2 + T + T_{\rm up}, t_1 + T), \\
s_1(t)=1,\, s_2(t)=1,\, h(t)=t - t_1 - T, & t \in [t_1 + T, t_1 + T + T_{\rm up}), \\
\ldots,\quad \ldots,\quad \ldots, & \ldots.
\end{array}$$

Clearly, in this case,

$$\begin{array}{ll}
s_1(t)=0,\, s_2(t)=0,\, h(t)=0, & t \in [t_2 + lT + T_{\rm up}, t_1 + lT), \\
s_1(t)=1,\, s_2(t)=0,\, h(t)=t - t_1 - kT, & t \in [t_1 + lT, t_1 + lT + T_{\rm up}), \\
s_1(t)=0,\, s_2(t)=0,\, h(t)=T_{\rm up}, & t \in [t_1 + lT + T_{\rm up}, t_2 + (l+1)T), \\
s_1(t)=0,\, s_2(t)=1,\, h(t)=t_2 + (l+1)T + T_{\rm up} - t, & t \in [t_2 + (l+1)T, t_2 + (l+1)T + T_{\rm up}), \\
\qquad l=1, 2, \ldots. &
\end{array}$$

Thus, m_2 is not starved (except for the initial interval $t \in [t_2, t_2 + T_{\rm up})$) and m_1 is never blocked.

Therefore, in both (a) and (b), $\mathrm{PR} = e$.

Case 3. $t_1 < t_2$. Similar arguments can be used. If $t_1 + T_{\rm up} \geq t_2$, we show that:

$$\begin{array}{ll} s_1(t)=0, s_2(t)=0, \ h(t)=0, & t \in [t_2+(l-1)T+T_{\rm up}, t_1+lT), \\ s_1(t)=1, s_2(t)=0, \ h(t)=t-t_1-lT, & t \in [t_1+lT, t_2+lT), \\ s_1(t)=1, s_2(t)=1, \ h(t)=t_2-t_1, & t \in [t_2+lT, t_1+lT+T_{\rm up}), \\ s_1(t)=0, s_2(t)=1, \ h(t)=t_2-t+lT+T_{\rm up}, & t \in [t_1+lT+T_{\rm up}, t_2+lT+T_{\rm up}), \\ \qquad l=1,2,\ldots. & \end{array}$$

Again, due to $N \geq T_{\rm down}$, machine m_1 is never blocked and m_2 is never starved. If $t_1 + T_{\rm up} < t_2$, the following hold:

$$\begin{array}{ll} s_1(t)=0, s_2(t)=0, \ h(t)=0, & t \in [t_2+(l-1)T+T_{\rm up}, t_1+lT), \\ s_1(t)=1, s_2(t)=0, \ h(t)=t-t_1-lT, & t \in [t_1+lT, t_1+lT+T_{\rm up}), \\ s_1(t)=0, s_2(t)=0, \ h(t)=T_{\rm up}, & t \in [t_1+lT+T_{\rm up}, t_2+lT), \\ s_1(t)=0, s_2(t)=1, \ h(t)=T_{\rm up}-t+lT+t_2, & t \in [t_2+lT, t_2+lT+T_{\rm up}), \\ \qquad l=1,2,\ldots. & \end{array}$$

Therefore, in both cases, PR $= e$.
Next, we repeat this analysis under the assumptions

$$\begin{array}{ll} (\alpha) s_i(0) = 1, & i = 1,2, \\ (\beta) s_1(0) = 1, & s_2(0) = 0, \\ (\gamma) s_1(0) = 0, & s_2(0) = 1, \end{array}$$

and show that after the initial period $t \in [0, T_{\rm up} + T_{\rm down})$, the line produces during the interval of the length $T_{\rm up}$ and does not produce during the interval of the length $T_{\rm down}$. Therefore, in all cases, PR $= e$. □

LEMMA 4.2 *Consider an M-machine line defined by assumption* (i)–(v) *with* $\mathrm{CV}_{{\rm up},i} = \mathrm{CV}_{{\rm down},i} = 0$, $i = 1, \ldots, M$, *and assume that* $T_{{\rm up},i} = T_{\rm up}$ *and* $T_{{\rm down},i} = T_{\rm down}$, $i = 1, \ldots, M$. *Then its production rate is given by*

$$\mathrm{PR} = e = \frac{T_{\rm up}}{T_{\rm up} + T_{\rm down}}.$$

Proof. From Lemma 4.1, machine m_2 is not starved except for a subinterval of $[0, T_{\rm up} + T_{\rm down})$. Thus, in the steady state, the two-machine line is up (producing) for $T_{\rm up}$ time units and down (not producing) for $T_{\rm down}$, which is equivalent to a single machine. Since $N \geq T_{\rm down}$, no blockage of m_2 can occur. Therefore, aggregating these two machines with the third one, we conclude that the three machines are again equivalent to a single machine. Continuing this process until all machines are aggregated, we obtain PR $= e$ for M-machine line when CV $= 0$ and $k \geq 1$. □

Proof of Theorem 4.1. Consider the production line, L, defined by assumptions (i)–(v) with arbitrary values of $T_{{\rm up},i}$ and $T_{{\rm down},i}$, but with $\mathrm{CV}_{{\rm up},i} = \mathrm{CV}_{{\rm down},i}=0$, $i = 1, \ldots, M$. Along with it, consider the production line, L', also defined by (i)–(v) but with identical machines and buffers given by:

$$
\begin{aligned}
T'_{\text{down}} &= \max_{i=1,\dots,M} T_{\text{down},i}, \\
e' &= \min_{i=1,\dots,M} \frac{T_{\text{up},i}}{T_{\text{up}} + T_{\text{down},i}} = \min_{i=1,\dots,M} e_i = e_{\min}, \\
T'_{\text{up}} &= \frac{e'}{1-e'} T'_{\text{down}}, \\
N' &= T'_{\text{down}}.
\end{aligned}
$$

As it follows from Lemma 4.2, the production rate, PR$'$, of line L' is $e_{\min}$. Due to (4.A.1) and the monotonicity property of production rate of serial lines with respect to machine and buffer parameters (see Shanthikumar and Yao, 1989), the production rate, PR, of line L satisfies the inequality

$$\text{PR} \geq \text{PR}'.$$

However, PR is limited by the least efficient machine in the system. Therefore,

$$\text{PR} = \text{PR}' = e_{\min}. \qquad \square$$

References

Altiok, T. (1985). Production lines with phase-type operation and repair times and finite buffers. *International Journal of Production Research*, 23:489–498.

Altiok, T. (1989). Approximate analysis of queues in series with phase-type service times and blocking. *Operations Research*, 37:601–610.

Altiok, T. (1997). *Performance Analysis of Manufacturing Systems.* Springer.

Altiok, T. and Ranjan, R. (1989). Analysis of Production Lines with general service times and finite buffers: A two-node decomposition approach. *Engineering Costs and Production Economics*, 17:155–165.

Le Bihan, H. and Dallery, Y. (2000). A robust decomposition method for the analysis of production lines with unreliable machines and finite buffers. *Annals of Operations Research*, 93:265–297.

Buxey, G.M., Slack, N.D., and Wild, R. (1973). Production flow line systems design—A review. *AIIE Transactions*, 5:37–48.

Buzacott, J.A. (1967). Automatic transfer lines with buffer stocks. *International Journal of Production Research*, 5:182–200.

Buzacott, J.A. and Hanifin, L.E. (1978). Models of automatic transfer lines with inventory banks: A review and comarison. *AIIE Transactions*, 10:197–207.

Buzacott, J.A. and Kostelski, D. (1987). Matrix-geometric and recursive algorithm solution of a two-stage unreliable flow line. *IIE Transactions*, 19:429–438.

Buzacott, J.A. and Shanthikumar, J.G. (1993). *Stochastic Models of Manufacturing Systems*. Prentice Hall.

Chiang, S.-Y., Kuo, C.-T., and Meerkov, S.M. (1998). Bottlenecks in Markovian production lines. *IEEE Transactions on Robotics and Automation*, 14:352–359.

Chiang, S.-Y., Kuo, C.-T., and Meerkov, S.M. (2000). DT-bottleneck in serial production lines: Theory and application. *IEEE Transactions on Robotics and Automation*, 16:567–580.

Chiang, S.-Y., Kuo, C.-T., and Meerkov, S.M. (2001). c-bottleneck in serial production lines: Identification and application. *Mathematical Problems in Engineering*, 6:543–578.

Choong, Y.F. and Gershwin, S.B. (1987). A decomposition method for the approximate evaluation of capacitated transfer lines with unreliable machines and random processing times. *IIE Transactions*, 19:150–159.

Dallery, Y. and Le Bihan, H. (1999). An improved decomposition method for the analysis of production lines with unreliable machines and finite buffers. *International Journal of Production Research*, 37:1093–1117.

Dallery, Y., David, R., and Xie, X.-L. (1988). An efficient algorithm for analysis of transfer lines with unreliable machines and finite buffers. *IIE Transactions*, 20:280–283.

Dallery, Y., David, R., and Xie, X.-L. (1989). Approximate analysis of transfer lines with unreliable machines and finite buffers. *IEEE Transactions on Automatic Control*, 34:943–953.

Dallery, Y. and Gershwin, S.B. (1992). Manufacturing flow line systems: A review of models and analytical results. *Queuing Systems*, 12:3–94.

Enginarlar, E., Li, J., Meerkov, S.M., and Zhang, R.Q. (2002). Buffer capacity to accommodating machine downtime in serial production lines. *International Journal of Production Research*, 40:601–624.

Gershwin, S.B. (1994). *Manufacturing Systems Engineering.* Prentice Hall.

Gershwin, S.B. and Berman, O. (1981). Analysis of transfer lines consisting of two unreliable machines and random processing times and finite storage buffers. *AIIE Transactions*, 13:2–11.

Gershwin, S.B. (1987). An efficient decomposition method for the approximate evaluation of tandem queues with finite storage space and blocking. *Operations Research*, 35:291–305.

Glassey, C.R. and Hong, Y. (1993). Analysis of behavior of an unreliable n-stage transfer line with $(n-1)$ inter-storage buffers. *International Journal of Production Research*, 31:519–530.

Hillier, F.S. and So, K.C. (1991a). The effect of the coefficient of variation of operation times on the allocation of storage space in production line systems. *IIE Transactions*, 23:198–206.

Hillier, F.S. and So, K.C. (1991b). The effect of machine breakdowns and internal storage on the performance of production line systems. *International Journal of Production Research*, 29:2043–2055.

Inman, R.R. (1999). De Koster Empirical evaluation of exponential and independence assumptions in queueing models of manufacturing systems. *Production and Operation Management*, 8:409–432.

Jacobs, D.A. and Meerkov, S.M. (1995a). A system-theoretic property of serial production lines: Improvability. *International Journal of System Science*, 26:95–137.

Jacobs, D.A. and Meerkov, S.M. (1995b). Mathematical theory of improvability for production systems. *Mathematical Problems in Engineering*, 1:95–137.

Jafari, M. and Shanthikumar, J.G. (1987). An approximate model of multistage automatic transfer lines with possible scrapping of workspieces. *IIE Transactions*, 19:252–265.

Koenigsberg, E. (1959). Production lines and internal storage—A review. *Management Sciences*, 5:410–433.

De Koster, M.B.M. (1987). Estimation of the line efficiency by aggregation. *International Journal of Production Research*, 25:615–626.

De Koster, M.B.M. (1988). An improved algorithm to approximate the behavior of flow lines. *International Journal of Production Research*, 26:691–700.

Li, J. and Meerkov, S.M. (2003). On the coefficients of variation of up- and downtime in manufacturing equipment. *Control Group Report CGR* 03-14, Forthcoming in: *IIE Transactions on Quality and Reliability Engineering.*

Lim, J.-T., Meerkov, S.M., and Top, F. (1990). Homogeneous, asymptotically reliable serial production lines: Theory and a case study. *IEEE Transactions on Automatic Control*, 35:524–534.

Liu, X.-G. and Buzacott, J.A. (1989). A zero-buffer equivalence technique for decomposing queueing networks with blocking. In: *Queueing Networks with Blocking* (H.G. Perros and T. Altiok, eds.), pages 87–104, North-Holland.

Papadopoulos, H.T. and Heavey, C. (1996). Queueing theory in manufacturing systems analysis and design: A classification of models for production and transfer lines. *European Journal of Operational Research*, 92:1–27.

Powell, S.G. (1994). Buffer allocation in unbalanced three-station lines. *International Journal of Production Research*, 32:2201–2217.

Sadr, J. and Malhame, R.P. (2003). Bottleneck identification and maximal throughput evaluation in unreliable transfer lines. *Proceedings of 4th Aegean International Conference on Analysis of Manufacturing Systems*, pages 41–49, Samos Island, Greece.

Sevast'yanov, S.T. (1962). Influence of storage bin capacity on the average standstill time of a production line. *Theory of Probability Applications*, 429–438.

Shanthikumar, J.G. and Yao, D.D. (1989). Monotonicity and concavity properties in cyclic queueing networks with finite buffers. In: *Queueing Networks with Blocking* (H.G. Perros and T. Altiok, eds.), pages 325–345, North-Holland.

Sheskin, T.J. (1976). Allocation of interstage storage along an automatic production line. *AIIE Transaction*, 8:146–152.

Soyster, A.L., Schmidt, J.W., and Rohrer, M.W. (1979). Allocations of buffer capacities for a class of fixed cycle production systems. *AIIE Transactions*, 11:140–146.

Tan, T. and Yeralan, S. (1997). A decomposition model for continuous materials flow production systems. *International Journal of Production Research*, 35:2759–2772.

Tempelmeier, H. (2003). Practical considerations in the optimization of flow production systems. *International Journal of Production Research*, 41:149–170.

Tempelmeier, H. and Burger, M. (2001). Performance evaluation of unbalanced flow line with general distributed processing times, failures and imperfect production. *IIE Transactions*, 33:293–302.

Terracol, C. and David, R. (1987). An aggregation technique for performance evaluation of transfer lines with unreliable machine and finite buffers. *Proceedings of IEEE International Conference on Robotics and Automation*, pages 1333–1338.

Vidalis, M.I. and Papadopoulos, H.T. (1999). Markovian analysis of production lines with coxian-2 service times. *International Transactions in Operational Research*, 6:495–524.

Viswanadham, N. and Narahari, Y. (1992). *Performance Modeling of Automated Manufacturing System*, Practice Hall.

Wijingaard, J. (1979). The effect of interstage buffer storage on the output of two unreliable production units in series with different production rates. *AIIE Transactions*, 11:42–47.

Yamashita, H. and Altiok, T. (1998). Buffer capacity allocation for a desired throughput of production lines. *IIE Transactions*, 30:883–891.

Chapter 5

SUPPLY CHAIN PRODUCTION PLANNING MODELING FACILITY LEAD TIME AND QUALITY OF SERVICE

Osman M. Anli
Michael C. Caramanis
Ioannis Ch. Paschalidis

Abstract We propose a decision support framework for the Supply Chain management of a manufacturing enterprise. It utilizes structured information sharing between a fluid approximation Master-Problem and facility specific Sub-Problems. We optimize weekly production schedules that minimize inventory and backlog costs subject to non-linear constraints on production imposed by weekly varying dynamic lead-times and inter-facility quality of service driven inventory hedging policies. Computational experience demonstrates that it is possible to achieve the same quality of service with significantly lower inventory and system times relative to static lead-time state of the art industry practice.

1. Introduction

1.1 Motivation and objectives

Modern manufacturing enterprises are becoming more global than ever. They encompass owned or contract manufacturing and transportation facilities, suppliers, distributors, and customer service centers scattered over the globe. Manufacturers are no longer the sole drivers of the *Supply Chain* (SC). A shift from a "push" to a "pull" environment is well on its way. Customer needs and preferences influence the SC's inner workings: product functionality, quality, speed of production, timeliness of deliveries, flexibility in adjusting to demand changes. In today's highly competitive marketplace, companies are challenged with achieving shorter order-to-delivery times while allowing customers to customize their orders. Manufacturers recognize the significance of short *Lead Times* and high *Quality of Service* (QoS) provisioning. Furthermore, time-based competition has had a significant impact on the design of production facilities (Product Cells) and their operation (Just In Time, Zero In Process Inventory, Lean Manufacturing etc). Finally, supplier-consumer information sharing has been looked upon as a means to reduce inventories needed to provide a desired service level. Although these efforts, together with the wider use of enterprise-wide transactions databases, have achieved remarkable productivity gains, further improvements in global SC lead times and QoS are critically required. The revolution in computational intelligence and communication capabilities assisted more recently by the emergence of sensor networks with dynamically reconfigurable topology, have brought these improvements within reach.

The lead time at each component of a SC contains information which is critical for effective coordination. Lead times change across weeks in the planning horizon. In fact, they vary non-linearly with load, production mix, lot sizes, detailed scheduling and other operational practices adopted during each week of the planning horizon. Nevertheless, widely used Material Requirements Planning (MRP) systems assume lead times are constant across the whole planning horizon to avoid the task of estimating and communicating variable lead time information. The use of limited information in the current state of the art industrial practice is responsible for inefficient planning and often chaotic and unstable operations hampered by chronic backlogs and widely oscillating inventories. Two major barriers preventing more extensive use of information are (i) the cost of collecting, processing and communicating the requisite information and (ii) computational and algorithmic challenges in using this information to plan and manage SCs optimally. We propose a time scale decomposition and information communication architecture framework that is capable to exploit sensor networks and overcome the communication barrier. We also propose an iterative decentralized coordination algorithm that provides proof of the concept that the computational barrier can be overcome as well.

1.2 Current industry practice

Whereas capacity is ignored and dynamics are modeled by constant lead times in the vanilla version of MRP models (Meal, Wachter and Whybark, 1987), advanced planning system (APS) approaches include adequate representation of material flow dynamics and detailed representation of effective (or expected) capacity. APS models rely on mathematical programming techniques and hierarchical decomposition (Hax and Meal, 1975; Bitran and Tirupati, 1993) to overcome combinatorial complexity explosion barriers while capturing the details of capacity restrictions. This task is particularly onerous in the face of discrete part integrality and complex production rules and constraints, which, together with uncertainty, render stochastic integer programming formulations computationally intractable.

Past approaches employed to bypass these hurdles include the theory of constraints, scheduling algorithms and fluid model approximations. The theory of constraints (Chen and Mandelbaum, 1991; Lambrecht and Decaluwe, 1988; Goldratt and Cox, 1984) approximates the model of the production system by a small number of bottleneck components that are modeled in great detail; production is scheduled around those components through constraint propagation over time. Two main shortcomings of the theory of constraints approach are: first the difficulty in identifying and modeling bottleneck components and second the fact that delays or lead time dynamics along part routes are nonlinear and difficult to model. The systematic modeling of individual facilities could be possibly used to alleviate the first shortcoming. However, it is very difficult to overcome the second shortcoming. A variety of scheduling algorithms ranging from mathematical programming and Lagrangian relaxation to genetic algorithms have been used, often effectively (Sharifnia, Caramanis and Gershwin, 1991; Caramanis and Liberopoulos, 1992; Liberopoulos and Caramanis, 1995; Kaskavelis and Caramanis, 1998; Khmelnitsky and Caramanis, 1998; Deleersnyder et al., 1992; Jain, Johnson and Safai, 1996; Goncalves et al., 1994; Brandimarte, Alfieri and Levi, 1998). Fluid model approximations have also been used extensively and with considerable success (Veatch and Caramanis, 1999; Chen and Yao, 1993; Bertsimas, Paschalidis and Tsitsiklis, 1994; Chen and Yao, 1992; Connors, Feigin and Yao, 1994; Feng and Leachman, 1996) but have not adequately addressed dynamic lead time modeling. It has been shown that deterministic fluid model approximations of stochastic discrete production networks can be employed to predict the qualitative nature of optimal scheduling rules (Caramanis and Liberopoulos, 1992; Liberopoulos and Caramanis, 1995; Veatch and Caramanis, 1999; Bertsimas, Paschalidis and Tsitsiklis, 1994) and to determine the stability and robustness of the approximated stochastic discrete networks (Sharifnia, 1994; Kumar and Meyn, 1995; Kumar and Seidman, 1990; Dai, 1995). The proposed algorithm exploits this line of research with particular emphasis on extending fluid network approximations to improve the dynamic lead time modeling capabilities.

Past efforts to model lead time in production planning are noteworthy (Graves, 1986; Lambrecht, Ivens and Vandaele, 1998) but are limited to static lead times estimated for average or typical production conditions.

1.3 Overview of proposed approach

The time scale driven decentralized information estimation and communication architecture that we propose in Section 2 enables coordination, planning and operational decisions of manufacturing cells, transportation activities, inventory and distribution facilities in a SC. We show that this can be achieved through *optimal* and *consistent* production targets and safety stock levels scheduled for each part type produced by each SC facility. We propose a framework of iterative information exchange between three decision making/performance evaluation layers that is indeed capable of achieving this coordination. The framework consists of a centralized *Planning Coordination* layer, a centralized QoS *Coordination* layer, and finally a decentralized *Performance Evaluation and Demand Information Layer.* The Planning layer determines facility specific production targets using performance and sensitivity information it receives from the decentralized performance evaluation and Information layer. The QoS layer combines interacting facility production capabilities and requirements (i.e. targets) to determine hedging inventory requirements that achieve exogenously specified QoS levels. The decentralized Performance Evaluation and Demand Information layer analyzes short term (hourly) stochastic dynamics of each facility to derive expected (weekly) work in process and safety stock inventory for each facility and their sensitivity w.r.t. planning level targets.

The major objective of the proposed framework is to capture second order effects of the steady state cell dynamics in order to model dynamic lead time effectively at the coarse (varying weekly) production planning dynamics layer. Weekly time averages are a statistic with relatively low variance due to law of large numbers effects and they can be effectively modeled as deterministic quantities within the planning layer. Furthermore, detailed information on machine specific queue and setup states is not globally available, hence, it is practical to share state information which is (i) time averaged to the coarse time scale, and (ii) grouped by facility. To this end, capacity, WIP and production requirements are facility specific aggregates. Production planning dynamics are thus constrained to satisfy minimum weekly average lead time requirements. Note that although facility lead times and inter facility hedging inventory requirements are averages over the fine (hourly) time scale dynamics modeled at the decentralized performance evaluation layer, they are dynamic relative to the coarse (weekly) time scale of the planning layer. Lead times and hedging inventory requirements are modeled as functions of production planning decisions (loading and mix). This constitutes the second order information which we have shown can be used (Caramanis, Pan and Anlı, 2001a) to significantly decrease inventory and backlog costs. Our planning coordination layer employs an iterative interaction of a single production planning master-problem on the one hand with the hedging policy QoS layer and the performance evaluation layer's multiple decentralized facility-specific sub-problems on the other.

The effectiveness of our planning layer model depends crucially on the quality with which the operational dynamics of the production facilities are modeled in the performance evaluation and information layer. To this end our framework relies on the following two building blocks:

1. Dynamic Lead Times Modeling. We rely on performance analysis results for stochastic queueing networks to accurately estimate average weekly lead times as functions of capacity utilization, production mix, production policies, and distributions of stochastic disturbances such as failure and repair times. This provides delivery requirements to upstream facilities and available supply to downstream facilities, which are necessary for efficient planning of production over a multi week horizon. These non-linear lead time functions, denoted by $\bar{g}(\cdot)$, are incorporated as weekly constraints on decision variables in production scheduling.

2. Provisioning of Quality of Service (QoS) guarantees. We introduce constraints that bound the *probability* of backlog at a SC facility. We believe that probabilistic constraints reflect *customer satisfaction* considerations and follow closely industry practice of providing QoS guarantees. We model these guarantees as non-linear constraints in our production scheduling framework, denoted by $\bar{h}(\cdot)$.

The main purpose of this article is to demonstrate that dynamic lead times and hedging inventory requirements can be modeled and included in the tractable determination of faster SC production plans while maintaining desired quality of service guarantees. Our aim is to provide a proof of concept regarding the feasibility of modeling dynamic lead times and quality of service guarantees as part of the production planning process. We do not claim that we are proposing a perfect model of reality, particularly as far as the decentralized queueing network sub-problem model is concerned. Many extensions in this direction are needed. We plan to address many of them in future research. The proof of concept presented here provides us with confidence that the remaining problems are tractable.

Section 2 introduces the proposed time scale driven data communication architecture. The SC problem and the performance evaluation, QoS, and planning layers are described in Section 3. Section 4 provides computational experience that shows the value of dynamic lead time and probabilistic QoS constraint information in the determination of a SC's coordinated production schedule. A three facility SC producing five different part types is used to develop various representative examples of SCs. Comparison to production schedules that are characteristic of current industry practice indicates that substantial improvements are possible. The impact of convexity of constraint feasible regions on algorithmic convergence is also discussed in Section 4. Examples of lead time feasible regions, ranging from convex to mildly non-convex and severely non-convex, are employed for illustrative purposes in a simpler two facility SC. Extensions of the proposed algorithm rendering it robust to non-convex constraints are finally presented.

2. Time scale driven decentralized data communication and decision support architecture

The multitude of strategic, planning, and operational decisions made routinely by SC participants are far too complex and the requisite information is far too large to handle in a centralized manner. Decentralized decision making has therefore been the norm. However, since the consequences of various decisions are interdependent, it follows that appropriate coordination can foster desirable efficiencies. Consider a decentralized decision making agent as "a decision node" in a network of communicating decision nodes. A key determinant of successful coordination is the systematic conversion of data available at a certain decision making node i to a compact representation of information "relevant" to the decision making process at node j. Relevant is construed here to mean *incorporating all information about the state, dynamics, and decision policies in node i that may contribute to efficient decision making in node j*. Compact representations of relevant information may take, for example, the form of a *statistic*: the time averaged lead time in a production system, the probability distribution and autocorrelation of a demand process, or a *performance target*, such as the desired weekly output of a manufacturing process. These compact representations provide key enabling efficiencies in both the estimation of the relevant information (which can be done in a decentralized distributed manner) as well as in its communication (the transmission of a statistic requires less bandwidth and energy that the time series it describes). Although several

Table 5.1. Example of time scale driven classification.

Scope / Functionality	**Enterprise**	**Factory**	**Cell**	**Process**
Resource Allocation	Hardware Investment	Group Technology	Layout; Overtime	Tools; Time
Contingency Planning	Risk Diversification	Outsourcing Safety Stock	Oper. Pol.; Perf. Eval.	SQC Maintenance
Sequencing	Plant Production Schedule	Cell Production Schedule	Machine Schedule	Real-Time Proc. Control

issues are still to be resolved, intelligent communicating mobile sensor networks have the potential to both estimate and communicate relevant information in ways that are superior to conventional alternatives in terms of cost, flexibility, and reliability.

We propose a time scale driven assignment of SC decisions to nodes that is suggestive of the "relevant" information exchange architecture. The idea of time scale driven decomposition is not new. In fact it has been widely used to great advantage in control theory (Saksena, O'Reilly and Kokotovic, 1984). The main idea here is the fact that decisions are characterized by a characteristic frequency and its corresponding time scale. For example, while machine operating decisions are made every few minutes, major resource acquisition decisions are made every few months or years. We further notice that supply chain decisions characterized by functionality (for example resource allocation, planning, sequencing) and scope (for example enterprise, plant, cell, process) are associated with a decreasing time scale as the scope narrows and the functionality changes from resource allocation to sequencing. Table 5.1 provides such a classification example where time scales decrease as the decision of interest moves to the south east.

The SC planning algorithm proposed here employs a decentralized decision making and information exchange architecture that is an instantiation of the time scale driven approach of Table 5.1. Figure 5.1 presents the information exchange architecture that supports factory scope production planning decisions, cell level performance evaluation, and process level operation control. Note that:

- The factory production planning node passes down weekly production targets to each cell.

- Each cell evaluates its performance during each week in the planning horizon, determines variability distributions and aggregates its hourly dynamics to weekly time averages of relevant performance measures such as Work In Process (WIP), Lead Time (LT), and their sensitivity with respect to weekly production targets passed down from the factory planning node.
- Contiguous cells coordinate horizontally to determine safety inventory of semi-finished and finished goods that assures desired quality of supply levels at each cell and quality of service to customers.
- Each cell communicates weekly averages and sensitivities up to the factory planning node and variability distributions horizontally to upstream and downstream cells.
- Using WIP, LT and sensitivity information, the factory planning node adjusts production targets so as to achieve material flows across cells that meet required WIP and safety stock levels while minimizing SC WIP and LT.

While the remainder of the paper elaborates on the algorithms that use the information flow described above to reach a stable and optimal production plan, we wish to emphasize that the proposed information architecture in addition to distributing computational effort (performance evaluation and handling of short time scale stochastic dynamics modeling is done in a decentralized manner at each cell), it also limits communication requirements to relevant information. For example, the factory planning node does not need to know the cell production details: labor and other resources available, machine capacities and manufacturing process specifics. It needs to know, however, and it does know, the weekly

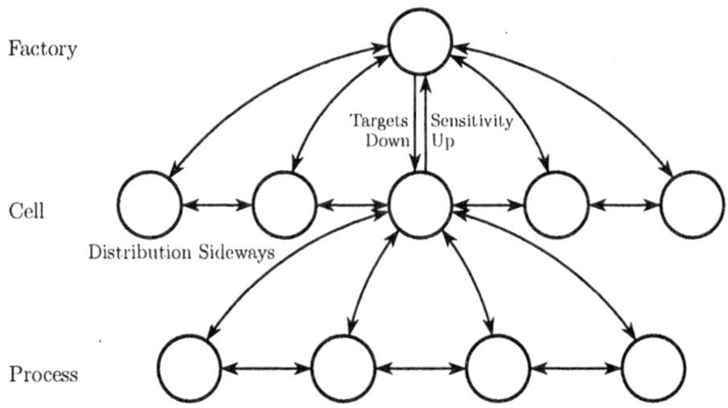

Figure 5.1. Example of information architecture.

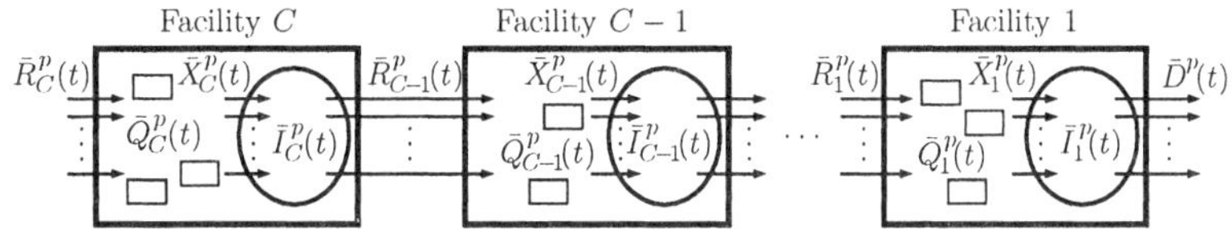

Figure 5.2. A multiclass SC with limited production capacity at each facility.

lead times at each cell and the hedging inventory between cells that are consistent with the production targets it sends to each cell. This is indeed the relevant information that enables superior production planning and SC speed up as we show in the computational results Section 4.

3. The SC management problem

3.1 SC problem overview

To describe our SC model and establish notation, we consider the system depicted in Figure 5.2 and the associated information exchange and decision layers are shown in Figure 5.3. Although a tree network of SC links or facilities can be modeled, we consider for ease of exposition, but without loss of generality, C production facilities connected in series. Production planning decisions and the resulting WIP and QoS hedging inventory requirements vary in the medium term (say across weeks) and the characteristic scale of their dynamics is called a period and denoted by $t \in \{1, 2, \ldots, T\}$. On the other hand, performance evaluation and demand dynamics vary many times within a period (say across hours) and their characteristic scale is called a time slot denote by the subscript k of t_k.

The QoS layer models random behavior of short term facility production capacity and final demand while the planning layer models expected values or time averages during a period (e.g., a week) under the underlying assumption that the period is long enough for the time slot stochastic process dynamics to reach steady state.

External demand is met from the available finished goods inventory in facility 1, and it is backordered if finished goods inventory (FGI) is not available. Every facility $c \in \{1, 2, \ldots, C\}$, produces a set of products and has a limited production capacity. Whereas facility C can draw from an infinite pool of inventory, production of facilities $(C-1, \ldots, 2, 1)$ is constrained by the production capacities of workstations and in addition by the work in process (WIP) in each facility. WIP is in turn constrained by (i) the FGI available upstream to release input into a facility that

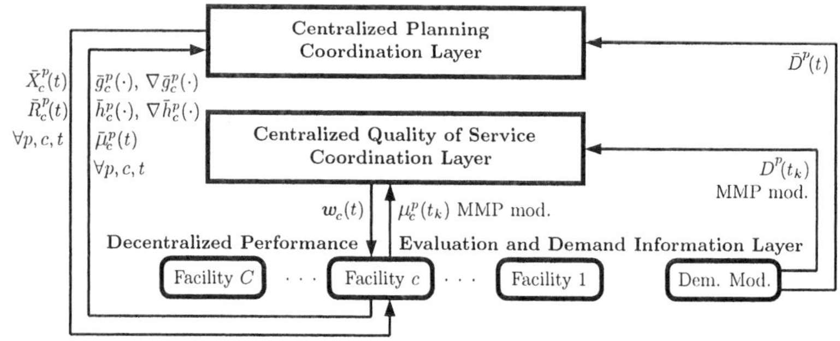

Figure 5.3. Information exchange amongst the coordination layers and the decentralized layer.

replenishes WIP and (ii) the production that depletes WIP. Again, for ease of exposition and without loss of generality, we assume all facilities process the same set of part types, $\{1, 2, \ldots, P\}$, and 1 part of part type p is required from an upstream facility $c+1$ to produce one unit of the same part type at facility c. The serial SC problem presented in Figure 5.2 retains the most salient features of the more general problem, particularly in terms of the general demand and service distributions we allow. As suggested by past experience in the literature (see Clark and Scarf, 1960; Lambrecht, Muckstadt and Luyten, 1984; Glasserman and Tayur, 1995), results for the simpler system can be routinely generalized to accommodate assembly/disassembly features.

$D^p(t_k)$ denotes the amount of external orders for product of class p arriving during time slot k of period t. $\mu^p_{c,m}(t_k)$ denotes the part type p production capacity in isolation of facility c workstation m during time slot k of period t. $X^p_c(t_k)$ denotes the number of type p parts facility c produces during time slot k of period t. $R^p_c(t_k)$ is the amount of WIP released into facility c from facility $c+1$ FGI. Finally, $Q^p_c(t_k)$ denotes the type p WIP at facility c, and $I^p_c(t_k)$ denotes the type p FGI at facility c available at time slot k of period t. Only at facility 1 we allow the FGI to take negative values to denote backordering. Following standard conventions, we use $\left(I^p_1(t_k)\right)^+$ and $\left(I^p_1(t_k)\right)^-$ to denote respectively $\max\{0, I^p_1(t_k)\}$ and $\max\{0, -I^p_1(t_k)\}$.

Since the period containing K time slots is the relevant time scale in the planning layer's dynamics, and since we assume it is long enough for the stochastic processes active at the time slot scale to reach steady state, we define the following time averaged variables: $\bar{R}^p_c(t) = (1/K)\sum_{k=1}^{K}$

$\mathbf{E}\left[R_c^p(t_k)\right]$, and similarly $\bar{X}_c^p(t)$, $\bar{Q}_c^p(t)$, $\bar{\mu}_c^p(t)$ and $\bar{I}_c^p(t)$. We use vector notation $\bar{\boldsymbol{X}}_c(t) = \left(\bar{X}_c^1(t), \ldots, \bar{X}_c^P(t)\right)$.

The SC management problem is implemented in three layers exchanging information as shown in Figure 5.3 and described below.

3.2 The performance evaluation and information layer

The performance evaluation and information layer shown in Figure 5.3 models the short term stochastic dynamics of production facilities at the operational level and develops the steady state or time averaged performance measure estimates of interest at the longer time scale of the planning layer. More specifically, by performance evaluation we mean:

1. The transformation of production targets in each period to estimates of minimum average WIP required during that period in each facility in order to meet the production targets set by the planning layer. This estimate will generally depend on production targets, $\bar{\boldsymbol{X}}_c(t)$, the probability distribution of all relevant random variables $\boldsymbol{P}_c(t)$, and other operational policies, $\boldsymbol{\pi}_c(t)$, during that period. The mapping of these inputs to the average WIP in facility c, $\bar{\boldsymbol{Q}}_c(t)$, is implicitly represented by function $\bar{g}_c^p\left(\bar{\boldsymbol{X}}_c(t), \boldsymbol{P}_c(t), \boldsymbol{\pi}_c(t)\right)$.

2. The estimation of sensitivities (or derivatives) of $\bar{g}_c^p(\cdot)$ with respect to production targets. This is needed for tractable representation of the highly non-linear relationship embodied in the $\bar{g}_c^p(\cdot)$ function.

3. The transformation of production targets from the planning layer, hedging inventory levels, $\boldsymbol{w}_c(t)$, from the QoS layer, and operational policies to the minimum average FGI required to meet the QoS constraint. The minimum average FGI requirements are represented by function $\bar{h}_c^p\left(\bar{\boldsymbol{X}}_c(t), \bar{\boldsymbol{X}}_{c-1}(t), \boldsymbol{P}_c(t), \boldsymbol{P}_{c-1}(t), \boldsymbol{w}_c(t), \boldsymbol{\pi}_c(t)\right)$. For purposes of demonstrating the concept of dynamic lead times associated with dynamic QoS guarantee provisioning, we consider here a limited pair-wise coupling of upstream and downstream facilities presented in Section 3.3.

4. The estimation of sensitivities (or derivatives) of the function $\bar{h}_c(\cdot)$ with respect to production targets. Again, to serve our proof of concept objective, we use relatively simple analytic approaches to the determination of $\bar{h}_c(\cdot)$ and its sensitivity requirement (see Section 3.3).

5. A representation of an aggregate probabilistic model of facility c short term capacity availability for use by the QoS layer. Whereas this can be in general a Markov Modulated Process model, we use here a simple, weighted bottleneck machine capacity exponential model, again in order to capture the correct factory physics and demonstrate the proof of concept in our planning layer algorithm. In practice, Markov Mod-

ulated Process models for production cells as well as for final demand can be estimated by analyzing possibly auto correlated production and shipment transaction databases (Paschalidis and Vassilaras, 2001).

The estimates produced by the performance evaluation algorithm are merely intended to demonstrate qualitatively appropriate behavior since our objective is to concentrate on an iterative planning layer. In real applications the performance evaluation and information layer can be implemented using more accurate approaches in a distributed/decentralized manner where efficiency and robustness is important but not crucial.

Production target decisions are realizable at the desired QoS level only if the requisite WIP and FGI are available at various facilities in a manner consistent with material conservation dynamics. Functions $\bar{g}_c^p(\cdot)$ and $\bar{h}_c(\cdot)$ establish the minimum WIP and FGI constraints employed at the planning layer discussed in Section 3.4.

The evaluation of $\bar{g}_c^p(\cdot)$ and $\bar{h}_c(\cdot)$ functions is a formidable task. Whereas we have used analytic models (Mean Value Analysis in this paper and the Queueing Network Analyzer elsewhere, see Caramanis, Pan and Anlı (2001a)) to study their behavior in an iterative decomposition algorithm and investigated convergence properties of the multilayer interactions depicted in Figure 5.3, Monte Carlo Simulation or complex analytical models involving Markovian or even more general stochastic decision processes are likely to be relevant in practice. Similar fluid approximation enhancements as those used in the deterministic algorithms of the planning layer, are also relevant in the context of stochastic models used at the decentralized layer (Caramanis, 1987; Kouikoglou and Phillis, 1991). These extensions are not trivial. For example, key events that are responsible for the efficiency of event driven simulation algorithms (e.g., a buffer fills or a buffer empties) proliferate (a buffer fills or empties partially with multiple partial full/empty states) requiring more sophisticated models (Caramanis, Wang and Paschalidis, 2003). The important advantage of fluid production stochastic models (whether simulation based or analytic) is their ability to provide sensitivity estimates more tractably than finite differencing of stochastic discrete production models.

Finally, the convexity of the feasible regions defined by the $\bar{g}(\cdot)$ and $\bar{h}(\cdot)$ functions is crucial to the convergence of the coordination layer. Figure 5.4 shows a realistic example of the feasible region boundaries for a two part type stochastic production network. More specifically, the maximum value of $\bar{X}_c^1(t)$ subject to $\bar{Q}_c^1(t) \geq \bar{g}_c^1(\bar{\boldsymbol{X}}_c(t), \boldsymbol{P}_c(t), \boldsymbol{\pi}_c(t))$ is plotted versus $\bar{Q}_c^1(t)$ and $\bar{X}_c^2(t)$.

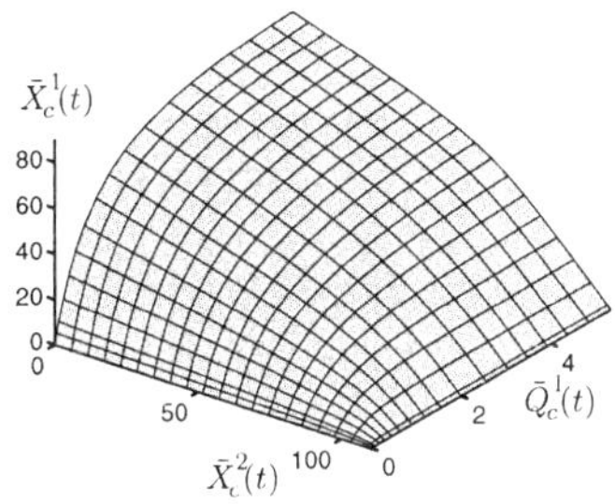

Figure 5.4. Feasible production $\bar{X}_c^p(t)$ as a function of WIP $\bar{Q}_c^p(t)$.

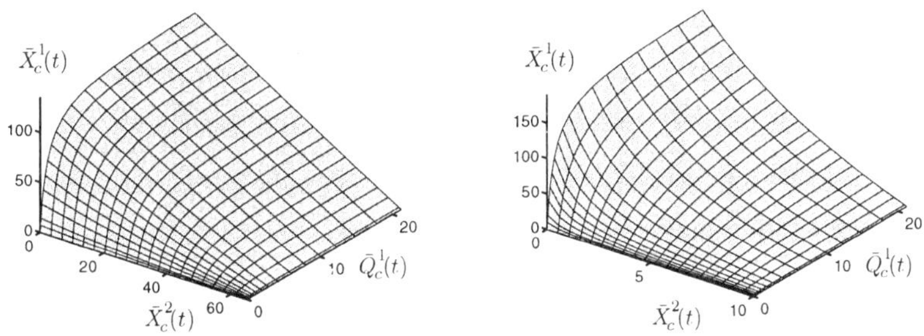

Figure 5.5. Examples of mildly non-convex and severely non-convex feasible regions.

Although the above constraints exhibit generally convex feasible regions, we have observed non-convex feasible regions that arise when either operational policies are flagrantly sub-optimal or facility designs are far from homogeneous (e.g., product classes impose diverse production time requirements on facility workstations, see Caramanis and Anlı, 1999b). Consider Figure 5.5 depicting mildly non-convex and severely non-convex feasible regions in contrast to the convex example in Figure 5.4. We have been able to construct robust iterative master-problem sub-problem algorithms that converge even under severe non-convexity conditions (Anlı, 2002). The impact of the non-convexities on optimization convergence of the proposed iterative algorithm is discussed in Section 4.2.

3.3 The Quality of Service coordination layer

The QoS layer interacts with other layers as shown in Figure 5.3. Its objective is to estimate a production policy that achieves the desired probabilistic QoS guarantees. Accurate modeling of QoS Coordination policies is an important research problem in itself that we are paying attention to Bertsimas and Paschalidis (2001); Paschalidis and Liu (2003); Paschalidis et al. (2004). A number of methodologies have been used including Multiclass Queuing Network Analysis (QNA), Monte Carlo simulation, Stochastic System Approximations, and large deviations asymptotics.

Given our stated objective to demonstrate the ability to construct a robust and efficient planning layer, we elected a simple but certainly near optimal (Liberopoulos and Caramanis, 1995) hedging policy that works as follows: Facility c produces at full capacity as long as the amount of work in its output buffers summed across all part types is below the hedging inventory level, $w_c(t)$, set for that week by the QoS layer. With the average release into the downstream cell from the FGI buffer of facility c equal to the downstream facility's production target, the hedging inventory level is selected by the QoS layer so that the probability of a stock out of the downstream facility does not exceed the desired level. We implement this idea using the following assumptions:

1. The desired stockout probability is selected according to a function shown in Figure 5.6 that is monotonically decreasing as the utilization of the downstream facility increases. This is a reasonable assumption since a production facility with a lot of extra capacity can withstand temporary stock outs more flexibly than a highly utilized facility.

2. The hedging inventory level is estimated from the stock out probability specified under 1 above by using an M/M/1 system where the arrival rate is equal to the downstream facility's production target while the processing rate equals to the weighted bottleneck of upstream facility work stations.

The hedging inventory policy described above achieves a constant QoS between facilities. Moreover, it decouples the implementation of operational decisions across facilities. Furthermore, we employ an aggregation to approximate multiple part types as a single part type and construct a fictitious workstation to represent the whole facility. A maximum tolerable starvation probability, $\epsilon_c(t)$, is empirically calculated as a function of workload in facility $c-1$ (a weighted average of utilizations of workstations). Examples of this empirical parametric function are plotted in Figure 5.6. The hedging inventory level that limits starvation to $\epsilon_c(t)$ is then obtained using the M/M/1 model described above. Specifically, noting that the level of the queue, $q_c(t_k)$, in the M/M/1

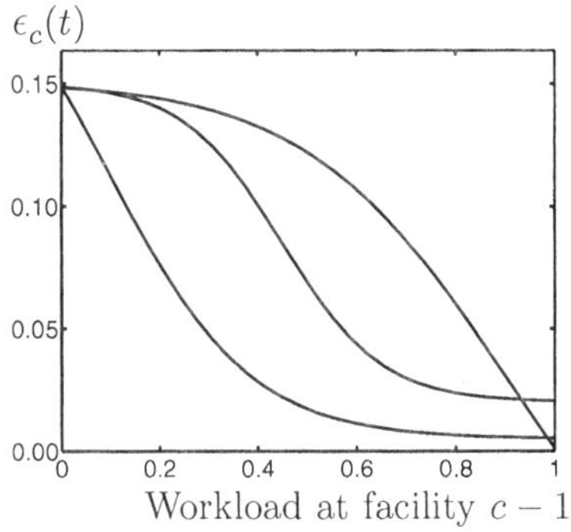

Figure 5.6. Maximum tolerable starvation probability as a function of the workload in facility $c-1$.

model is equal to the negative of the FGI in the real system, ranging from -$\boldsymbol{w}_c(t)$ to $+\infty$. We then find a hedging point level -$\boldsymbol{w}_c(t)$ at which the departure process (corresponding in the model to the upstream facility's production capacity) is starved in such a manner that $\text{Prob}\left(q_c(t_k) > 0\right) = \epsilon_c(t)$. The average hedging inventory is then evaluated as $\sum_{i=-\boldsymbol{w}_c(t)}^{0} |i| \, \text{Prob}\left(q_c(t_k) = i\right)$, and denoted by the non-linear function $\bar{h}_c\left(\bar{\boldsymbol{X}}_c(t), \bar{\boldsymbol{X}}_{c-1}(t), \boldsymbol{P}_c(t), \boldsymbol{P}_{c-1}(t), \boldsymbol{w}_c(t), \boldsymbol{\pi}_c(t)\right)$.

This allows us to express the FGI constraint function in terms of work by calculating the average aggregate FGI level and requiring a minimum weighted average to be met by the planning problem's FGI variables, $\bar{I}_c^p(t)$, $\forall p$. Figure 5.7 shows an instance of the FGI constraint function described above. Three identical machines with production capacity of 50 parts per time period have been assumed in the upstream and downstream facilities.

3.4 The planning coordination layer

The planning layer shown in Figure 5.3 and its role in the collaborative framework where the Master and Sub-problem layers interact in an iterative algorithm that produces the optimal production plan are described next.

3.4.1 The master-problem optimization algorithm. Given the longer (weekly) time scale of the planning layer, we use a Linear Programming (LP) based fluid model approximation of the discrete part production planning problem. Moreover, we extend the fluid model to

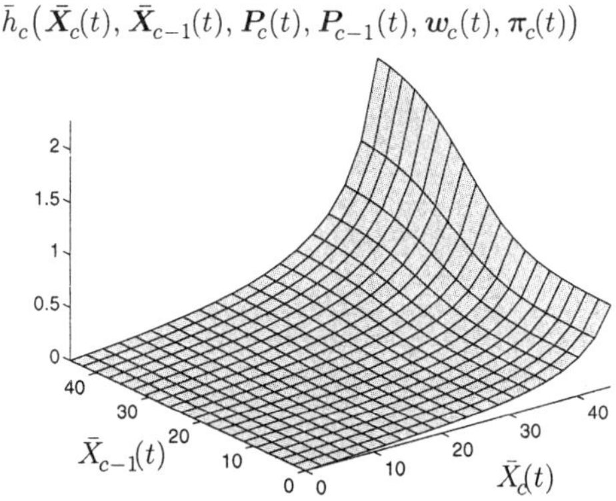

Figure 5.7. Horizontal coordination constraint function, $\bar{h}_c(\bar{\mathbf{X}}_c(t), \bar{\mathbf{X}}_{c-1}(t), \bar{\mathbf{P}}_c(t), \bar{\mathbf{P}}_{c-1}(t), \mathbf{w}_c(t), \boldsymbol{\pi}_c(t))$, for a single part type system.

represent WIP and FGI driven dynamic lead times through the non-linear constraint surfaces $\bar{g}_c^p(\cdot)$ and $\bar{h}_c(\cdot)$ defined in Section 3.2 and Section 3.3. These two constraints are key components of the planning layer model (see equations (5.6) and (5.7) in Exhibit 1).

The planning layer's objective is to determine targets of weekly production and release that minimize weighted WIP, FGI, and backlog costs over the horizon T, subject not only to the capacity and material conservation constraints, but also the non-linear constraints on weekly average WIP and hedging inventory that capture the essence of faster time scale stochastic dynamics driven by QoS provisioning and other operational policies that may be in place at some facility.

The Planning Layer solves the optimization problem given in Exhibit 1 where minimization takes place over weekly production target, $\bar{X}_c^p(t)$, and release, $\bar{R}_c^p(t)$, decision variables. $\bar{\alpha}_c^p$, β_c^p and γ_1^p denote medium term WIP holding, FGI holding and backorder cost rates[1], respectively. The optimization problem of Exhibit 1 is subject to the usual positivity, initial condition, capacity (5.5), material conservation (5.2), (5.3), and (5.4) and non-linear constraints (5.6) and (5.7) that capture lead time and QoS dynamics. $P_c(t)$ and $\pi_c(t)$ are as defined in Section 3.2. $\eta_c(t)$ is the maximum allowed utilization level for workstations in fa-

[1]Note that the costing of Backorders at the planning layer does not contradict the probabilistic QoS guarantees provided by the hedging point policy implemented through production policy thresholds $w_c(t)$. The planning layer's observation of constraint (5.3) assures that weekly production targets are consistent with probabilistic QoS requirements.

$$\min_{\bar{X}^p_c(t),\bar{R}^p_c(t)} \sum_{p,t} \left[\sum_{c=2}^{C} \left(\bar{\alpha}^p_c \bar{Q}^p_c(t) + \bar{\beta}^p_c \bar{I}^p_c(t)\right) + \bar{\beta}^p_1 \left(\bar{I}^p_1(t)\right)^+ + \bar{\gamma}^p_1 \left(\bar{I}^p_1(t)\right)^- \right] \quad (5.1)$$

subject to:

$$\bar{Q}^p_c(t+1) = \bar{Q}^p_c(t) + \bar{R}^p_c(t) - \bar{X}^p_c(t) \qquad \forall p, c, t \quad (5.2)$$

$$\bar{I}^p_c(t+1) = \bar{I}^p_c(t) + \bar{X}^p_c(t) - \bar{R}^p_{c-1}(t) \qquad \forall p, c \geq 2, t \quad (5.3)$$

$$\bar{I}^p_c(t+1) = \bar{I}^p_c(t) + \bar{X}^p_c(t) - \bar{D}^p(t) \qquad \forall p, c = 1, t \quad (5.4)$$

$$\sum_{p=1}^{P} \frac{\bar{X}^p_c(t)}{\overline{\mu}^p_{c,m}(t)} \leq \eta_c(t) \qquad \forall m, c, t \quad (5.5)$$

$$\bar{Q}^p_c(t) \geq \bar{g}^p_c\left(\bar{\boldsymbol{X}}_c(t), \boldsymbol{P}_c(t), \boldsymbol{\pi}_c(t)\right) \qquad \forall p, c, t \quad (5.6)$$

$$\sum_{p=1}^{P} \tau^p_c(t) \bar{I}^p_c(t) \geq \bar{h}_c\left(\bar{\boldsymbol{X}}_c(t), \bar{\boldsymbol{X}}_{c-1}(t), \boldsymbol{P}_c(t), \boldsymbol{P}_{c-1}(t), \boldsymbol{w}_c(t), \boldsymbol{\pi}_c(t)\right) \qquad \forall p, c \geq 2, t \quad (5.7)$$

Exhibit 1: Planning layer optimization problem.

cility c during time period t. Vector $\boldsymbol{w}_c(t)$ is determined at the QoS layer. It denotes the hedging Inventory implemented by facility c to control its production schedule without causing downstream stock out with frequency exceeding the desired probabilistic QoS constraint. The coefficients $\tau^p_c(t)$ in constraint (5.7) are the weights that convert each unit of FGI, $\bar{I}^p_c(t)$, to units of work, namely the units of the right hand side, $\bar{h}_c(\cdot)$. In this instance, work is the processing time needed at the upstream facility to produce the required amount of inventory. We use $\tau^p_c(t)$ equal to the reciprocal of the minimum production capacity for part type p of all machines in facility c, $\tau^p_c(t) = \left[\min_{\forall m \in M_c} \mu^p_{c,m}(t)\right]^{-1}$.

3.4.2 Iterative algorithm for layer collaboration and coordination. An iterative single master-problem (Centralized Planning Coordination Layer) multiple sub-problem (Decentralized Performance and Information Layer) algorithm has been developed to model the non-linear constraints and derive optimal production plans that explicitly account for variable lead time and QoS constraints (see Figure 5.3).

The efficient representation of constraints 5.6 and (5.7) in Exhibit 1 requires point estimates and sensitivity estimates so as to approximate the non-linear constraint boundaries. In fact, the iterative fine-tuning of a finite number of appropriately selected local approximations leads to convergence under mild convexity or quasi convexity conditions (Caramanis and Anlı, 1999b). We summarize below this iterative employment of master and sub-problems:

1. The Planning Layer calculates tentative production and material release schedules $\bar{X}_c^p(t)$, $\bar{R}_c^p(t)$ $\forall p, c, t$, and conveys them to facility sub-problems along with the WIP and FGI levels $\bar{Q}_c^p(t)$, $\bar{I}_c^p(t)$, that result from the production schedule through the material flow constraints (5.2), (5.3), and (5.4) in Exhibit 1.

2. Each facility performs local planning conditional on the production and release targets obtained from the Planning Layer. The Quality of Service Layer proceeds to calculate hedging points $\boldsymbol{w}_c(t)$ that can achieve probabilistic QoS constraints at each facility in the presence of upstream and downstream production targets, and production capacities at each pair of upstream/downstream facilities. Each facility determines in a decentralized manner production policies that can best attain planning layer production targets. Finally each facility evaluates its performance and provides for each time period the following feedback to the planning layer:

(i) the required WIP and hedging inventory ($\bar{g}_c^p(\cdot)$, $\bar{h}_c(\cdot)$), and

(ii) their sensitivities ($\nabla \bar{g}_c^p(\cdot)$, $\nabla \bar{h}_c(\cdot)$) w.r.t. the production targets. Each facility evaluates its performance and provides the Planning Layer with hyper planes tangent to functions ($\bar{g}_c^p(\cdot)$ and $\bar{h}_c(\cdot)$) at the most recent iteration's production target values $\bar{X}_c^p(t)$. Using the tangent hyper planes accumulated over past iterations, the Planning Layer solves again the master-problem as a Linear Program to produce a new set of tentative targets that are associated with additional tangent hyperplanes added to the master-problem for use in its next iteration. Note that the representation accuracy of the non-linear constraint surfaces increases monotonically with the addition of tangent hyper planes. As the problem converges, most tangent hype planes are added in the vicinity of the optimal solution rendering the accuracy error arbitrarily small.

We next employ an example to introduce the basic idea of lead time dynamics at the coarse time scale planning layer and its significance in yielding a superior production plan. Consider a simple manufacturing facility consisting of a 10 workstation transfer line. Each workstation has stochastic processing time and average production capacity of 10 parts per week. This is also the capacity of the facility.

Figure 5.8 shows the average weekly facility production rate, $\bar{X}$, that is achievable as a function of the average available WIP, denoted by $\bar{Q}$, under three, successively more accurate, fluid model approximations (FMAs). The *Naive* FMA does not impose any average WIP constraints, and the production rate is simply constrained by the average production capacity. The *certainty equivalent* FMA considers deterministic workstation capacities equal to the average capacity availability. As a result it disregards queueing delays. It underestimates the average WIP needed for sustaining a certain average production rate, setting it equal to the sum of the 10 machine utilizations. This is a linear constraint involving Production and WIP. Finally, the *Optimal Open Loop Feedback Controller* (OOLFC) employs a stochastic model that captures the fact that effective workstation processing times are random variables. This results in the correct estimate of average WIP, and hence of average Lead Time, which reflects the average queue levels in front of work stations in addition to the average number of parts being processed.

In the context of the SC planning coordination layer, the OOLFC model approximation is implemented via the inclusion of non-linear constraints on production $\bar{X}_c^p(t)$ represented by the WIP and FGI functions, $\bar{g}_c^p(\cdot)$ and $\bar{h}_c(\cdot)$, introduced in Sections 3.2 and 3.3. We have successfully implemented several approaches to handle these constraints, so far based on iterative tangent or piece-wise linear approximations (Caramanis, Pan and Anlı, 2001a) that allow LP optimization.

The non-linear constraints render the production plans generated not only feasible but also superior to plans obtained by alternative deterministic approaches based on certainty equivalent formulations (Connors, Feigin and Yao, 1994; Caramanis and Anlı, 1999a,b). The introduction

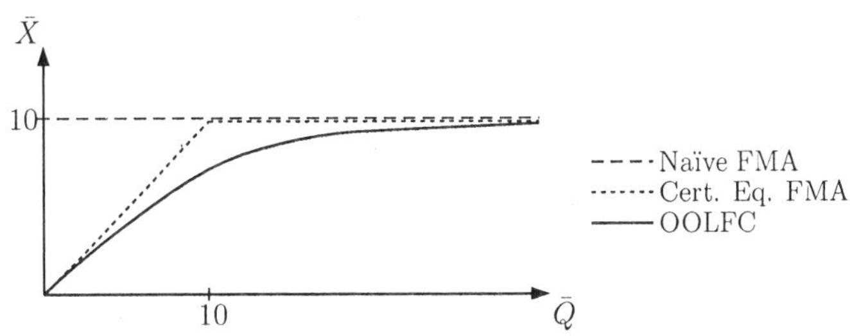

Figure 5.8. Feasible production as a function of WIP with various deterministic fluid and stochastic model approximations.

of the non-linear constraints is equivalent to formulating and solving the open loop optimal controller known to dominate the certainty equivalent controller (Bertsekas, 1995). The certainty equivalent controller, possibly adjusted to model a larger, worst case, constant lead time delay in each activity, represents today's industry practice, and, as such, it is compared to the proposed open loop optimal controller in Section 4.

4. Algorithmic implementation and numerical evaluation of benefits

In this section we report computational experience in the following two areas:

Firstly, we discuss computational experience of the proposed dynamic lead time and QoS hedging Inventory modeling algorithm that we refer to as Optimal Open Loop Feedback Controller (OOLFC). We compare OOLFC performance to two algorithms that represent industry practice in order to explore the value that one may attribute to the dynamic lead time and QoS hedging Inventory information employed by our OOLFC approach.

Secondly, we present a robust control extension to the basic OOLFC algorithm that enables it to handle difficult cases in which a facility's operational control is mismanaged or the part type family assigned to a facility is poorly designed to consist of severely non-similar part types. In these cases, the non-linear lead time constraints are non-convex and do not allow outer linearization. The proposed algorithmic extension is shown to overcome the complication introduced by this type of convexity failures.

4.1 Computational experience and the value of dynamic lead time and QoS hedging inventory information

Whereas we have previously reported production planning algorithms that model explicitly dynamic lead times (Caramanis, Pan and Anlı, 2001a,b; Caramanis and Anlı, 1999a,b; Caramanis, Paschalidis and Anlı, 1999; Caramanis and Anlı, 1998), we present here for the first time results reflecting probabilistic QoS policies.

Numerical experience from 6 demand scenarios applied to a 5 part type 3 facility SC shown in Figure 5.9 follows. Facilities 3, 2, 1 have 5, 4, and 6 workstations, respectively, with average production capacities as shown in Table 5.2, where part types correspond to columns and workstations to rows. Facilities are allowed a maximum workstation utilization of $\eta_c(t) = 0.9$. Initial levels of WIP and FGI are set to zero in all facilities. The cost coefficients are shown in Table 5.3.

The first 2 demand scenarios consist of a 23 week planning horizon with constant demand for each of the five part types during weeks 5–19 equal to vectors of $(14, 7, 5, 6, 7)$ and $(11, 11, 8, 3, 7)$. Weeks 1–4 and 20–23 were on purpose assigned zero demand to allow the algorithm to fill

Table 5.2. Average workstation production capacities.

		p.t.1	p.t.2	p.t.3	p.t.4	p.t.5
Facility 3	$m_{3,1}$	70	75	75	85	70
	$m_{3,2}$	70	65	65	70	85
	$m_{3,3}$	55	60	75	90	80
	$m_{3,4}$	55	70	75	80	75
	$m_{3,5}$	55	55	80	65	90
Facility 2	$m_{2,1}$	55	55	55	65	70
	$m_{2,2}$	55	55	60	55	70
	$m_{2,3}$	55	65	55	70	65
	$m_{2,4}$	55	55	65	70	65
Facility 1	$m_{1,1}$	40	40	60	55	65
	$m_{1,2}$	35	50	50	65	65
	$m_{1,3}$	40	40	55	65	70
	$m_{1,4}$	40	35	50	55	70
	$m_{1,5}$	35	50	60	55	65
	$m_{1,6}$	35	50	55	65	70

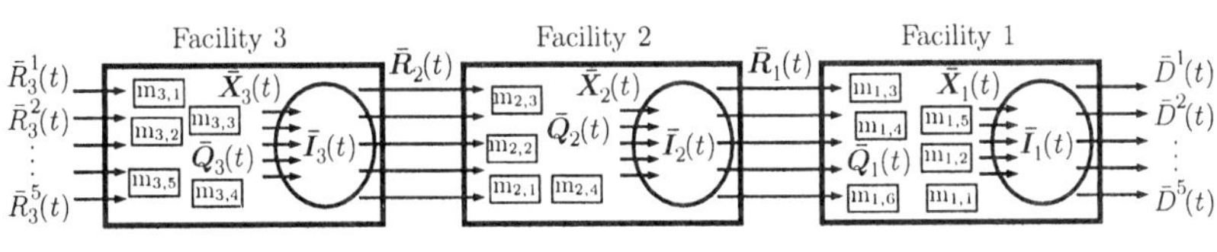

Figure 5.9. An example supply chain with 3 facilities producing 5 part types.

Table 5.3. WIP, FGI holding and FGI backlog cost coefficients.

		p.t.1	p.t.2	p.t.3	p.t.4	p.t.5
Facility 3	WIP	20	15	10	10	5
	FGI^+	27	20	12	23	8
Facility 2	WIP	30	25	15	30	10
	FGI^+	55	40	40	42	26
Facility 1	WIP	60	50	50	45	30
	FGI^+	150	110	120	100	70
	FGI^-	3000	2200	2400	2000	1400

Table 5.4. Weekly demand for scenarios 3, 4, 5, and 6; weeks 6 through 18.

		6	7	8	9	10	11	12	13	14	15	16	17	18
Demand Scenario 3	p.t. 1	23	3	5	13	27	18	0	14	24	13	0	4	5
	p.t. 2	7	6	23	16	2	5	23	12	0	8	7	17	19
	p.t. 3	0	10	2	2	1	5	8	3	3	2	12	4	2
	p.t. 4	0	23	5	6	6	11	7	6	9	12	25	10	4
	p.t. 5	10	8	4	2	3	4	4	7	5	9	8	8	13
Demand Scenario 4	p.t. 1	14	2	7	25	46	23	0	24	26	13	0	5	2
	p.t. 2	4	4	33	29	4	6	37	21	0	8	8	21	8
	p.t. 3	0	8	3	3	2	6	13	6	3	2	14	5	1
	p.t. 4	0	17	8	11	10	13	12	10	10	12	31	11	2
	p.t. 5	6	6	6	3	5	5	6	11	6	9	9	10	6
Demand Scenario 5	p.t. 1	0	30	13	20	17	23	20	17	17	20	53	20	3
	p.t. 2	12	2	6	23	41	21	6	10	17	6	6	14	2
	p.t. 3	4	17	9	7	9	8	8	8	8	8	9	19	6
	p.t. 4	2	9	3	3	2	1	1	4	12	10	5	10	4
	p.t. 5	2	2	14	12	2	3	3	3	0	9	10	14	4
Demand Scenario 6	p.t. 1	0	47	20	31	27	23	14	12	12	14	37	14	2
	p.t. 2	15	2	7	28	51	21	4	7	12	4	4	10	1
	p.t. 3	6	23	13	10	13	8	6	6	6	6	6	13	4
	p.t. 4	3	16	5	5	3	1	1	3	8	7	4	7	3
	p.t. 5	3	3	19	17	3	3	2	2	0	6	7	10	3

and empty the system optimally without disadvantaging the sub-optimal industry practice alternative planning algorithms considered. The last 4 demand scenarios also have planning horizons of 23 weeks, with demand during weeks 1–5 and 19–23 set to zero. The first and the last 5 weeks have zero demand for reasons similar to those stated. Demand scenarios 3, 4, 5 and 6 during weeks 6–18 are given in Table 5.4. Part types correspond to rows.

Facility loading required to meet demand in a just in time manner is quite even in scenario 3 while production mix varies more to reflect product specific volatility even under level overall loading. Scenarios 4, 5, and 6 represent increasingly more costly situations for the SC where, for some intermediate time periods in the planning horizon, namely periods (12–19), (11–20), and (8–20), respectively, the cumulative demand load exceeds the cumulative production capacity.

To compare the proposed SC planning algorithm to state of the art practice in industry we distinguish between three planning approaches:

(i) The Optimal Open Loop Feedback Controller approach (OOLFC) which corresponds to the algorithm that we propose in this article as it is described in Section 3.4. It models the dynamics of required lead

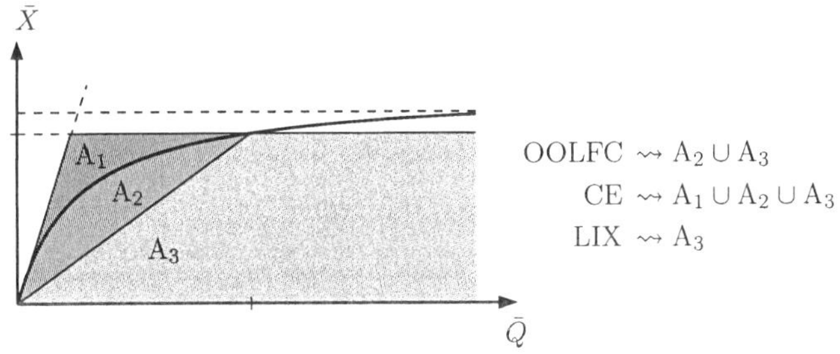

Figure 5.10. OOLFC, CE, and LIX.

times and QoS hedging inventory by employing the non-linear constraint hyper plane outer linearizations. For a one dimensional case, this can be shown symbolically as corresponding to use as feasible decision variable region the area equivalent to the union of areas A_2 and A_3 in Figure 5.10. Little's law can be easily employed to convert the WIP, denoted by $\bar{Q}$, versus Production, denoted by $\bar{X}$ space to a Lead time versus Production space.

(ii) The Certainty Equivalent (CE) approach that uses an optimistic feasible region which corresponds to the assumption that facility lead times consist of processing times alone and that there are no queueing delays. Symbolically, the feasible decision space can be shown by the union of areas A_1, A_2, and A_3 in Figure 5.10. We implement this approach by solving the production planning problem with the optimistic fixed lead time constraint and no QoS constraints. This gives a production plan that violates the actual lead time as well as QoS constraints. The raw material release trajectory in this infeasible plan is then used as a constraint and the best feasible production plan is found whose raw material release trajectory reflects the exceedingly optimistic short lead time assumptions described above. As a result, the CE plan, almost surely leads to sizable shortfalls and its performance is disadvantaged by the associated backlog costs. A more realistic representation of industry practice that is still based on a fixed lead time assumption but allows for queueing delays is also considered below.

(iii) The Limited Information Exchange (LIX) approach relies on a constant-hence limited information exchange of master and sub-problems-lead time and QoS hedging inventory. However, these constant lead time and QoS hedging inventory constraints are set conservatively equal to a linear interpolation between zero when capacity utilization is zero and the lead time and hedging inventory needed when capacity utilization equals the highest allowable level (this is $\eta_c(t)$ in the problem definition of

Exhibit 1 and Section 3.4). To make sure that a feasible production plan is generated, the non-linear lead time and QoS constraints are imposed in addition to these linear conservative constraints. Symbolically, the feasible decision space can be shown as the area which is the union A_3 of Figure 5.10. The LIX approach should be expected to give excellent schedules under stable conditions when the production facilities are level loaded, whereas when demand varies over time the resulting production plans will be associated with higher than necessary WIP levels. Some combination of the LIX conservative constant lead time approach and the optimistic one under CE is quite representative of industry practice. Cost comparisons may therefore be interpreted accordingly.

Figures 5.11, 5.12, and 5.13 present the optimal solution time trajectories for facilities 1, 2 and 3 of the proposed OOLFC approach under demand scenario 4. Part type specific quantities are shown from top to bottom. The release rate of parts into each facility, average WIP in each facility, facility specific production rates, and FGI (facility 1) or hedging inventory (facilities 2 and 3) trajectories are shown. Demand trajectories are superimposed for easy reference on FGI trajectories as a bar graph on the same scale. We observe that for part types 1, 2, and 5 production plans at facility 1 are characterized by significant pre-production during time periods 1 through 12, with demand fulfilled in a just-in-time manner and with relatively low FGI during the rest of the planning horizon. Examining the cost coefficient and production capability structure of this production system, we notice that part types 1 and 2 are associated with relatively higher cost and capacity, and hence are the as most likely candidates for build up of FGI in advance of the time demand occurs. The justification for high production rates of part type 5 can be sought in the examination of the solution details associated with facilities 2 and 3 in Figures 5.12 and 5.13.

In these two figures the last column depicts the hedging inventory after the associated facility. The aggregate nature of the FGI constraints (5.7) allows the optimization algorithm to free capacity by building inventory in one part type so that, in the event of a short fall in an other part type, it can assign its free capacity to it. Computational experience indicates that hedging inventory is built primarily for one part type –the one with the lowest inventory holding cost– to satisfy constraint (5.7) in the least costly manner. An example of this is seen in Figure 5.12 associated with facility 2 where part type 4 hedging inventory is built during time periods 1 through 8, and part type 5 hedging inventory is built during time periods 6 through 16.

In Figures 5.14, 5.15, and 5.16 we compare the production targets of facility 1 obtained by each of the three approaches. Solid, dashed and

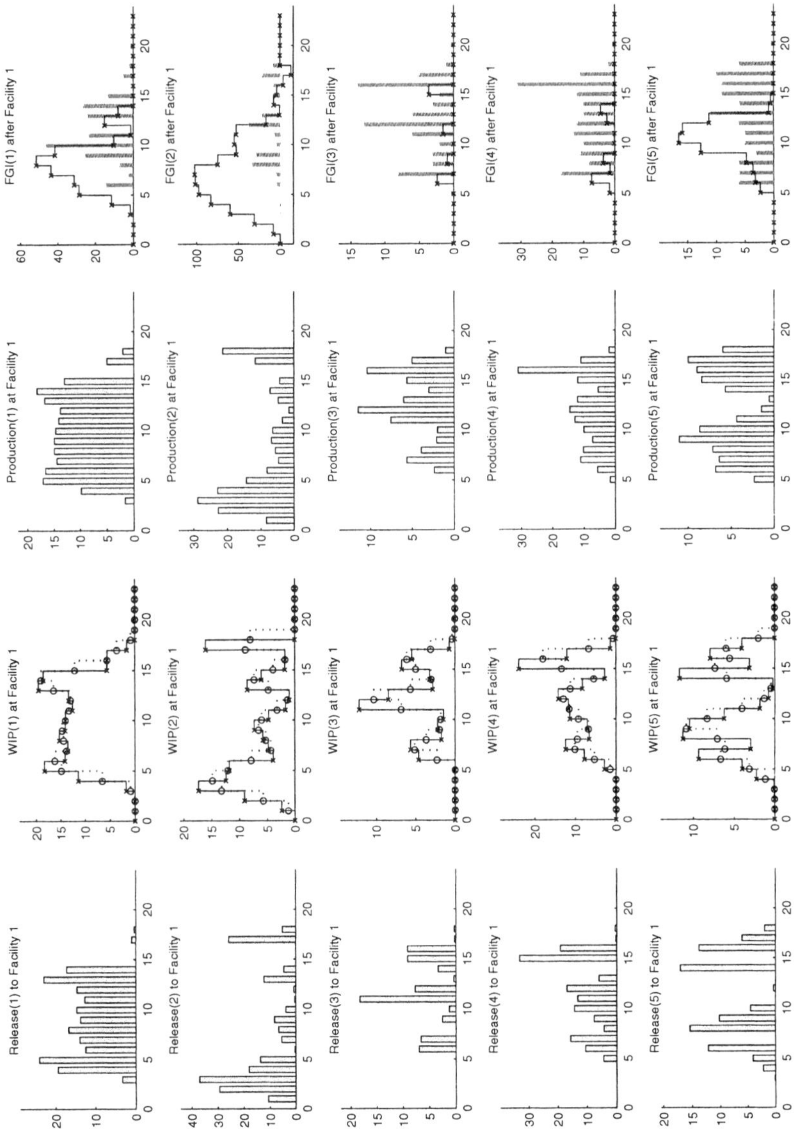

Figure 5.11. Facility 1 optimal solution details of OOLFC under demand scenario 4.

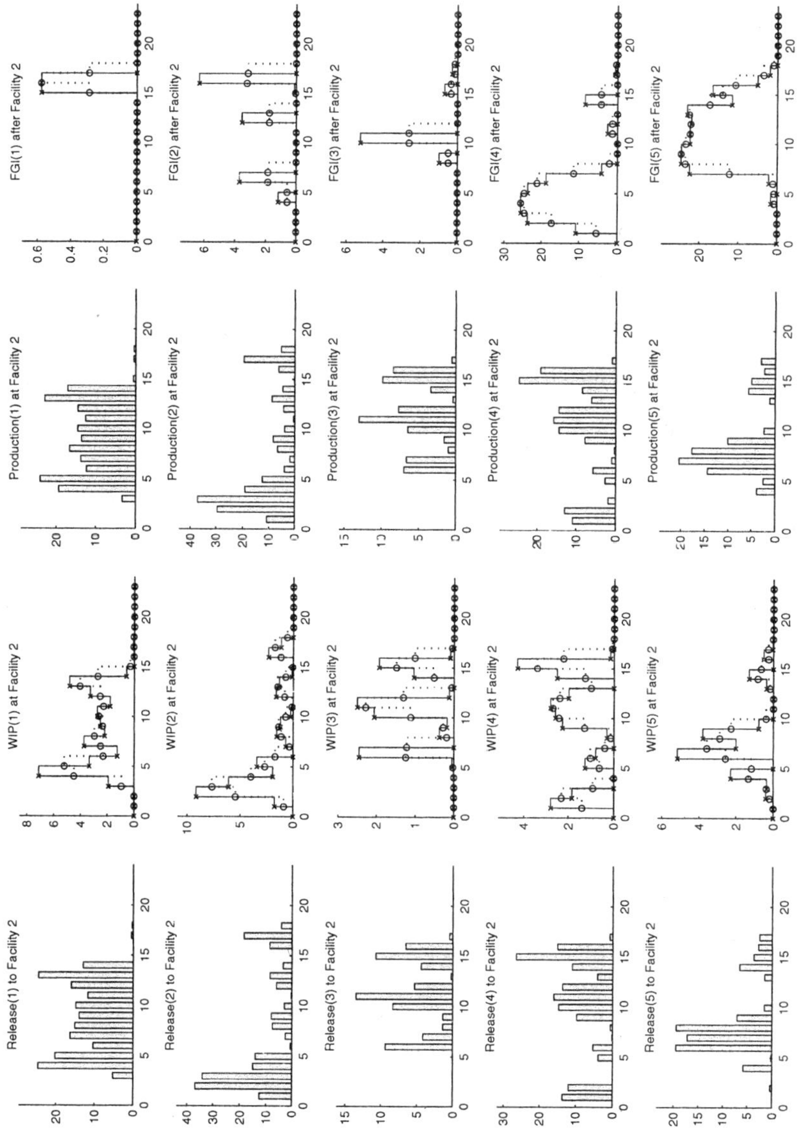

Figure 5.12. Facility 2 optimal solution details of OOLFC under demand scenario 4.

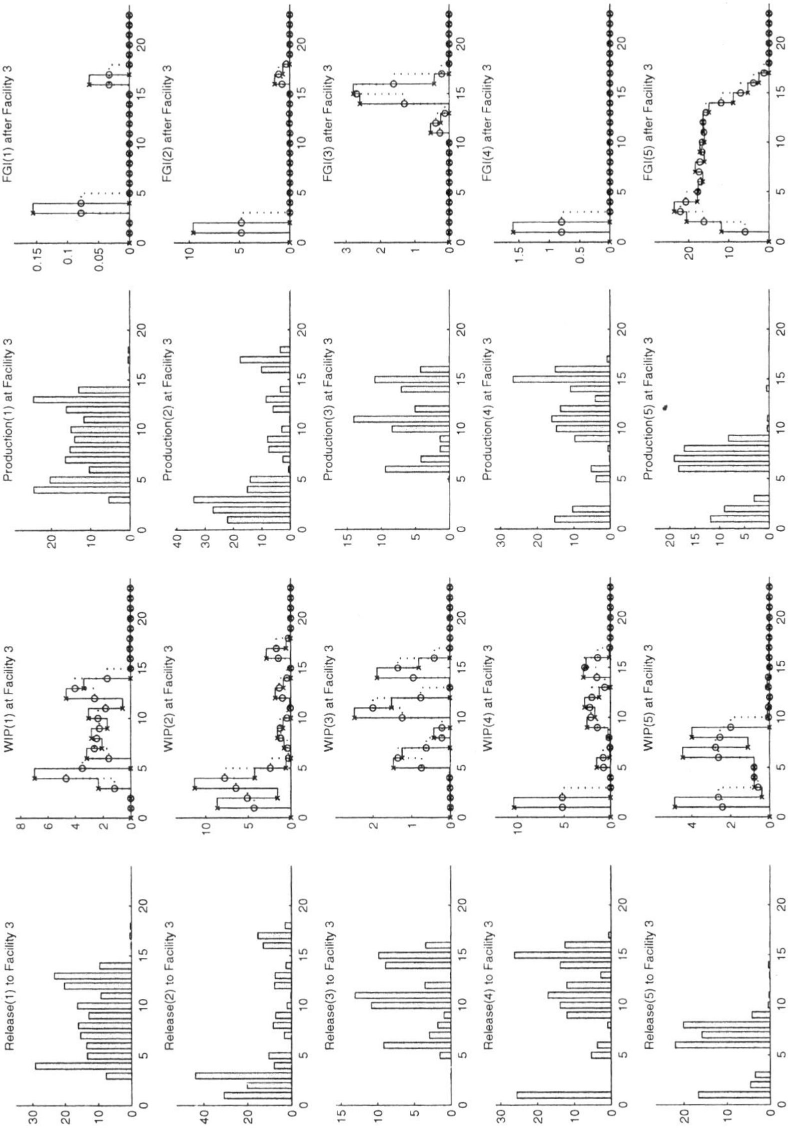

Figure 5.13. Facility 3 optimal solution details of OOLFC under demand scenario 4.

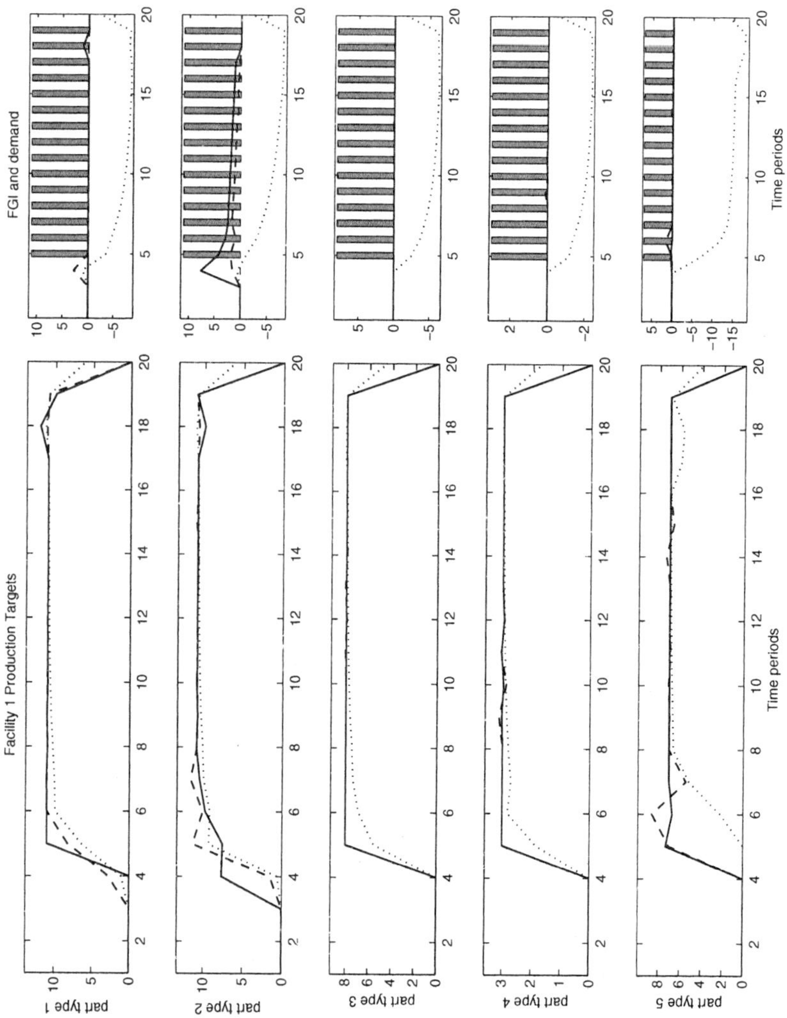

Figure 5.14. Facility 3 production targets and FGI levels comparison among OOLFC, LIX, and CE approaches under demand scenario 2.

dotted line styles are used to distinguish OOLFC, LIX and CE approach weekly production target trajectories for three different demand scenarios. Part type 1 through 5 production is shown from top to bottom respectively.

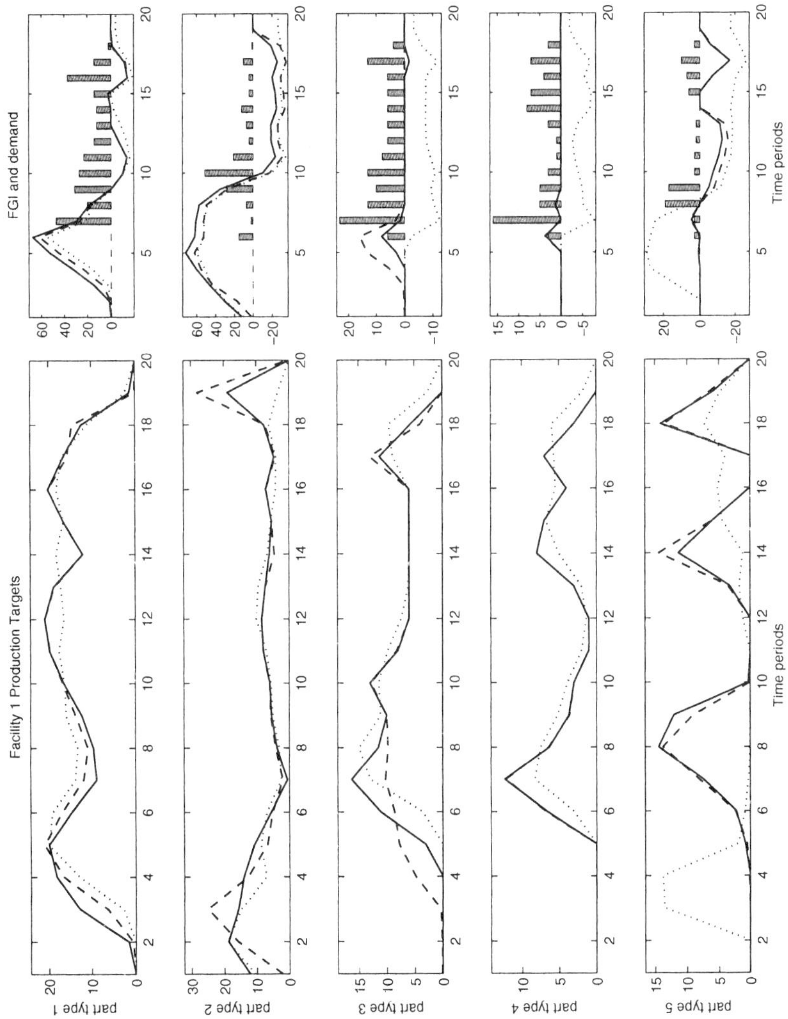

Figure 5.15. Facility 3 production targets and FGI levels comparison among OOLFC, LIX, and CE approaches under demand scenario 6.

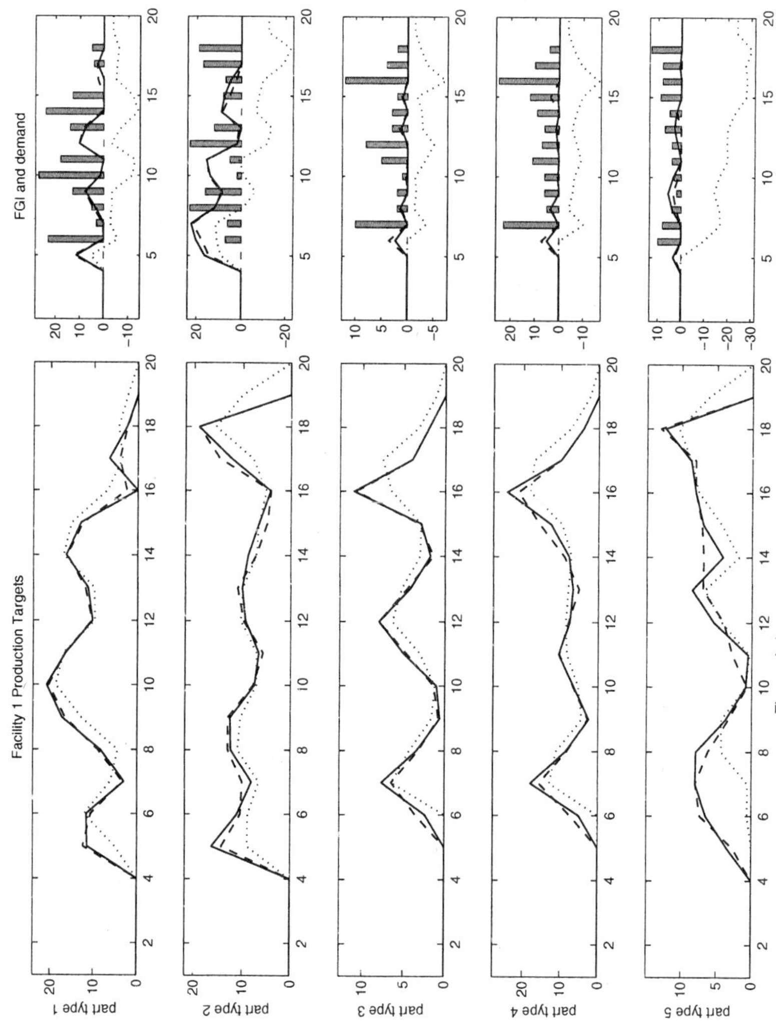

Figure 5.16. Facility 3 production targets and FGI levels comparison among OOLFC, LIX, and CE approaches under demand scenario 3.

Demand scenario 2 results shown in Figure 5.14 correspond to a constant workload of about 80% on facility 1 and constant part type demand mix. Since this load is close to the 90% for which the constant lead time and hedging inventory constraint used in the LIX approach has been calculated, one would expect a relatively small cost difference in the performance of the LIX approach relative to the optimal OOLFC approach. Indeed, in a production system processing a single part type or equivalently a family of perfectly similar part types, level loading that is close to the maximum allowed utilization $\eta_c(t)$ yields minimal performance differences (Anlı, 2003). The considerable performance differences observed here (40% as reported in Table 5.5) are due to the low production target levels that are needed during time periods at the beginning and the end of the planning horizon when the WIP and hedging inventory levels are "ramped up" from zero levels and "ramped down" back to zero levels. The level loading-constant production mix observed in Figure 5.14 verifies the proximity of LIX and OOLFC during the middle portion of the planning time horizon. The remaining differences are due to the non-linear lead time and hedging inventory requirements during the transition period at the beginning and end of non-zero demand activity.

Figure 5.15 shows the production target and FGI differences among the three approaches under scenario 6. Although the overall horizon workload is feasible, demand scenario 6 contains periods of very high demand. Solution trajectories for each of the three approaches attempt to balance pre-production and shortages in different ways. There are multiple "ramp-up" and "ramp-down" trajectory segments during which the LIX approach results follow the OOLFC approach results with a delay. Comparison of OOLFC and LIX plans for demand scenarios 2 and 6 (see Table 5.5) indicates that the LIX approach performs better under demand scenario 6 than it does under demand scenario 2 (20% higher than OOLFC versus 40%). This is due to the overwhelming importance of backlogs under demand scenario 6 where backlog costs are in absolute terms much higher than the WIP costs under scenario 2. Figure 5.16 compares the different approaches under demand scenario 3. Note that LIX solution trajectories deviate even more from OOLFC trajectories reported in Figure 5.15.

As mentioned already, while OOLFC corresponds to the proposed modeling of dynamic lead times through the incorporation of non-linear constraints (5.6) and (5.7) of Exhibit 1, we use the LIX approach as a representative of the current practice of planning with a fixed Lead Time and Safety Stock assumptions, usually corresponding to level loading conditions at the maximum allowed efficiency (worst case analysis).

Finally, the CE approach is representative of an over optimistic, i.e. too low, lead time allowance that assumes no queueing delays. Obviously, CE schedules are hampered by shortfalls and backlogs, rendering the CE approach one order of magnitude worst that the current industry practice representative (LIX). Table 5.5 summarizes the potential benefits of the proposed OOLFC framework. Note that the Total Cost (T.C.) column shows the value of the optimal objective function for OOLFC and the relative percentage of the CE and LIX approaches, while the remaining columns represent a decomposition of the total cost into its WIP, positive FGI, and backlog components. The superiority of the OOLFC dynamic lead time approach is clear. In terms of computational effort, all three approaches under all the six scenarios converged in no more than 12 master-problem sub-problem iterations except for CE requiring 17 iterations with scenario 2.

Table 5.5. OOLFC, CE, and LIX Comparison.

	Approach	T.C.	WIP	FGI^+	FGI^-
Demand Scenario 1	OOLFC	27,173	85%	15%	0%
	LIX	161%	129%	32%	0%
	CE	4397%	72%	14%	4311%
Demand Scenario 2	OOLFC	33,477	78%	22%	0%
	LIX	140%	109%	31%	0%
	CE	3756%	62%	12%	3681%
Demand Scenario 3	OOLFC	60,165	42%	58%	0%
	LIX	113%	54%	58%	0%
	CE	2660%	29%	24%	2607%
Demand Scenario 4	OOLFC	210,815	15%	67%	18%
	LIX	128%	19%	62%	47%
	CE	733%	11%	47%	675%
Demand Scenario 5	OOLFC	398,267	9%	39%	51%
	LIX	115%	12%	35%	68%
	CE	381%	7%	33%	340%
Demand Scenario 6	OOLFC	831,031	5%	12%	83%
	LIX	122%	6%	11%	105%
	CE	259%	3%	10%	246%

4.2 Convex and non-convex lead time feasible regions

When lead time feasible regions are convex, namely when constraint functions are convex, the sub-problem linearization task generates a hyper plane tangent to the constraint boundary that constitutes a supporting plane for the feasible region of the lead time constraint. All linearizations, past and current, are added and retained as linear inequalities and each master-problem solution is simply the solution of a linear program. The number of constraints of the Linear Program master-problem increases with the progress of iterations. With the accumulation of constraints, the local representation accuracy of the non-linear lead time constraints increases arbitrarily till a convergence criterion is met. This procedure is known as the Generalized Benders' decomposition algorithm (Caramanis, 1987).

When the lead time feasible region is not convex, tangent hyper planes can no longer provide an outer linearization of the feasible region. In fact, certain tangent hyper planes, particularly those generated at long past iterations and associated with a tentative production target that was far from the optimal, may cut into the feasible region and exclude some of it from consideration. To avoid this exclusion of feasible region subsets, the iterative algorithm is converted to a modified Generalized Benders' algorithm consisting of the following two phases:

- *Phase* 1 accumulates all constraints generated at each iteration and the algorithm proceeds until a feasible solution is obtained. Since at least some lead time constraint feasible regions are not convex, the linear polyhedron corresponding to the tangent hyper planes will be likely to have excluded a subset of the feasible region where the optimal solution lies. The successive LP master-problems will converge to a solution, but the solution will not necessarily be a local optimum. Figure 5.17 provides an illustrative example for a one part type facility where the WIP required to achieve a production level forms a non-convex feasible set.

In this example the LP solution of the master problem with linear constraints returns the point $({}^{1}\bar{Q}, {}^{1}\bar{X})$ which is then associated with the point $({}^{1}\hat{\bar{Q}}, {}^{1}\hat{\bar{X}})$ on the constraint surface. The tangent hyper plane generated at this point eliminates a potentially superior solution such as $({}^{3}\hat{\bar{Q}}, {}^{3}\hat{\bar{X}})$ to which the OOLFC algorithm can converge only after the hyper plane constraint at $({}^{1}\hat{\bar{Q}}, {}^{1}\hat{\bar{X}})$ is removed. If this constraint is not removed, the algorithm may converge to an inferior solution such as $({}^{2}\bar{Q}, {}^{2}\bar{X})$.

• *Phase* 2 aims at proceeding to a solution that is at least locally optimal. This is achieved by discarding hyper planes generated in earlier iterations, keeping only a subset of constraints generated during a fixed number of the most recent iterations. The fixed number of iterations is selected to equal a fraction of the number of iterations needed to converge in phase 1. During phase 2, provided that convergence is achieved, all of the hyper plane constraints retained are tangent to points on the surface of the lead time constraint functions that are arbitrarily close to each other. Therefore, convergence indicates that the problem is locally convex and the solution is locally optimal. A constrained gradient method has been used repeatedly following the convergence of phase 2 without being able to improve the cost of the solution. This is consistent with the conclusion that the solution obtained is indeed a local optimum.

Since the two phase algorithm can only guarantee a local optimum when some lead time feasible regions are not convex, the trajectory of points on the lead time constraint surface where hyper planes are generated is important as it can lead to a different local optimum. Recall that LP generated tentative production targets are generally infeasible, i.e. they violate the true lead time constraint. In terms of Figure 5.4, this can be visualized as a production target and work in process pair that specifies a point that lies *above* the surface of the $\bar{g}(\cdot)$ function. Such a point is associated in the subsequent iteration to a point *on* the surface of the $\bar{g}(\cdot)$ function and a new tangent hyper plane generated at that point. To assure the convergence of the master-problem, this association must guarantee that the tangent hyper plane that passes through the point on the function surface excludes the point above the surface on its infeasible side. We use an association where the tentative $\bar{\boldsymbol{X}}_c(t)$

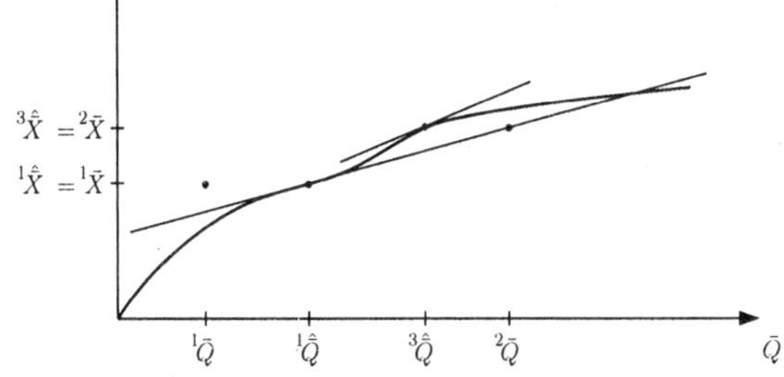

Figure 5.17. Example of a non-convex lead time constraint feasible region.

Table 5.6. Average workstation production capacities.

	Type 1		Type 2		Type 3	
	$m_{c,1}$	$m_{c,2}$	$m_{c,1}$	$m_{c,2}$	$m_{c,1}$	$m_{c,2}$
p.t. 1	150	110	135	65	190	10
p.t. 2	90	220	135	65	190	10

Table 5.7. WIP, FGI holding and FGI backlog cost coefficients.

		p.t. 1	p.t. 2
Facility 2	WIP	10	10
	FGI^+	15	15
Facility 1	WIP	15	15
	FGI^+	20	20
	FGI^-	200	200

value is retained unchanged while the $\bar{Q}_c(t)$ value is adjusted as needed to create a point on the surface of the $\bar{g}(\cdot)$ function.

tab7

A simple two identical facility supply chain where each facility processes two part types on two workstations with FCFS queue protocol exponential processing times (Figure 5.18) is used for simplicity and brevity. Numerical experience on larger systems such as the one used in Section 4.1 is qualitatively comparable. Three lead time constraint function types are used: type 1 referring to unambiguously convex feasible regions for all facilities and for all part types, type 2 referring to some mildly non-convex feasible regions, and type 3 referring to some severely non-convex feasible region brought about by using a First Come First Served (FCFS) queue protocol when part type processing times differ

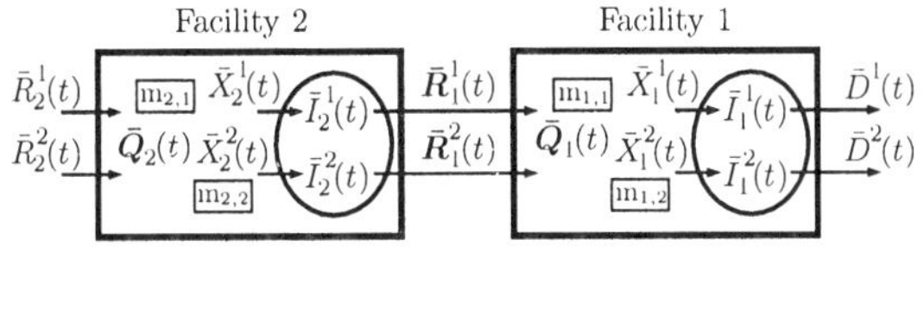

Figure 5.18. Example 2 facility supply chain producing 2 part types.

Table 5.8. Demand scenarios with the weekly demand expressed as a percentage of the part type's production capacity.

		1	2	3	4	5	6	7	8	9	10	11	12	13	14	15
Demand Scenario 7	p.t. 1	60	20	60	40	60	40	40	20	40	40	60	40	60	0	0
	p.t. 2	40	70	50	50	70	40	40	50	50	50	40	20	50	0	0
Demand Scenario 8	p.t. 1	0	0	60	40	60	40	40	20	80	60	80	40	60	0	0
	p.t. 2	0	0	50	50	70	50	50	70	50	80	50	20	50	0	0
Demand Scenario 9	p.t. 1	0	0	80	80	80	60	40	40	40	40	40	40	40	0	0
	p.t. 2	0	0	20	20	20	50	50	50	50	80	80	80	80	0	0
Demand Scenario 10	p.t. 1	0	0	40	40	60	100	80	60	60	20	40	40	40	0	0
	p.t. 2	0	0	40	40	70	80	80	80	70	50	40	40	20	0	0

significantly. Moreover, four demand scenarios were used to test algorithmic performance more effectively.

The cost coefficients used in all lead time constraint function types and demand scenarios are given in Table 5.7 as WIP holding cost, FGI holding cost, and FGI backlog cost, respectively. Columns are associated with part types and rows with facilities. Table 5.6 shows production rate capacities of part types (denoted as p.t. 1 and p.t. 2) at workstations 1 and 2 (denoted as $m_{c,1}$ and $m_{c,2}$) for each of the three types of lead time constraint functions considered.

Table 5.8 shows the demand scenarios used. The demand for each part type is expressed as a percentage of its production capacity, actual values vary under the three lead time constraint function types. Initial WIP and FGI is assumed to be zero at both facilities. All demand scenarios have approximately the same total horizon demand. Scenario 7, however, has a demand activity during periods 1 and 2 as well.

Figures 5.19 through 5.21 show the performance of the two phase algorithm for the convex, mildly and severely non-convex lead time feasible

		d.s. 7	d.s. 8	d.s. 9	d.s. 10
Type 1	phase 1	8	8	6	6
	phase 2	7	7	4	4
Type 2	phase 1	8	7	7	5
	phase 2	7	6	6	5
Type 3	phase 1	8	8	7	9
	phase 2	5	5	5	6

Table 5.9. Number of iterations needed to converge in phase 1 and the number of most recent constraint sets retained in phase 2.

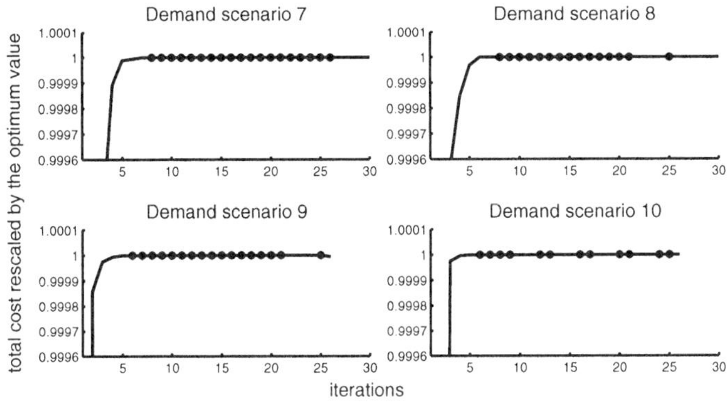

Figure 5.19. Convex lead time feasible region.

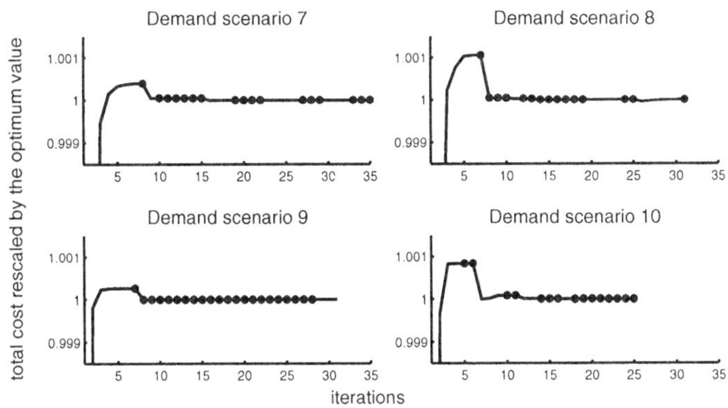

Figure 5.20. Mildly non-convex lead time feasible region.

regions. For each feasible region type, the objective function value is plotted against the iteration number and expressed as a proportion of the lowest feasible solution cost obtained for the given demand scenario. phase 2 is applied immediately after phase 1. Table 5.9 shows the number of iterations needed to converge in phase 1 and the number of most recent constraint sets kept in phase 2 for the three constraint function types under the four demand scenarios. The first few iterations are omitted to allow better resolution in the vicinity of interest. Iterations that produced feasible solutions are denoted by filled circle markers on the trajectories.

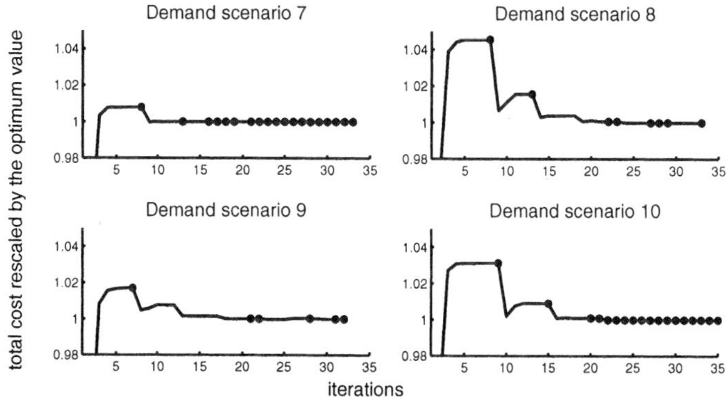

Figure 5.21. Severely non-convex lead time feasible region.

For the convex and mildly non-convex feasible regions, convergence is achieved in all demand scenarios. In fact, for the convex type, phase 1 is sufficient. Once the support hyper planes that were generated far from the optimal solution are discarded the algorithm converges to the optimum and reaches the lower cost feasible solution. In Figure 5.21 we observe an even more drastic drop in the cost function value and loss of feasibility after we switch to phase 2. For the severely non-convex case we must maintain an even smaller number of recently generated constraints to avoid excluding critical portions of the feasible region. When phase 2 takes over, there are not enough linear constraints in the vicinity of the optimal solution to characterize the curvature of the nonlinear constraint surface. This results in a sharp decline of the cost function at the expense of feasibility. As the iterations proceed, however, the necessary nonlinear curvature information is represented via subsequent linearizations, feasibility is enforced again and the algorithm converges to the optimal solution.

5. Conclusion

We presented an efficient and robust approach for SC coordination and planning that is able to reduce lead time substantially, in many cases by as much as a factor of 2 or better, and guarantee probabilistic quality of service performance at low WIP and backlog costs. Indeed, since average production rates are comparable, lead time delays are proportional to WIP and FGI comparisons where the OOLFC production schedule is at least 10 percent better and in most cases more than 50 percent better.

The strong indication of the superior feasibility, efficiency, robustness and tractability of the proposed production planning approach is sufficient proof of concept and motivation to continue with the investigation of ways to incorporate sophisticated QoS and probabilistic multi echelon QoS strategies. Ongoing research is focusing on improving QoS layer algorithms, enhancing stochastic fluid Monte Carlo simulation models for effective sub-problem modeling, and including decentralized facility resource planning in our SC coordination framework. This research is particularly important in view of the emergence of sensor networks that promise to render the collection of relevant information that enables cost efficient and reliable modeling of interesting stochastic dynamics such as demand variability and autocorrelation, raw material delivery variations, and manufacturing process reliability. The contribution of this paper is indeed the proof of the concept that reliable and cost efficient information about production, shipping, transportation, receiving, warehousing and retail activity stochastic dynamics can be translated to significant efficiency gains in SC management.

Acknowledgments

NSF Grant DMI–0300359 is acknowledged for partial support of research reported here.

References

Anlı, O.M. (2002). Iterative approximation selection algorithms for supply chain production planning. *Working paper*, Boston University Center for Information and Systems Engineering.

Anlı, O.M. (2003). Supply chain production planning: Modeling dynamic lead times and efficient inter-cell inventory policies. Boston University Center for Information and Systems Engineering, *Technical Report and Ph.D. dissertation.*

Bertsekas, D.P. (1995). *Dynamic Programming and Optimal Control*, Volumes I and II, Athena Scientific, Belmont, MA.

Bertsimas, D. and Paschalidis, I.Ch. (2001). Probabilistic service level guarantees in make-to-stock manufacturing systems. *Operations Research*, 49(1):119–133.

Bertsimas, D., Paschalidis, I.Ch., and Tsitsiklis, J.N. (1994). Optimization of multiclass queueing networks: Polyhedral and nonlinear characterizations of achievable performance. *The Annals of Applied Probability*, 4(1):43–75.

Bitran, G.R. and Tirupati, D. (1993). Hierarchical production planning. In: *Logistics of Production and Inventory* (S.C. Graves, A.H.G. Rinnooy Kan, and P.H. Zipkin, eds.), Chapter 10, volume 4, Amsterdam, North Holland.

Brandimarte, P., Alfieri, A., and Levi, R. (1998). LP-based heuristics for the capacitated lot sizing problem. *CIRP Annals — Manufacturing Technology*, 47(1):423–426.

Caramanis, M. (1987). Production system design: A discrete event dynamic system and generalized Benders' decomposition approach. *International Journal of Production Research*, 8(25):1223–1234.

Caramanis, M. and Anlı, O.M. (1998). Manufacturing supply chain coordination through synergistic decentralized decision making. In: *Proceedings of Rensselaer's International CAICIM* (Oct. 7-9).

Caramanis, M. and Anlı, O.M. (1999a). Dynamic lead time modeling for JIT production planning. In: *Proceedings of the IEEE Robotics and Automation Conference*, 2:1450–1455, Detroit, MI.

Caramanis, M. and Anlı, O.M. (1999b). Modeling work in process versus production constraints for efficient supply chain planning: Convexity issues. In: *Proceedings of the IEEE CDC*, pages 900–906, Phoenix, AZ.

Caramanis, M. and Liberopoulos, G. (1992). Perturbation analysis for the design of flexible manufacturing system flow controllers. *Operations Research*, 40(6):1107–1125.

Caramanis, M., Pan, H., and Anlı, O.M. (2001a). A closed-loop approach to efficient and stable supply-chain coordination in complex stochastic manufacturing systems. In: *Proceedings of the American Control Conference*, pages 1381–1388, Arlington, VA.

Caramanis, M., Pan, H., and Anlı, O.M. (2001b). Is there a trade off between lean and agile manufacturing? A supply chain investigation. In: *Proceedings of the Third Aegean International Conference on Design of Manufacturing Systems*, Tinos Island, Greece.

Caramanis, M., Paschalidis, I.Ch., and Anlı, O.M. (1999). A framework for decentralized control of manufacturing enterprises. In: *Proceedings of the DARPA-JFACC Symposium on Advances in Enterprise Control*, pages 99–109, San Diego, CA.

Caramanis, M., Wang, J., and Paschalidis, I.Ch. (2003). Enhanced gluid approximation models discrete part dynamics while enabling Monte

Carlo simulation-based. I.P.A. *Working paper*, Boston University Center for Information and Systems Engineering.

Chen, H. and Mandelbaum, A. (1991). Discrete flow networks: Bottleneck analysis and fluid approximations. *Operations Research*, 16(2): 408–447.

Chen, H. and Yao, D.D. (1992). A fluid model for systems with random disruptions. *Operations Research*, 40(2):239–247.

Chen, H. and Yao, D.D. (1993). Dynamic scheduling of a multiclass fluid network. *Operations Research*, 41(6):1104–1115.

Clark, A.J. and Scarf, H. (1960). Optimal policies for a multi-echelon inventory problem. *Management Science*, 6:475–490.

Connors, D., Feigin, G., and Yao, D.D. (1994). Scheduling semiconductor lines using a fluid network model. *IEEE Transactions on Robotics and Automation*, 10(2):88–98.

Dai, J.G. (1995). On the positive Harris recurrence for multiclass queueing networks: A unified approach via fluid models. *The Annals of Applied Probability*, 5:49–77.

Deleersnyder, J., Hodgson, T., King, R., O'Grady, P., and Savva, A. (1992). Integrating Kanban type pull systems and MRP type push systems: Insights from a Markovian model. *IIE Transactions*, 24(3): 43–56.

Feng, Y. and Leachman, R. (1996). A production planning methodology for semiconductor manufacturing based on iterative simulation and linear programming calculations. *IEEE Transactions on Semiconductor Manufacturing*, 9(3):257–269.

Glasserman, P. and Tayur, S.R. (1995). Sensitivity analysis for base-stock levels in multiechelon production-inventory systems. *Management Science*, 45(2):263–281.

Goldratt, E. and Cox, J. (1984). *The Goal: Excellence in Manufacturing.* 262 pages, North River Press.

Goncalves, J.F., Leachman, R.C., Gascon, A., and Xiong, Z. (1994). A heuristic scheduling policy for multi-item, multi-machine production systems with time-varying, stochastic demands. *Management Science*, 40(11):1455–1468.

Graves, S.C. (1986). A tactical planning model for a job shop. *Operations Research*, 34(4):522–533.

Hax, A.C. and Meal, H.C. (1975). Hierarchical integration of production planning and scheduling. In: *Logistics, TIMS Studies in the Management Sciences* (M.A. Geisler, ed.), pages 53–69. North Holland.

Jain, S., Johnson, M., and Safai, F. (1996). Implementing setup optimization on the shop floor. *Operations Research*, 43(6):843–851.

Kaskavelis, C. and Caramanis, M. (1998). Efficient Lagrangian relaxation algorithms for real-life-size job-shop scheduling problems. *IIE Transactions on Scheduling and Logistics*, 30(11):1085–1097.

Khmelnitsky, E. and Caramanis, M. (1998). One-machine n-part-type optimal set-up scheduling: Analytical characterization of switching surfaces. *IEEE Transactions on Automatic Control*, 43(11):1584–1588.

Kouikoglou, V.S. and Phillis, Y.A. (1991). An exact discrete-event model and control policies for production lines with buffers. *IEEE Transactions on Automatic Control*, 36(5):515–527.

Kumar, P.R. and Meyn, S.P. (1995). Stability of queueing networks and scheduling policies. *IEEE Transactions on Automatic Control*, 40(2):251–260.

Kumar, P.R. and Seidman, T.I. (1990). Dynamic instabilities and stabilization methods in distributed real-time scheduling of manufacturing systems. *IEEE Transactions on Automatic Control*, 35(3):289–298.

Lambrecht, M. and Decaluwe, L. (1988). JIT and constraint theory: The issue of bottleneck management. *Production and Inventory Management Journal*, 29(3):61–66.

Lambrecht, M.R., Ivens, P.L., and Vandaele, N.J. (1998). ACLIPS: A capacity and lead time integrated procedure for scheduling. *Management Science*, 44(11):1548–1561.

Lambrecht, M.R., Muckstadt, J.A., and Luyten, R. (1984). Protective stocks in multi-stage production systems. *International Journal of Production Research*, 22(6):1001–1025.

Liberopoulos, G. and Caramanis, M. (1995). Dynamics and design of a class of parameterized manufacturing flow controllers. *IEEE Transactions on Automatic Control*, 40(6):1018–1028.

Meal, H.C., Wachter, M.H., and Whybark, D.C. (1987). Material requirements planning in hierarchical production planning systems. *International Journal of Production Research*, 25(7):947–956.

Paschalidis, I.Ch. and Liu, Y. (2003). Large deviations-based asymptotics for inventory control in supply chains. *Operations Research*, 51(3):437–460.

Paschalidis, I.Ch. and Vassilaras, S. (2001). On the estimation of buffer overflow probabilities from measurements. *IEEE Transactions on Information Theory*, 47(1):178–191.

Paschalidis, I.Ch., Liu, Y., Cassandras, C.G., and Panayiotou, C. (2004). Inventory control for supply chains with service level constraints: A synergy between large deviations and perturbation analysis. *The Annals of Operations Research (Special Volume on Stochastic Models of Production-Inventory Systems)*, 126:231–258.

Saksena, V.R., O'Reilly, J., and Kokotovic, P.V. (1984). Singular perturbations and time-scale methods in control theory: Survey 1976–1983. *Automatica*, 20(3):273–293.

Sharifnia, A. (1994). Stability and performance of distributed production control methods based on continuous-flow models. *IEEE Transactions on Automatic Control*, 39(4):725–737.

Sharifnia, A., Caramanis, M., and Gershwin, S.B. (1991). Dynamic setup scheduling and flow control in manufacturing systems. *Discrete Event Dynamic Systems: Theory and Applications*, 1(2):149–175.

Veatch, M. and Caramanis, M. (1999). Optimal average cost manufacturing flow controllers: Convexity and differentiability. *IEEE Transactions on Automatic Control*, 44(4):779–783.

II

LARGE SCALE OR MULTI AGENT SYSTEM PROBLEMS

Chapter 6

PROVIDING QOS IN LARGE NETWORKS: STATISTICAL MULTIPLEXING AND ADMISSION CONTROL

Nikolay B. Likhanov
Ravi R. Mazumdar
François Théberge

Abstract In this paper we consider the problem of providing statistical Quality of Service (QoS) guarantees defined in terms of packet loss when independent heterogeneous traffic streams access a network router of high capacity. By using a scaling technique we show how this problem becomes tractable when the server capacity is large and many traffic streams are present. In particular we show that we can define an effective bandwidth for the sources that allows us to map the model onto a multirate loss model. In particular we show several insights on the multiplexing problem as the capacity becomes large. We also provide numerical and simulation evidence to show how the largeness of networks can be used to advantage in providing very simple admission control schemes. The techniques are based on large deviations, local limit theorems, and the product-form associated with co-ordinate convex policies.

1. Introduction

Quality of Service (QoS) guarantees are going to be distinct features of the services that users will obtain from next generation high-speed networks. In the emerging networks, the QoS issue will be much more complicated since the QoS requirements will differ from user to user. Indeed networks will need to offer heterogeneous QoS. This is an important issue due to the fact that QoS based pricing structures are increasingly being advocated.

One of the challenging problems in networks is to characterize the admissible region of the numbers of connections or flows that can be admitted into the network in order to guarantee a given level of Quality

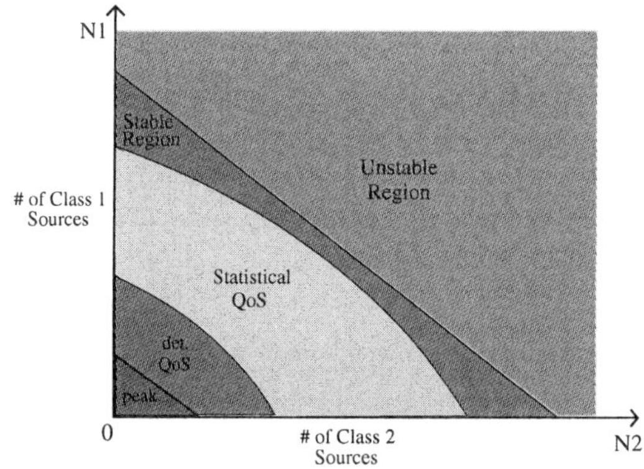

Figure 6.1. Admissible region when using different criteria for the case of 2 classes.

of Service (QoS). QoS is usually specified by loss probability constraints or bounds on the delays incurred by the bits as they traverse the network from source to destination. There are two approaches: providing deterministic or statistical guarantees.

Deterministic guarantees are hard guarantees and the analysis is usually based on a worst-case analysis. When traffic streams are shaped or their sample paths forced to conform to a given envelope, a powerful approach called *network calculus* (Chang, 1998; LeBoudec, 1998) has been developed. This approach allows us to consider the end-to-end problem but yields conservative results due to the fact that it is essentially a worst-case approach. Providing statistical QoS is much more efficient in terms of resource utilization (in this context being able to support a larger number of flows) and when networks are large they lead to *economies of scale* (Duffield and O'Connell, 1995). This is the phenomenon of *statistical multiplexing.* The maximum number of flows is limited by the stability requirement that the average total rate of the flows using a particular resource must be less than the capacity of that resource.

Figure 6.1 shows the typical scenario when different criteria are selected for admitting users into a network assuming that there are 2 classes of users with a high level of burstiness.

One of the key issues in providing statistical QoS is our ability to estimate and/or measure packet loss at a network element under general traffic assumptions. This is intractable in general except in a few simple cases. However knowing that the statistical performance requirements are fairly stringent, i.e. the probability of packet loss to be in the range $10^{-9} - 10^{-5}$, implies that we are concerned with the tail probability distribution that naturally leads to the study of asymptotics. This makes the problem more tractable. There are basically two types of asymptotics of interest: 1) The large buffer asymptotic when there are a few traffic streams which share a resource and a given stream can consume a significant amount of the bandwidth. This scenario occurs when there is a substantial amount of non delay sensitive traffic, and 2) The many sources asymptotics when many small sources share the resource. This scenario is of importance in the core of the network or in MPLS (Multipath Label Switching) when virtual routes are established within a network and a flows between the origin-destination (O-D) are aggregated and switched along them. This latter asymptotic is in fact an instance of the so-called statistical multiplexing and is the scenario we consider.

Large deviations asymptotics have been studied for a number of scenarios and by now there is quite a lot of know-how from the analytical standpoint. For single buffers based on First In First Out (FIFO) disciplines they can be found in A. Botvich and Duffield (1995); Choe and Shroff (1999); Courcoubetis and Weber (1996); Likhanov and Mazumdar (1999). For multi-buffered systems large buffer asymptotics for a priority model can be found in O'Connell (1998) while the many-sources case with HOL priority schemes are reported in Delas et al. (2002). For fair queueing or GPS disciplines large buffer results under restrictive assumptions on the sources can be found in Massoulie (1999); Bertsimas et al. (1998) while the many sources case can be found in Kotopoulos (2000). In the FIFO case the delay distributions are readily obtained from the so-called buffer overflow results. In the HOL and GPS cases it can be shown the delay distributions when a large number of sources are present are also closely related to the loss rate asymptotics (Shakkottai and Srikant, 2001; Delas et al., 2002). A recent monograph (Ganesh et al., 2004) is quite comprehensive in discussing the various asymptotics mainly in the context of single queues.

In spite of the successes in analyzing the single node case in both the large buffer and many sources contexts, there has been only limited success at identifying the asymptotics for network models. In general, only when the input and output rate functions have *linear geodesics*[1], the end-to-end analysis is feasible by an iterative procedure. This however is not very useful from the point of gaining insights and to perform admission control. In Ganesh and O'Connell (1998); O'Connell (1997),

[1]This corresponds to the situation where the most likely path of the buffer occupancy process to an extreme value is a straight line.

it is shown that in queues with more than one input, the departure process need not have the linear geodesic property and thus identifying the rate functions of the outputs explicitly in terms of the inputs is not easy. However, there has been some partial success in the many sources case. This is due to the fact that in the many sources context, the resulting measure is a convolution and the buffer occupancy probability goes to zero even for very small buffers. This result has been exploited in Wischik (1999) where it is shown that the moment generating function (m.g.f.) for a single input does not change as it passes through a node when multiplexed with many similar inputs. Since the m.g.f. is the necessary information required to determine the overflow asymptotics, one can analyze large in-tree networks where each node receives only a small number of inputs from a large number of other independent nodes. However, such a situation is of limited scope and not valid in networks where many flows utilize many links. In Eun and Shroff (2003) this result is extended by considering a two-stage queueing system where the first node serves many flows, of which a certain fixed finite set arrive to the second node. In this case, the first node can be ignored for calculating the overflow probability of the second node as the number of flows of the first node (N) increases. But this result does not hold when the number of output flows arriving at the downstream node is of $O(N)$. This situation is however of interest when trying to estimate the end-to-end QoS in the general scenario described in the beginning. Although overflow probability still goes to zero at the downstream node, determining the rate of this decay depends on the particular time scale and on the sample path characteristics of the output flows from the upstream nodes.

When the buffers are small, it has been shown that the instantaneous values of the total input rate determine the asymptotics (Likhanov and Mazumdar, 1999; Mandjes and Kim, 2001).

This small buffer scenario is actually of much interest in today's networks where buffers are small (in comparison to the server capacity). This is the essence of the so-called **rate envelope multiplexing** in networks (see Roberts, 1998) where buffers are small to absorb local fluctuations but essentially the network can be modeled by bufferless nodes. Moreover it is shown in Ozturk et al. (2004) that asymptotically the admissible region can be computed by only knowing the input characteristics and thus from the point of view of admission control it is enough to consider a single buffer scenario.

One of the basic issues associated with the connection acceptance phase is the determination of the bandwidth associated with a given connection. When the sources are CBR (constant bit rate) this is relatively easy to do because the bandwidth is fairly constant modulo jitter introduced in the network. However, when source bandwidth varies randomly over the duration of the connection this is much more difficult given the QoS constraints. Allocating the peak rate for such a connection will negate the gains achievable by statistical multiplexing (see Figure 6.1) while the mean rate would be very poor with regard to packet loss. This issue has been addressed in the context of *effective bandwidths* in Hui (1988) for unbuffered models and more recently in an excellent survey paper (Kelly, 1996). Packet loss is directly related to the number of connections being carried and thus there is a tight coupling between the connection acceptance and packet level phenomena. One major issue is the great difference in time-scales and due to the extremely high bit rates feedback information is not very useful to control bit loss while feedback is the basic means at the connection level. This coupling leads to a bootstrapping between open loop and closed loop control i.e. packet level phenomena impacts the number of connections allowable for which the a priori statistical information of bit flow must be used and the number of connections in turn alter bit loss. In this paper we address this issue and provide a framework in which this procedure can be done based on apriori information about the arrival rates of connections or sessions.

The basic aim of this paper is to show how the largeness of network capacity and the multiplexing of many independent flows can be used to advantage. It yields simple closed-form results and a very simple admission procedure based on the notion of an *effective bandwidth* of a connection mentioned earlier. This just amounts to approximating the boundary of the acceptance region by a hyperplane constructed at a particular point defined by the equilibrium distribution of the set of connections that lead to a violation of the QoS constraints. Moreover, we show that this is completely determined by knowing their arrival rates into the network leading to a more robust procedure since in the classical effective bandwidth idea the effective bandwidth of sources changes with the connection mix. We then show some asymptotic properties in that as the size of the system increases, the effective bandwidths of sources converge to their mean rates. Finally we show the connection between the admission control strategy and the problem of estimating connection blocking which is also an important design parameter called the Grade of Service (GoS) used in defining so-called Service Level Agreements (SLAs). In particular, we show that the effective bandwidth as defined in this paper provides the consistent mapping that maps packet level phenomena to the connection level phenomena in that the most likely equilibrium state for the blocking coincides with the most likely equilibrium state for loss.

The organization of the paper is as follows. In Section 2 we formulate the problem and then use recent results from local limit large deviations to obtain an estimate of connection acceptance region. In Section 3 we develop the notion of the most likely loss configuration and show some of its properties. Section 4 stitches together the results developed in the previous sections for the CAC scheme. In Section 5 we illustrate the application of the CAC procedure on an example with ON-OFF sources and compare the analytical results with simulations.

2. Problem formulation and acceptance region

Consider a link of capacity C units of bandwidth which is accessed by M types of independent stationary, ergodic connections or flows. It is assumed that there are M heterogeneous classes of connections and a connection of type $i; i \in \{1, 2, \ldots, M\}$ arrives according to a Poisson process with intensity λ_i. A source when connected has a certain bit rate, say, $a_i(t) \in [0, \Pi_i]$ where Π_i denotes the peak rate of the source. Let $r_i = \mathbb{E}[a_i(t)]$ denote the mean bit rate.

We assume that the link has a given configuration of the number of connections of each type being carried at a given time which we assume is held invariant. We will obtain the bit loss[2] probability for a given fixed configuration.

More precisely, suppose that the given link has $\mathbf{m} = (m_1, m_2, \ldots, m_M)$ number of connections being carried at time t where m_i denotes the number of connections of type i. Let $X_i(t)$ denote the instantaneous load on the link due to source i then by definition $X_i(t) = \sum_{n=1}^{m_i} a_{n,i}(t)$ where $a_{n,i}(t)$ are i.i.d. with the same distribution as $a_i(t)$ and $0 \leq X_i(t) \leq m_i \Pi_i$.

We assume that $\sum_{i=1}^{M} \Pi_i m_i > C$ since otherwise there can be no loss. Furthermore, since we are interested in very small bit loss probabilities we assume that the average load is less than C i.e. $\sum_{i=1}^{M} m_i r_i < C$. The other situations are not of interest although in principle the development carried out below can still be done.

Let $X(t) = \sum_{i=1}^{M} X_i(t)$ denote the instantaneous load on the system. Then the number of bits lost during an interval of length T is just given by: $N(T) = \int_0^T (X(t) - C)^+ dt$.

Let $N_1(T) = \int_0^T \sum_{i=1}^{M} m_i a_i(t)\, dt$ denote the total number of bits which are offered to the system in an interval of length T and $a_i(t)$ is the instantaneous rate of type i calls which is a r.v. with values in $[0, \Pi_i]$. Then $N(T)/N_1(T)$ denotes the fraction of bits lost. Since the processes

[2]Although in this paper we refer to the flow in bits, the granularity can be taken to be in packets in which case we have a packet loss measure.

are stationary and ergodic, by the strong law of large numbers, the stationary bit loss probability is given by:

$$\mathbb{P}(\text{bit loss}) = \lim_{T\to\infty} \frac{N(T)}{N_1(T)} = \frac{\mathbb{E}[(X-C)^+]}{\sum_{i=1}^M m_i r_i} \tag{6.1}$$

where $x^+ = \max\{x, 0\}$.

Now to compute $\mathbb{E}[(X-C)^+]$ we need to determine the tail distribution $\mathbb{P}(X > x)$ and this is given by a convolution measure since the connections are independent. This is extremely intractable in general.

In the sequel we will show that in fact when the size of the system is large then one can exploit very elegant results from the theory of sums of independent random variables (termed local limit theorems, which can be found in Petrov (1975) or Korolyuk et al. (1985)) to obtain explicit analytical results that are $O(1)$ in complexity.

The notion of a large system can be viewed in many ways. A particularly attractive way is to view a large system as a scaled version of a nominal system. More precisely, we scale both the capacity C and the number of sources $\{m_i\}$ by a factor N i.e. $C(N) = NC$ and $m_i(N) = Nm_i$; $i = 1, 2, \ldots, M$. This scaling keeps the ratio of the number of sources to the capacity constant. This is a purely analytic device which will allow us to determine the accuracy of our results.

Let us now assume the scaled version. Let $P_N(\mathbf{m})$ denote the bit loss probability given by:

$$P_N(\mathbf{m}) = \frac{\mathbb{E}[(X^N - NC)^+]}{N\sum_{i=1}^M r_i m_i} = \frac{\int_{NC}^{\sum Nm_i\Pi_i} dF^{(N)}(x)}{N\sum_{i=1}^M m_i r_i}$$

where $F^{(N)}(x)$ is the distribution of $X^N(t) = \sum_{i=1}^M X_i^N(t)$ and $X_i^N(t) = \sum_{j=1}^{Nm_i} a_{i,j}(t)$.

We begin with the following result which is a simple consequence of sums of independent r.v's. The proof is trivial and so we omit it.

LEMMA 6.1 *Define $\eta = \sum_{i=1}^M \xi_i$ where ξ_i are independent r.v's with distribution the same as $\sum_{j=1}^{m_i} a_{i,j}$. Let $\{\eta_i\}$ be an independent collection of r.v's with the same distribution as η defined above. Then the random variables $\sum_{i=1}^N \eta_i$ and $\sum_{i=1}^M X_i$, where X_i are independent r.v's with distribution as $\sum_{j=1}^{Nm_i} a_{i,j}$, have the same distribution.*

REMARK 11 The importance of this result is to convert the summation with respect to the number of types to a summation in the scale N. Then noting that the probability distribution of sums of N .i.d. r.v's can be written as the N-fold convolution of the common distribution we can construct useful estimates based on measure changes and local limit theorems.

As a result of the above Lemma we can write $F^{(N)}(dx) = \mu^{*N}(dx)$ where $\mu(dx)$ is the measure of η. We can now obtain estimates for the required loss by now invoking the Bahadur-Rao theorem (Bahadur and Rao, 1960) from local limit large deviations. We state the theorem below as well as a theorem on local limit large deviations for densities due to Petrov (1975).

PROPOSITION 6.1 *Let* $X^N = \sum_{j=1}^{Nm} X_j$ *where* $\{X_j\}_{j=1}^{Nm}$ *are i.i.d. r.v's with moment generating function* $\phi(h)$.

Then as $N \to \infty$, uniformly for any $u > 0$,

Petrov. For all $u \in (-\infty, +\infty)$

$$p^{X^N}(Nu)\,du = \mathbb{P}(X^N \in [Nu, Nu + du)) = \frac{e^{-NI(u)}}{\sqrt{2\pi\sigma^2 N}} du\left(1 + O\left(\frac{1}{N}\right)\right) \qquad (6.2)$$

Bahadur-Rao. For $u > E[X_i]$:

$$\mathbb{P}\{X^N \geq Nu)\} = e^{-NI(u)}\frac{1}{\tau\sqrt{2\pi\sigma^2 N}}\left(1 + O\left(\frac{1}{N}\right)\right) \qquad (6.3)$$

where

$$I(u) = u\tau - m\log\left(\phi(\tau)\right) \qquad (6.4)$$

$\tau(u)$ is the unique solution to

$$m\frac{\phi'(\tau)}{\phi(\tau)} = u \qquad (6.5)$$

and

$$\sigma^2 = m\left(\frac{\phi''(\tau)}{\phi(\tau)} - \left(\frac{\phi'(\tau)}{\phi(\tau)}\right)^2\right). \qquad (6.6)$$

$I(u)$ is referred to in large deviations theory as the rate function.

We apply the above result by taking $\mu = \mu_\eta$ where μ_η is the distribution corresponding to the r.v. η and $u = C$. By the definition of η the moment generating function is given by

$$\phi_\eta(t) = \prod_{k=1}^{M} \left(\phi_k(t)\right)^{m_k} \qquad (6.7)$$

where $\phi_k(t)$ is the moment generating function of a_k.

Therefore

$$I(C) = C\tau_c - \sum_{i=1}^{M} m_i \ln\left(\phi_i(\tau_c)\right) \tag{6.8}$$

where τ_c is the unique (since $\sum_{i=1}^{M} m_i \Pi_i > C$ and $\sum_{i=1}^{M} m_i r_i < C$ by assumption) solution of

$$\sum_{i=1}^{M} \frac{m_i \phi_i'(\tau_c)}{\phi_i(\tau_c)} = C. \tag{6.9}$$

Using the Bahadur-Rao theorem (Bahadur and Rao, 1960) and the result of Petrov (1975), we can then show the following result for the bit loss probability necessary to characterize the acceptance region. The proof can be found in Likhanov and Mazumdar (1999); Likhanov et al. (1996).

PROPOSITION 6.2 *Consider an unbuffered system of capacity NC which carries Nm_i $(1 \le i \le M)$ independent stationary, ergodic sources of type i, where a source of type i has instantaneous rate $a_i(t)$, a r.v. which takes values in $[0, \Pi_i]$ with mean $r_i = \mathbb{E}[a_i(t)]$. Under the hypotheses that $\sum_{i=1}^{M} m_i \Pi_i > C$ and $\sum_{i=1}^{M} m_i r_i < C$ the stationary bit loss probability is given by:*

$$P(\text{bit loss}) = \frac{e^{-NI(C)}}{\tau_c^2 C \rho \sqrt{2\pi\sigma^2 N^3}} \left(1 + O\left(\frac{1}{N}\right)\right) \tag{6.10}$$

where:

(i) $\tau_c > 0$ *is the unique solution to*

$$\sum_{i=1}^{M} \frac{m_i \phi_i'(\tau_c)}{\phi_i(\tau_c)} = C$$

where $\phi_i(t)$ is the moment generating function of a_i.

(ii) $I(C)$ *is the rate function given by:*

$$I(C) = C\tau_c - \sum_{i=1}^{M} m_i \ln\left(\phi_i(\tau_c)\right)$$

(iii) σ^2 *is given by*

$$\sigma^2 = \sum_{i=1}^{M} m_i \left(\frac{\phi_i''(\tau_c)}{\phi_i(\tau_c)} - \left(\frac{\phi_i'(\tau_c)}{\phi_i(\tau_c)}\right)^2 \right)$$

and $\rho = \sum_{i=1}^{M} r_i m_i$.

REMARK 12 If the bit overflow probability is used as the QoS parameter, then the bound is given by the Chernoff bound which is just $e^{-NI(C)}$. This is the starting point of the approach in Hui (1988); Kelly (1996).

EXAMPLES We now give explicit relations for some commonly used traffic sources in applications.

ON-OFF sources. These are the most commonly used source models to represent variable bit rate (VBR) traffic. The importance of these models is that they serve as *worst case* traffic for a given set of traffic models characterized by burst length, peak and mean rates as shown in Doshi (1995); Guillemin et al. (2002). Given their importance we provide detailed expressions for the required quantities.

By definition an ON-OFF source has an instantaneous rate $a_i \in \{0, \Pi_i\}$ i.e. it is either OFF (and therefore with rate 0) or ON at peak rate Π_i. Let p_i denote the stationary probability that a source is ON, then $r_i = \Pi_i p_i$. (Alternatively p_i is obtained from the mean rate specification by the preceding relation).

For this particular case the instantaneous load on the system is completely specified by the number of connections which are in their ON state at a given time. Given the independence assumption on the sources we obtain:

$$P\Big(\sum_{i=1}^{M} X_i = k\Big) = \sum_{\mathbf{n}\in\mathbf{A}(k)} \prod_{i=1}^{M} \binom{m_i n_i}{p}_i^{n_i} (1-p_i)^{m_i - n_i} \tag{6.11}$$

where

$$\mathbf{A}(k) = \Big\{\mathbf{n}\colon \sum_{i=1}^{M} n_i \Pi_i = k\Big\}. \tag{6.12}$$

In spite of the above explicit form the computational complexity remains for large systems. For this model the quantities necessary to compute the bit loss probability are:

(i) $\phi_\eta(t) = \prod_{i=1}^{M} (p_i e^{t\Pi_i} + 1 - p_i)^{m_i}$.

(ii) τ_c is the unique solution to

$$\sum_{i=1}^{M} \frac{m_i p_i \Pi_i e^{\tau_c \Pi_i}}{p_i e^{\tau_c \Pi_i} + 1 - p_i} = C.$$

(iii) The rate function $I(C)$ is given by:

$$I(C) = C\tau_c - \sum_{i=1}^{M} m_i \ln(p_i e^{\tau_c \Pi_i} + 1 - p_i).$$

(iv) σ^2 (variance under the changed distribution)

$$\sigma^2 = \sum_{i=1}^{M} \frac{m_i p_i \Pi_i^2 e^{\tau_c \Pi_i}(1 - p_i)}{(p_i e^{\tau_c \Pi_i} + 1 - p_i)^2}.$$

Uniform sources. These correspond to sources when there is complete lack of information regarding the bit flow. In particular the probability of the instantaneous bit flow is equally weighted between all the states i.e. $F_i(x) = x/\Pi_i$ for $x \in [0, \Pi_i]$.

In this case

(i) $\phi_i(t) = 1/\Pi_i(1 - e^{t(\Pi_i)}/1 - e^t)$.

(ii) $r_i = \Pi_i/2$.

(iii) τ_c and $I(C)$ can be calculated knowing $\phi_i(t)$.

Markov modulated sources. This is also a commonly used source model where the instantaneous rate a_i corresponds to the state of the underlying Markov chain. In this case $p_{i,j} = \pi_i(j)$ where $\pi_i(.)$ denotes the stationary distribution of the Markov chain defined on $\{0, 1, \ldots, \Pi_i\}$ for source of type i.

Thus we see that the stationary bit loss probability can easily calculated once the underlying model for the rate process is specified.

In the following subsection we discuss the accuracy of the estimates obtained for the ON-OFF source model by comparing the results with those obtained by the commonly used Gaussian approximations as well as simulations.

2.1 Accuracy of estimates

We now demonstrate the accuracy of the estimate given by Proposition 6.2 by comparing it with simulation results as well as the a Gaussian approximation based on a central limit approximation for the convolution measure.

It is readily seen that the validity of the Gaussian approximation is completely out of the range of bit loss probabilities we are interested in i.e. the Gaussian approximation is only useful for relatively large values of bit loss (in our context when the scale is small). The results reported are for bit loss probabilities of the order 10^{-6} since below this level it is very difficult to obtain any reasonable confidence in simulations. But even at this level the accuracy of the method proposed is obvious.

In the following example, we set the capacity $C = 20$ with two classes of traffic, i.e. $M = 2$. We use the following data

$$m_1 = 20, \quad m_2 = 10, \quad p_1 = .275, \quad p_2 = .8, \quad \Pi_1 = 2, \quad \Pi_2 = 1$$

and we use different values for the multiplier N.

Results in Table 1.1 are given as base 10 logarithms, so the losses are of the orders $10^{-6} - 10^{-3}$ We note that Proposition 2.2 is very precise when bit loss is small, which is not the case for the Gaussian approximation. We have also given the results based on a simple application of the Chernoff bound.

Table 6.1. Accuracy of bit loss probabilities.

N	Simulation (99% conf. int.)	Chernoff bound (overflow prob.)	Gaussian	Prop. 2.2
60	(-3.5,-3,3)	-1.4	-4.4	-3.2
80	(-4.0,-3.7)	-1.7	-5.0	-3.6
100	(-4.3,-4.0)	-2.0	-5.6	-4.0
120	(-4.7,-4.4)	-2.3	-6.1	-4.4
140	(-5.0,-4.7)	-2.6	-6.7	-4.7
160	(-5.3,-5.0)	-2.8	-7.2	-5.1
180	(-5.4,-5.3)	-3.1	-7.7	-5.4
200	(-5.7,-5.8)	-3.4	-8.2	-5.7

Let us now note the importance of the bit loss probability estimates obtained above. For convenience, with regards to Table 1.1 above, suppose that ε the bound on the loss probability is required to be $\sim 10^{-5}$. The simulations indicate that with a capacity of 3200 (i.e. $N = 160$), the system can handle 3200 connections of type 1 and 1600 connections of type 2. For this configuration the Gaussian estimates suggest bit loss of the order 10^{-7} implying more connections can be admitted while in fact admitting more connections can only drastically reduce performance. Thus for using the Gaussian approximation is too optimistic for bit loss. Keeping the number of connections of type 1 at 3200, calculations based on the Gaussian estimate for loss probabilities of the order 10^{-5} gives the number of type 2 connections to be 1680. For this configuration the simulation results with 99% confidence give bit loss estimates of the order

10^{-3} which is clearly out of the acceptance region. The corresponding results using the results of Proposition 2.2 fall within the margin of error for the simulations. From the values for bit loss, using the Chernoff bound is readily seen to be too conservative.

2.2 Acceptance region

In the development so far we assumed that the configuration of the number of connections of each type i.e. $\{m_i\}_{i=1}^M$ is known. We then obtained the approximation to the stationary bit loss probability for a given configuration. In reality, the arrivals of connections and whether they are admitted or not implies that the configuration is in fact a random variable (which depends on the admission strategy).

Let $P_L^N(m_1, m_2, \ldots, m_M)$ denote the bit loss probability for the given configuration of $(Nm_1, Nm_2, \ldots, Nm_M)$ viewed as a function of the connection configuration. This will allow us to determine the so-called acceptance region since this mapping defines the bit loss probabilities over all possible connections. The QoS requirements on the bit loss probability are typically of the order $10^{-6} - 10^{-9}$. Let ε denote the QoS bound on the bit loss.

Define:

$$\Omega_\varepsilon = \{\mathbf{m} \colon P_L^N(\mathbf{m}) \leq \varepsilon\} \tag{6.13}$$

where $\mathbf{m} = (m_1, m_2, \ldots, m_M)^t$.

Then Ω_ε defines all possible connection configurations which meet the QoS constraints and is referred to as the *acceptance region.*

Let us define the boundary of the acceptance region as

$$\partial\Omega_\varepsilon = \{\mathbf{m} : P_L^N = \varepsilon\}. \tag{6.14}$$

Let us see an important property associated with the acceptance region. This is the property of *coordinate convexity* whose definition we recall below:

DEFINITION 6.1 *Let* $\mathbf{S}$ *be the set of all possible configurations. Then* $\mathbf{S}$ *is said to be coordinate convex if for* $\mathbf{m} \in \mathbf{S}$ *implies that the vector* $\mathbf{m} - e_k \in \mathbf{S}$ *for all* $k = 1, 2, \ldots, M$ *such that* $m_k > 0$*, where* e_k *is the unit vector in dimension* M *with a* 1 *in the* k*th row and 0 elsewhere.*

Coordinate convexity implies that an arriving connection is accepted if and only if the new configuration obtained after the addition of the new connection remains in the set $\mathbf{S}$ after admittance. The importance of the coordinate convexity is that the equilibrium distribution of the configuration has a so-called *product-form* (Ross, 1995) which we will exploit in the following section.

Let us now return to the properties of the acceptance region. First of all it is clear that the acceptance region is coordinate convex under the mapping of the true bit loss probability. This follows directly since:

$$\sum_{j=1}^{m_k} a_{j,k} \geq \sum_{j=1}^{m_k - 1} a_{j,k}$$

where $\{a_{j,k}\}$ are i.i.d. with the same distribution as a_k.

Therefore $X = \sum_{i=1}^{M} X_i$ stochastically dominates $Y = \sum_{i=1, i \neq k}^{M} X_i + X_k'$ where X_k' corresponds to one less connection of type k. Hence $P(X > a) \geq P(Y > a)$ for all a from which it readily follows that the corresponding bit loss probabilities will dominate since they are specified by the complementary distribution.

We now show that if the result of Proposition 6.2 is used to define the acceptance region then the resulting region is coordinate convex.

Before proceeding with the proof we let us note a few interesting properties associated with the size of the system i.e. scaling. For $\mathbf{m} \in \partial\Omega_\varepsilon$ a little reflection shows that as the scaling increases, keeping the QoS constraint fixed at ε implies that the corresponding m_i used in the unscaled system must decrease which implies that τ_c decreases. In fact with some further analysis it can be shown that the τ_c associated with the measure change goes to 0 at a rate of the order $O(1/\sqrt{N})$. This is important in the sequel in which we retain so-called significant or dominating terms.

PROPOSITION 6.3 *For a given* QoS *constraint specified by* ε*, the acceptance region* Ω_ε *obtained by using the result of Proposition* 2.2 *for the bit loss probability is coordinate convex for large systems.*

Proof. To prove this result it is sufficient to show that the following monotonicity result holds:

Let $N\mathbf{m}$ and $N\mathbf{m} - e_k$ be two configurations such that $m_k > 0$, then $P(\text{bit loss} \mid N\mathbf{m}) > P(\text{bit loss} | N\mathbf{m} - e_k)$ for $k \in \{1, 2, \ldots, M\}$ for which $m_k > 0$.

Note for the scaled variables i.e. Nm_k the perturbation e_k is of order $1/N$ implying it is small and therefore the above result will be shown by treating the integer variables $\mathbf{m}$ as continuous and then showing that the partial derivative of the bit loss w.r.t. Nm_k is positive (which implies monotonicity).

Neglecting the $O(1/N)$ term in Proposition 6.2 the partial derivative can be shown to be:

$$\frac{\partial P(\text{bit loss})}{\partial N m_k}$$
$$= P(\text{bit loss})\Big[-\Big(\frac{\partial I(C)}{\partial m_k}+\frac{\partial \tau_c}{N\partial m_k}\Big)-\frac{1}{N\text{Den}}\frac{\partial \text{Den}}{\partial m_k}\Big] \quad (6.15)$$

In the above the term *Den* refers to the denominator of (6.10) and is given by:

$$\text{Den} = \sqrt{2\pi N\sigma\tau_c^2}\sum_{i=1}^{M} m_i r_i.$$

From the definition of $I(C)$ and τ_c it can be readily seen that:

$$\frac{\partial}{\partial m_k}(I(C)+\tau_c) = -\ln\big(\phi_k(\tau_c)\big)+\frac{\partial \tau_c}{\partial m_k}.$$

Noting that $\tau_c > 0$ this implies that $\ln\big(\phi_k(\tau_c)\big) > 0$.

Now from the definition of τ_c it can be shown that $\partial\tau_c/\partial m_k$ is given by the solution to:

$$\sigma^2\frac{\partial \tau_c}{\partial m_k} = -\frac{\phi_k'(\tau_c)}{\phi_k(\tau_c)}$$

and since $\sigma^2 > 0$ it implies that $\partial\tau_c/\partial m_k < 0$.

From the definition of Den we have:

$$\frac{1}{N\text{Den}}\frac{\partial Den}{\partial m_k} = \frac{1}{N\sigma}\frac{\partial\sigma}{\partial m_k}+2\frac{1}{N\tau_c}\frac{\partial\tau_c}{\partial m_k}+\frac{r_k}{N\sum_{i=1}^{M} m_i r_i}.$$

In the expression above the first term is $O(1/N\sigma^4)$ (from the definition of σ^2) and hence under the condition $\sum_{i=1}^{M} m_i r_i < C$ is bounded by a constant divided by N. The third term can also be bounded by a constant divided by N while the second term can contribute significantly when τ_c is small since it is of order $O(1/\sqrt{N})$. Hence, for N large we can write:

$$\frac{\partial P(\text{bit loss})}{\partial m_k} = P(\text{bit loss})\Big[\ln\big(\phi_k(\tau_c)\big)+\frac{1}{N\sigma^2}\Big(1+\frac{2}{\tau_c}\Big)\frac{\phi_k'(\tau_c)}{\phi_k(\tau_c)}-O\Big(\frac{1}{N}\Big)\Big]$$

where the $O(1/N)$ term above is positive but smaller in magnitude in comparison to the first two terms for large N. This implies that the positive terms dominate implying that:

$$\frac{\partial P(\text{bit loss})}{\partial m_k} > 0$$

for all $m_k > 0$; $k = 1, 2, \ldots, M$ and hence the proof is done. □

REMARK 13 If the Chernoff bound is used as the approximation then the coordinate convexity is immediate.

Having established that for large systems the acceptance region is coordinate convex (as a function of the number of connections) when the approximation formula is used for bit loss we now are in a position to further develop the CAC for unbuffered models.

3. Most likely bit loss configuration and its role

Let us recall the model under consideration. A link of capacity NC is accessed by M classes of independent, stationary, ergodic sources. A connection of type i arrives at Poisson rate $N\lambda_i$; $i = 1, 2, \ldots, M$ and a connection once admitted holds the resources for a random time of unit mean in duration. Once the connection is established, the bit flow is has a random rate $a_i(t)$ as discussed above. The sources are assumed to be mutually independent.

In the previous section, given a configuration $(Nm_1, Nm_2, \ldots, Nm_M)$ we gave an $O(1)$ (in complexity) approximation to compute the stationary bit loss probability. Throughout this section and the following sections we assume that the formula given in Proposition 2.2 is used to compute the bit loss probability. We also saw that for large N the acceptance region specified by the QoS denoted by Ω_ε is coordinate convex.

For the model above, it is well known that for coordinate convex state-space the joint distribution of the number of connections under stationarity is given by the following product-form distribution which is insensitive to the actual holding time distribution (see Labourdette and Hart, 1992, for example):

$$\Pi(\mathbf{m}) = \frac{1}{G}\prod_{i=1}^{M}\frac{(N\lambda_i)^{m_i}}{m_i!} \tag{6.16}$$

where G is the normalizing constant given by:

$$G = \sum_{\mathbf{m}\in\Omega_\varepsilon}\frac{(N\lambda_i)^{m_i}}{m_i!}$$

and $\mathbf{m}$ is the vector of the number of connections being held of each type.

We now restrict ourselves to the set $\partial\Omega_\varepsilon$ which we denote as the boundary states. By definition, $\partial\Omega_\varepsilon$ is the subset of states in which the bit loss meets the constraints exactly and thus correspond to the allowable states with the maximal bit loss permissible. We now isolate amongst these states the state with the highest probability of occurring which we define to be the *most likely bit loss configuration* i.e.

DEFINITION 6.2 *The state(s)* $\mathbf{m}^* \in \partial\Omega_\varepsilon$ *given by*

$$\mathbf{m}^* = \text{argmax}_{\mathbf{m}\in\partial\Omega_\varepsilon}\Pi(\mathbf{m}) \tag{6.17}$$

is (are) said to be the most likely bit loss configuration(s).

Let $P_L(\mathbf{N}m)$ denote the stationary bit loss probability for the configuration $(Nm_1, \ldots, Nm_M)$. Then the most likely bit loss state can be computed by from the constrained nonlinear optimization problem:

$$\text{Max}\,\Pi(\mathbf{N}m) \quad \text{subject to} \quad P_L(\mathbf{N}m) = \varepsilon.$$

The above problem is a constrained nonlinear integer optimization problem. However, due to the size, unit increments are of negligible relative order and hence we can treat it as a constrained nonlinear optimization problem over non-negative reals. Even so the problem as posed is formidable. However, we can exploit the fact that for N large we can approximate the terms by using the Stirling approximation.

Let us multiply the numerator and denominator (i.e. G) in (6.16) by $\prod_{i=1}^{M} e^{-N\lambda_i}$. Now neglecting the normalizing factor G (since it is a constant) we can approximate the numerator using Stirling's approximation for Nm_i by:

$$\prod_{i=1}^{M} \frac{e^{-N\lambda_i}(N\lambda_i)^{Nm_i}}{Nm_i!} = \prod_{i=1}^{M} \frac{e^{-N\lambda_i f(\beta_i)}}{\sqrt{2\pi N m_i}} \tag{6.18}$$

where

$$\beta_i = \frac{m_i}{\lambda_i} \tag{6.19}$$

and

$$f(x) = x\ln(x) - x + 1. \tag{6.20}$$

Hence, for large N the optimization problem can be written as:

$$\text{Max}\prod_{i=1}^{M} \frac{e^{-N\lambda_i f(\beta_i)}}{\sqrt{2\pi N\lambda_i\beta_i}} \quad \text{subject to} \quad P_L(\mathbf{N}\lambda\beta) = \varepsilon.$$

By introducing the Lagrange multiplier $\mathbf{y}$ we can convert this problem to an unconstrained minimization problem as:

$$\min \sum_{i=1}^{M} N\lambda_i f(\beta_i) + \mathbf{y}\big(\ln P_L(\lambda\beta) - \ln(\varepsilon)\big) - \sum_{i=1}^{M} \ln(\sqrt{2\pi N\beta_i\lambda_i}) \tag{6.21}$$

and we note that

$$\ln\big(P_L(\lambda\beta)\big) = -\big(NI(C)+\tau_c\big) - \ln(\sqrt{2\pi N^3}) \\ -2\ln\big(\sqrt{\sigma}(\tau_c\big) - \ln(\sum_{i=1}^{M} N\lambda_i\beta_i r_i\big).$$

The reason for writing the equality constraint in the variational form in terms of logarithms is now obvious i.e. to make both terms of the same order.

For performing the optimization we only retain terms involving $\mathbf{m}$.

Therefore, with the above observation we define $\mathbf{m}^*$ as the vector which minimizes (neglecting the constant terms i.e. not depending on $\mathbf{m}$):

$$J(\mathbf{N}m) = \sum_{i=1}^{M}\Big(N\lambda_i f(\beta_i) - \frac{1}{2}\ln(N\beta_i\lambda_i)\Big) \\ - \mathbf{y}\big[\ln\big(P_L(N\lambda\beta)\big) - \ln(\varepsilon)\big]. \quad (6.22)$$

From standard nonlinear optimization theory, see Luenberger (1984) for example, the necessary first-order conditions that $\mathbf{m}^*$ satisfies are:

$$\frac{\partial\big(\lambda_i f(\beta_i)\big)}{\partial m_j}\delta_{i,j} - \mathbf{y}\frac{\partial P_L(N\lambda\beta)}{N\partial m_j}\bigg|_{\mathbf{m}=\mathbf{m}^*} = 0; \quad i,j = 1,2,\ldots,M \quad (6.23)$$

where the Lagrange multiplier $\mathbf{y}$ is such that the equality constraint is achieved.

Now using the expression for the partial derivative of $P_L(N\lambda\beta)$ we obtain the result that the most likely state $N\mathbf{m}^*$ satisfies:

$$Nm_j^* = N\lambda_j\big(\phi_j(\tau_c)\big)^{\mathbf{y}}\exp\left\{\frac{\mathbf{y}}{N\sigma^2}\left[(1+\frac{2}{\tau_c})\frac{\phi_j'(\tau_c)}{\phi_j(\tau_c)}\right]\right\} \quad (6.24)$$

and τ_c is the solution to:

$$\sum_{i=1}^{M}\frac{m_i^*\phi_i'(\tau_c)}{\phi_i(\tau_c)} = C \quad (6.25)$$

which gives $M+1$ equations to compute $\mathbf{m}$ and τ_c as functions of the given parameters and the Lagrange multiplier $\mathbf{y}$.

Finally the Lagrange multiplier $\mathbf{y}$ is chosen to satisfy the constraint:

$$P_L(N\mathbf{m}^*) = \varepsilon \quad (6.26)$$

thus giving us $M+2$ equations for computing the $M+2$ unknowns given the source and arrival characteristics $\phi_i(.)$, $\{\lambda_i\}_{i=1}^M$, C as well as the scaling factor N and the QoS constraint ε.

By computing the Hessian at $N\mathbf{m}^*$ it can be shown that the Hessian is positive definite and thus the solution $N\mathbf{m}^*$ is regular. We omit it for sake of brevity.

REMARK 14 From above it follows that the assumption that the states are of order $O(N)$ as assumed in Section 1 is satisfied and the bit loss probability approximation as given in Proposition 6.2 is valid.

Let us now study the role of the most likely bit loss configuration above.

LEMMA 6.2 *Let $N\mathbf{m}^*$ be the most likely bit loss configuration and $N\mathbf{m}$ be any other configuration in $\partial\Omega_\varepsilon$. Let $\Pi(\mathbf{m})$ denote the stationary distribution for $\mathbf{m}$.*

Then:

$$\frac{\Pi(N\mathbf{m})}{\Pi(N\mathbf{m}^*)} \sim O(e^{-N}) \tag{6.27}$$

Proof. First note that for N large:

$$\Pi(N\mathbf{m}) = \frac{1}{\prod_{i=1}^M \sqrt{2\pi N m_i}} e^{-N\sum_{i=1}^M \lambda_i f(\beta_i)}$$

where β_i is defined by (6.19) and $f(x)$ by (6.20).

Let $N\mathbf{m} \in \partial\Omega_\varepsilon$. Now from the fact that $N\mathbf{m}^*$ is the unique minimizer of $N\sum_{i=1}^M \lambda_i f(\beta_i)$ for $\mathbf{m} \in \partial\Omega_\varepsilon$ we see that

$$\frac{\Pi(N\mathbf{m})}{\Pi(N\mathbf{m}^*)} = \sqrt{\prod_{i=1}^M \frac{m_i^*}{m_i}} e^{-N\sum_{i=1}^M \lambda_i (f(\beta_i) - f(\beta_i^*))}$$

and

$$\sum_{i=1}^M \lambda_i (f(\beta_i) - f(\beta_i^*)) > 0$$

for $\beta \neq \beta^*$ which proves the result for $\mathbf{m} \in \partial\Omega_\varepsilon$. □

On the other hand if $\mathbf{m} \in interior(\Omega_\varepsilon)$ the above estimates hold for the bit loss probabilities. Indeed, for a configuration $\mathbf{m} \in$ interior(Ω_ε) from the definition of the rate function the corresponding rate function denoted by $\mathcal{I}(C)$ is strictly larger than $I(C)$. This follows from the fact that the partial derivative with respect to $\mathbf{m}$ is negative. In this case,

$$\frac{P_L(\mathbf{m})}{P_L(\mathbf{m}^*)} \sim O\big(e^{-N(\mathcal{I}(C)-I(C))}\big)$$

which gives the ratio of bit loss probabilities this time of $O(e^{-N})$.

The importance of the above result is that, in so far as we consider the boundary states which correspond to the maximum bit loss permissible, the contribution of the other boundary configurations with respect to the most likely bit loss configuration is exponentially negligible as the size of the system becomes large. This important property will be utilized in defining the connection acceptance control which is addressed in the next section.

4. Effective bandwidths, CAC and connection blocking

In this section we develop the connection acceptance control strategy building upon the results in the previous sections.

First note that once we characterize Ω_ε a given connection request is admitted if the new configuration with the connection request added is within Ω_ε. In order to do so one would have to compute the region Ω_ε which in light of the expression for the bit loss probability is a daunting task. Moreover as mentioned in the introduction, the calculation of the connection blocking probability is needed for bandwidth allocation for a given VP in the MPLS context for a given Grade of Service (GoS). This is defined as the ratio of the number of arriving connections which cannot be accepted over the total number of arriving requests. This involves the computation of $\sum_{\mathbf{m}\in\partial\Omega_\varepsilon} \Pi(\mathbf{m})$ which is also computationally heavy since it involves calculating the boundary. However, as we have seen in the previous section, the most likely bit loss configuration dominates the bit loss. To avoid this the notion of *effective bandwidths* is introduced (see Kelly, 1996) by which the nonlinear boundary surface is approximated by a tangent hyperplane. The relative slopes of this hyperplane then define the effective bandwidths. Clearly because of the nonlinear nature of the boundary, the slopes of the hyperplanes can differ substantially depending on where the hyperplane is constructed. This amounts to saying that effective bandwidths are not intrinsically associated with the sources but the mixtures. However, as we now show, in large systems the existence of the most likely loss state where the loss is dominant leads to a natural point to construct the hyperplane. Moreover the slopes are completely characterized in terms of the connection arrival rates and the individual rate functions.

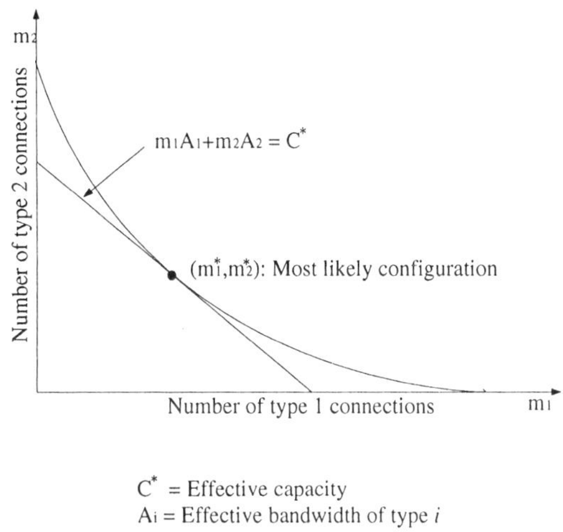

Figure 6.2. Effective bandwidths via hyperplane approximation.

The basic idea is that since $N\mathbf{m}^*$ lies on $\partial\Omega_\varepsilon$ we construct the tangent hyperplane to $\partial\Omega_\varepsilon$ at $N\mathbf{m}^*$. The slope of the hyperplane then defines the relative contributions in terms of the necessary incremental bandwidth requirements of the various types of connections. This is what we identify as the *effective bandwidths* of the various types of sources.

Figure 6.2 illustrates the idea of the effective bandwidths.

Define:

$$a_j = \ln\left(\phi_j(\tau_c)\right) + \frac{1}{N\sigma^2}(1 + \frac{2}{\tau_c})\frac{\phi_j'(\tau_c)}{\phi_j(\tau_c)} \tag{6.28}$$

Let $a_{\min} = \min(a_1, a_2, \ldots, a_M)$ and then define:

$$A_j = \frac{a_j}{a_{\min}}. \tag{6.29}$$

Note since a_j is just the ratio of the partial derivative of the bit loss to the bit loss probability it represents the sensitivity of the bit loss probability (normalized with respect to the minimum of a_j) and thus represents the change in bit loss as the connection is increased and can be identified with the supplementary bandwidth associated with the connection when we try to admit one more.

Then define C^* as

$$C^* = \sum_{i=1}^{M} m_i^* A_i. \tag{6.30}$$

Then the interpretation of $\{A_i\}_{i=1}^M$ and C^* is that the A_i denote the *effective bandwidths* of the connection (with the smallest connection assigned a unit bandwidth) and NC^* the *effective capacity* of the Virtual Path.

With the above terms defined the tangent hyperplane to $\partial\Omega_\varepsilon$ can be approximated as:

$$\mathbf{T} = \left\{ \mathbf{m} : \sum_{i=1}^{M} m_i A_i = NC^* \right\} \tag{6.31}$$

Therefore the Connection Acceptance Control strategy can now be formalized as follows:

1. Compute A_j for a given incoming request of type j.
2. If $A_j + \sum_{i \in \text{ongoing}} m_i A_i \leq NC^*$ then admit the connection, else reject the request.

Thus the Connection Acceptance Control strategy involves computing the available bandwidth at the instant of arrival and seeing whether the effective bandwidth of the incoming request is less than the available bandwidth. This linear truncation now allows us to compute the blocking probability for a given connection request. Before doing so let us see some properties of the effective bandwidths and how good the tangent hyperplane approximation is for large systems.

Throughout the following discussion we assume that $N\mathbf{m} \in \partial\Omega_\varepsilon$ i.e. the bit loss probability is held fixed. We will study the behavior of $A_j(N)$, $C^*(N)$, and the tangent hyperplane $T(N)$ (where the explicit dependence of these quantities on the size specified by N is noted) as $N \to \infty$.

Let $N\mathbf{m}(N)$ be a configuration corresponding to bit loss probability ε where $\mathbf{m}(N)$ is made to depend on N.

PROPOSITION 6.4 *Let $P_L(N\mathbf{m}(N))$ be the bit loss probability for configuration $N\mathbf{m}(N)$ which is assumed to be held constant at ε. Then as $N \to \infty$ the following properties hold:*

1. *$\mathbf{m}(N)$ converges to $\mathbf{m}^o$ such that $\sum_{i=1}^M m_i^o r_i = C$ where r_i is the mean rate of a connection of type i.*
2. *$A_j(N)$ converges to $r_j/r_{\min}$.*
3. *$C^*(N)$ converges to $C/r_{\min}$.*
4. *The tangent hyperplane $T(N)$ coincides with $\partial\Omega_\varepsilon$.*

Proof.

Proof of 1. First note that the bit loss probability $P_L(.)$ can be completely characterized by $\big(N, \mathbf{m}(N), \tau_c(N)\big)$. Keeping it fixed and increasing N implies that $\tau_c(N)$ must go to zero since $I(C) \to 0$ and τ_c satisfies:

$$\sum_{i=1}^{M} m_i(N) \frac{\phi_i'\big(\tau_c(N)\big)}{\phi_i\big(\tau_c(N)\big)} = C.$$

Therefore:

$$\lim_{N\to\infty} \frac{\phi_i'\big(\tau_c(N)\big)}{\phi_i\big(\tau_c(N)\big)} = \frac{\phi_i'(0)}{\phi_i(0)} = r_i$$

and hence

$$\lim_{N\to\infty} \mathbf{m}(N) \to \mathbf{m}^o$$

where $\mathbf{m}^o$ satisfies

$$\sum_{i=1}^{M} m_i^o r_i = C.$$

Proof of 2. From the definition of A_j we have:

$$A_j(N) = \frac{\ln\big(\phi_j\big(\tau_c(N)\big)\big) + \alpha\big(\tau_c(N), N\big)\frac{\phi_j'(\tau_c(N))}{\phi_j(\tau_c(N))}}{\ln(\phi_{\min}(\tau_c(N))) + \alpha\big(\tau_c(N), N\big)\frac{\phi_{\min}'(\tau_c(N))}{\phi_{\min}(\tau_c(N))}}$$

where $\alpha\big(\tau_c(N), N\big)$ denotes the multiplying factor in the definition of A_j.

Now noting that as $\tau_c(N) \to 0$, $\ln(\phi_j\big(\tau_c(N)\big) \to 0$ we have

$$\lim_{N\to\infty} A_j(N) \to \frac{r_j}{r_{\min}}.$$

Proof of 3. This follows directly from the definition of C^* and the two results above.

Proof of 4. This follows directly from 1., 2. and 3. since the properties hold for any $N\mathbf{m}(N) \in \partial\Omega_\varepsilon$. □

REMARK 15 The importance of the above result is that for virtual paths with extremely large capacity in MPLS architectures that can be associated with a very large scaling factor N, the problem becomes completely decoupled with the bandwidth assignment for a given class associated with its mean rate and the effective capacity equal to the given capacity. In this case the only problem is to design capacity allocation to meet GoS requirements using the standard loss model.

REMARK 16 From the result above we see that the boundary of the acceptance region is a hyperplane only in the limit as N goes to infinity. When the system is very large then the linearity of the acceptance boundary implies that the overall performance would be completely insensitive to the choice of the point where the hyperplane is taken and thus in the case of very large systems one would expect very little difference from the results in Elwalid et al. (1995); Kelly (1996).

We now obtain the expression for the connection blocking probability which is needed for the GoS determination.

Let us recall the basic assumptions: connection arrivals are Poisson with a connection of type i having intensity $N\lambda_i$. It is assumed without loss of generality that the connections hold the assigned bandwidth for a random amount of time of unit mean. Connections are assumed to be independent.

From the above, a connection of type i is assigned the effective bandwidth A_i which in the MPLS context corresponds to the number of VC's required. For convenience we assume all quantities are integer valued i.e. we re-define the quantities:

$$A_i \leftarrow \text{int}(A_i), \quad m_i^* \leftarrow \text{int}(m_i^*)$$

where $\text{int}(x)$ is the smallest integer $\geq x$.

As a consequence C^* will also be integer valued. From a practical viewpoint the integer valued effective bandwidths will then denote the number of VC's required and NC^* will be the total effective capacity of the VP in terms of number of VC's.

With the above parameters, the consequence of the CAC rule specified by the linear rule renders the model as a classical multi-rate loss model with rates A_i for a connection of type i. Once again using the fact that the system is large on account of the scaling factor N we can use the approximations for the blocking probabilities for the multi-rate model given in Gazdicki et al. (1993); Mitra and Morrison (1994), noting that by definition the smallest effective bandwidth is 1 and hence the GCD (greatest common divisor) of $\{A_i\}_{i=1}^M$ is 1. We quote the result below:

PROPOSITION 6.5 *Consider a multirate loss model specified by arrival rates $N\lambda_i$, effective bandwidths A_i and effective capacity NC^*.*

Then, the following expressions for the blocking probability are obtained:

Light load: *If $\sum_{i=1}^M \lambda_i A_i < C^*$, then the connection blocking probability for class k; $k = 1, 2, \ldots, M$ is given by:*

$$P_k(N)$$

$$= \exp\big(-I(C^*)\big)\frac{1}{\sqrt{2\pi N}\sigma}\left(\frac{1-\exp(t_{c^*}A_k)}{1-\exp(t_{c^*})}\right)\left(1+O\left(\frac{1}{N}\right)\right) \quad (6.32)$$

Critical load: *If* $\sum_{i=1}^M \lambda_i A_i = C^*$ *then*

$$P_k(N) = \sqrt{\frac{2}{\pi N}}\frac{A_k}{\sigma}\left(1+O\left(\frac{1}{\sqrt{N}}\right)\right) \quad (6.33)$$

Heavy Load: *If* $\sum_{i=1}^M \lambda_i A_i > C^*$ *then*

$$P_k(N) = \big(1-\exp(t_{c^*}A_k)\big)\left(1+O\left(\frac{1}{N}\right)\right) \quad (6.34)$$

where the quantities t_{c^*}, $I(C^*)$, *and* σ *are given by:*

(i) t_{c^*} *is the unique solution to the equation* $\sum_{i=1}^M \lambda_i A_i \exp(t_{c^*}A_i) = C^*$

(ii) $I(C^*) = C^* t_{c^*} - \sum_{i=1}^M \lambda_i\big(\exp(t_{c^*}A_i) - 1\big)$

(iii) $\sigma^2 = \sum_{i=1}^M \lambda_i A_i^2 \exp(t_{c^*}A_i)$.

Thus, once the most likely bit loss configuration is identified from the source characteristics and the corresponding values of C^* and $\{A_i\}_{i=1}^M$ are determined the connection blocking probability can be determined from above.

It is worth remarking one point with regard to the definition of the effective capacity C^*. In Labourdette and Hart (1992), it is shown that the most likely state at capacity for the multi-rate loss model i.e. the most likely configuration $\mathbf{m}$ such that $\sum_{i=1}^M m_i A_i = C^*$ is given by:

$$m_i^* = \lambda_i \alpha^{A_i}$$

where α is the unique solution to

$$C^* = \sum_{i=1}^M \lambda_i A_i \alpha^{A_i}.$$

From the definition of the most likely bit loss configuration above and the definition of A_i the effective bandwidths it is readily seen that in fact the most likely bit loss configuration corresponds to the most likely configuration for the loss model with effective capacity NC^*. To see this, define $\alpha = e^{a_{\min}\mathbf{y}}$. Then by definition $m_i^* = \lambda_i \alpha^{A_i}$ and $C^* = \sum_{i=1}^M \lambda_i A_i \alpha^{A_i}$.

This basically shows that in mapping the bit loss phenomena to the connection level multirate loss model we remain consistent i.e. the most likely bit loss configuration remains invariant since it corresponds to the most likely configuration at capacity for the multirate loss model.

Thus the basic problem of CAC to meet a given QoS specified by ε is equivalent to transforming the model to a multi-rate loss model where the bit level phenomena is mapped into the connection level through the definition of the effective bandwidth and effective capacity.

Let us now recapitulate all the results concerning the CAC and connection blocking.

Assume that the following data is given: connection arrival rates $\{\lambda_i\}_{i=1}^M$, bit flow characteristics $\{\phi_i(t)\}_{i=1}^M$, link capacity C, QoS parameter ε and scaling factor N.

Step 1. Determine the parameters τ_c, $\{m_i^*\}_{i=1}^M$, and Lagrange multiplier $\mathbf{y}$ from the following set of $(M+2)$ equations:

$$m_j^* = \lambda_j(\phi_j(\tau_c))^{\mathbf{y}} \exp\left(\frac{\mathbf{y}}{N\sigma^2}\left[1+\frac{2}{\tau_c}\right]\frac{\phi_j'(\tau_c)}{\phi_j(\tau_c)}\right),$$

$$C = \sum_{i=1}^M m_i^* \frac{\phi_i'(\tau_c)}{\phi_i(\tau_c)},$$

$$\varepsilon = P_L(N\mathbf{m}^*).$$

Step 2. Having obtained τ_c and m_i^* above compute the effective bandwidth A_i and then the effective capacity C^*.

Step 3. Depending upon the loading condition compute the blocking probabilities $P_k(N)$ using appropriate form from Proposition 6.5 with multirate parameters λ_i, A_i and C^*

In the next section we go through an example of the above procedure and compare the results obtained with simulations.

5. Numerical example of QoS and GoS with proposed admission control

In this example, we use the same parameters as in Table 1.1. We set capacity $C = 20$ with two classes of traffic such that

$$\lambda_1 = \lambda_2 = 14, \quad p_1 = .275, \quad p_2 = .8, \quad \Pi_1 = 2, \quad \Pi_2 = 1$$

and we use multiplier $N = 100$. The bit loss constraint is set to $\varepsilon = 10^{-4}$. This value is larger than reality to allow for accurate simulations to be performed. Recall that we suppose that each class of traffic is modeled as an ON-OFF process and therefore,

$$\phi_j(\tau_c) = p_j \exp(\tau_c \Pi_j) + 1 - p_j.$$

We numerically solve Step 1 given at the end of section 3 to find m_1^*, m_2^*, τ_c and $\mathbf{y}$. We obtain the following solution

$$m_1^* = 14.156, \quad m_2^* = 14.217, \quad \tau_c = .06215, \quad \mathbf{y} = .2245.$$

With these values, we go to Step 2 of the procedure and we compute $a_1 = .04948$, $a_2 = .06852$ which we normalize to get $A_1 = 1.0$, $A_2 = 1.385$ and $C^* = 33.847$.

In Step 3 of the procedure, we use the approximation formulae of Proposition 6.5 to evaluate call blocking probabilities. We first notice that $\sum_j A_j \lambda_j < C^*$ so we are in the light load case. The blocking probabilities are reported in Table 1.2.

Table 6.2. Connection blocking probabilities.

Technique	Class 1 blocking	Class 2 blocking
Simul. (95 % conf. int.)	.00427-.00501	.00631-.00724
Proposition 4.2	.00479	.00661

This procedure defines an acceptance region of the form

$$\sum_j A_j m_j \leq NC^*.$$

To justify the use of such an acceptance region, we used simulation to evaluate bit loss at a number of points on the boundary of the acceptance region. Recall that the objective was to keep bit loss below $10^{-4} = \varepsilon$. The results are given in Table 1.3 where we see that our linear acceptance region is conservative and very close to the true (unknown) acceptance region.

Concluding remarks

In this paper we have proposed a framework for addressing the problem of admission control to offer connections statistical guarantees on their QoS requirements based on the loss probability. This can be readily extended to delay distributions. We have shown how large system size can be used to advantage via the ideas of scaling and moreover we have shown that there is a natural definition for the effective bandwidths

Table 6.3. Bit loss probabilities.

Number of Class 1 calls	Number of Class 2 calls	Base 10 logarithm of 95% conf. int. for bit loss
500	2083	(-4.13,-4.03)
1000	1722	(-4.19,-4.11)
1416	1422	(-4.16,-4.09)
1500	1361	(-4.30,-4.23)
2000	1000	(-4.25,-4.17)

that gives a consistent mapping from the fast bit time scale to the slow connection time scale that allows for GoS dimensioning. The approach can be readily extended to more complicated scheduling disciplines provided one can analytically obtain estimates for the loss probabilities or delay distributions. We have also shown how the effects of multiplexing lead to a reduction of effective bandwidths and shown other scaling properties of large systems showing loss concentrates at certain points on the boundary of the acceptance region in equilibrium.

Acknowledgments

This research has been supported in part by grants 0087404–ANI and 0099137–SPN from the National Science Foundation, and by a grant from France Telecom through the CTI program.

References

Botvich, A. and Duffield, N.G. (1995). Large deviations, economies of scale and the shape of the loss curve in large multiplexers. *Queueing Systems*, 20:293–320.

Bahadur, R.R. and Ranga Rao, R. (1960). On deviations of the sample mean. *Annals of Mathematical Statistics*, 31:1015–1027.

Bertsimas, D., Paschalidis, I.C., and Tsitsiklis, J.N. (1998). Asymptotic buffer overflow probabilities in multiclass multiplexers: An optimal control approach. *IEEE Transactions on Automatic Control*, 43(3):315–335.

Chang, C.-S. (1998). On deterministic traffic regulation and service guarantees: A systematic approach by filtering. *IEEE Transactions on Information Theory*, 44(3):1097–1110.

Choe, J. and Shroff, N.B. (1999). On the supremum distribution of integrated stationary Gaussian processes with negative linear drift. *Advances in Applied Probability*, 31(1):135–157.

Courcoubetis, C. and Weber, R.R. (1996). Buffer overflow asymptotics for a switch handling many traffic sources. *Journal of Applied Probability*, 33:886–903.

Delas, S., Mazumdar, R.R., and Rosenberg, C.P. (2002). Tail asymptotics for HOL priority queues handling a large number of independent stationary sources. *Queueing Systems: Theory and Applications*, 40(2):183–204.

Doshi, B.T. (1995). Deterministic rule based traffic descriptors for broadband isdn: Worst case behavior and connection acceptance control. *International Journal of Communication Systems*, 8:91–109.

Duffield, N.G. and O'Connell, N. (1995). Large deviations and overflow probabilities for the general single-server queue with applications. *Mathematical Proceedings of the Cambridge Philosophical Society*, 118:363–374.

Elwalid, A., Mitra, D., and Wentworth, R.H. (1995). A new approach for allocating buffers and bandwidth to heterogenous regulated traffic in an atm node. *IEEE Journal on Selected Areas in Communications*, 13(6):1115–1127.

Eun, D.Y. and Shroff, N. (2003). Simplification of network analysis in large-bandwidth systems. *Proceedings of IEEE INFOCOM'03*, San Francisco, CA.

Ganesh, A.J. and O'Connell, N. (1998). The linear geodesic property is not generally preserved by a FIFO queue. *Annals of Applied Probability*, 8(1):98–111.

Ganesh, A.J., O'Connell, N., and Wischik, D. (2004). *Big Queues*. Springer-Verlag, N.Y.

Gazdicki, P., Lambadaris, I., and Mazumdar, R.R. (1993). Blocking probabilities for large multirate Erlang loss systems. *Advances in Applied Probability*, 25(4):997–1009.

Guillemin, F., Likhanov, N., Mazumdar, R., and Rosenberg, C. (2002). Extremal traffic and bounds on the mean delay of multiplexed regulated traffic streams. In: *Proceedings of the INFOCOM* '2002, N.Y.

Hui, J.Y. (1988). Resource allocation in broadband networks. *IEEE Journal of Selected Areas in Communications*, 6:1598–1608.

Kelly, F.P. (1996). Notes on effective bandwidths. In: *Stochastic Networks* (F.P. Kelly, S. Zachary, and I. Zeidins, eds.). Oxford University Press.

Korolyuk, V., Portenko, N., Skorohod, A., and Turbin, A. (1985). *Handbook on Probability Theory and Mathematical Statistics* (in Russian). Nauka, Moscow.

Kotopoulos, K. (2000). Asymptotics of multibuffered systems with GPS service. *Ph.D dissertation*, University of Essex.

Labourdette, J.F. and Hart, G. (1992). Blocking probabilities in multirate loss systems: Insensitivity, asymptotic behavior and approximations. *IEEE Transactions on Communications*, COM-8:1239–1249.

LeBoudec, J.-Y. (1998). Application of network calculus to guaranteed service networks. *IEEE Transactions Informormation Theory*, 44(3):1087–1096.

Likhanov, N. and Mazumdar, R.R. (1999). Cell loss asymptotics in buffers fed with a large number of independent stationary sources. *Journal of Applied Probability*, 36(1):86–96.

Likhanov, N., Mazumdar, R.R., and Théberge, F. (1996). Calculation of cell loss probabilities in large unbuffered multiservice systems. In: *Proceedings of the IEEE ICC*, Montréal.

Luenberger, D. (1984). *Linear and Nonlinear Programming*, 2nd, Edition. Addison-Wesley Publ. Co.

Mandjes, M. and Kim, J.H. (2001). Large deviations for small buffers: an insensitivity result. *Queueing Systems*, 37(4):349–362.

Massoulie, L. (1999). Large deviations estimates for polling and weighted fair queueing service systems. *Advances in Performance Analysis.*

Mitra, D. and Morrison, J.A. (1994). Erlang capacity and uniform approximations for shared unbuffered resources. *IEEE/ACM Transactions on Networking*, 2:558–570.

O'Connell, N. (1997). Large deviations for departures from a shared buffer. *Journal of Applied Probability*, 34(3):753–766.

O'Connell, N. (1998). Large deviations for queue lengths at a multi-buffered resource. *Journal of Applied Probability*, 35(1):240–245.

Ozturk, O., Mazumdar, R., and Likhanov, N. (2004). Many sources asymptotics in networks with small buffers. *Queueing Systems (QUESTA)*, 46(1-2).

Petrov, V.V. (1975). *Sums of Independent Random Variables*. Springer-Verlag, Berlin.

Roberts, J.W. (1998). *COST* 242: *Methods for the Performance Evaluation and Design of Broadband Multiservice Networks*. Elsevier N.V.

Ross, K.W. (1995). *Multiservice Loss Networks for Broadband Telecommunication Networks*. Springer-Verlag, N.Y.

Shakkottai, S. and Srikant, R. (2001). Many-sources delay asymptotics with applications to priority queues. *Queueing System*, 39(2-3):183–200.

Wischik, D. (1999). Output of a switch, or, effective bandwidths for networks. *Queueing Systems*, 32:383–396.

Chapter 7

COMBINED COMPETITIVE FLOW CONTROL AND ROUTING IN NETWORKS WITH HARD SIDE CONSTRAINTS

Rachid El Azouzi
Mohamed El Kamili
Eitan Altman
Mohammed Abbad
Tamer Başar

Abstract We consider in this paper the problem of combined flow control and routing in a noncooperative setting, where each user is faced with a multi-criteria optimization problem, formulated as the minimization of one criterion subject to constraints on others. We address here the basic questions of existence and uniqueness of equilibrium. We show that an equilibrium indeed exists, but it may not be unique due to the multi-criteria nature of the problem. We are able, however, to obtain uniqueness in some weaker sense under appropriate conditions; we show in particular that the link utilizations are uniquely determined at equilibrium and the normalized Nash equilibrium is unique.

1. Introduction

Flow control and routing are two components of resource and traffic management in today's high-speed networks, such as the Internet and the ATM. Flow control is used by best-effort type traffic in order to adjust the input transmission rates (the instantaneous throughput of a connection) to the available bandwidth in the network. Routing decisions are taken to select paths with certain desirable properties, such as minimum delays. In real time applications, however, an application may have several criteria for quality of service. It might be sensitive to delays and losses, or it might seek to minimize some cost imposed on the use of network resources. In the presence of several users each with several objectives, who determine the routes for flows they control, this gives rise to a noncooperative multicriteria game. As is often the case in today's networks, quality of service of an application is often given through an upper bound on some performance measure (delay, loss rate or jitter, see e.g. ATM forum, 1999)). An appropriate framework for modeling this situation is that of noncooperative game theory.

Traditional noncooperative games combining flow and routing decisions have been studied in the past; see, for example, Haurie and Marcotte (1985); Patriksson (1994), and references therein. In particular, it is well known that, when the cost function of each player is the sum of link costs minus a reward which is a function of its throughput, then the underlying game can be transformed into one involving only routing decisions. Other recent papers that consider a combined flow control and routing game are Rhee and Konstantopoulos (1998, 1999), where the utility of each player is related to the sum of powers over the links. (The power criterion is the ratio between some function of the throughput and the delay.) The part of the utility in Rhee and Konstantopoulos (1998, 1999) that corresponds to the delay is given by the sum of all link capacities minus all link flows, and in Rhee and Konstantopoulos (1998) it is further multiplied by some entropy function. Thus, the utility in this case does not directly correspond to the actual expected delay, but it has the advantage of leading to a computable Nash equilibrium in the case of parallel links. Yet another reference, Altman, Başar and Srikant (1999), on the other hand, deals with the actual power criterion, i.e., the ratio between (some increasing function of) the total throughput of a user and the average delay experienced by traffic of that user. The paper introduces approximate Nash equilibrium corresponding to the case when the number of players is very large, establishes the existence of such an equilibrium and a characterization for it, and further shows that the solution needs not be unique.

In this paper, we consider such constraints which can be expressed as bounds on the sum of some functions (such as delays or costs) along all links constituting a path from an origin to a destination. In mathematical terms, such problems are games where the users strategy sets are not independent but coupled. Games of this kind are called coupled-constraint games (see Rosen, 1965), and we call the corresponding solution concept the *constrained (or coupled) Nash equilibrium.*

In El Azouzi and Altman (2003), we have investigated the special topology of parallel links without flow control. We showed through a simple example that there may exist several equilibria, although in the absence of side constraints there would be a single equilibrium Orda,

Rom and Shimkin (1993). We then showed the uniqueness of normalized equilibrium *a la* Rosen (Rosen, 1965) for the constrained parallel links problem, and discussed an application of this to pricing. Our objective here is to extend these results to the case where the global throughput of each user may also be controlled. This approach allows us to establish the uniqueness of normalized Nash equilibrium in the combined flow control and routing game with constraints of quality of service in a network of parallel links, and to obtain several qualitative characterizations of this equilibrium.

In the general topology case, it is known that even without flow control and constraints there may be several equilibria; counter examples that show nonuniqueness have been given in Orda, Rom and Shimkin (1993). Yet in two special cases, uniqueness of Nash equilibrium has been established in general topology networks in the absence of side constraints, Orda, Rom and Shimkin (1993):

1. In the symmetric case, there is a unique Nash equilibrium which is itself symmetric.

2. Consider two equilibria each with the property that whenever a player sends a positive amount of flow to some link then all other players also do so. We call this *assumption of "all positive flow"*. Then, the total link flow under the two equilibria are equal.

We focus in this paper on these cases, with flow control and additional side constraints, and show that the type of uniqueness results that hold for the unconstrained case extends also to the constrained normalized equilibrium.

These specific results in this paper extend some previous results in Rosen (1965) on the uniqueness of the Nash equilibrium in the context of noncooperative control of flow and routing without constraints.

The structure of the paper is as follows. In the next section we introduce the model and assumptions. In Section 3, we establish the existence of coupled Nash equilibria and normalized Nash equilibria for general topology networks and motivate its use for decentralized pricing. In Section 4, we consider the case of two nodes connected by a set of parallel links and study the uniqueness of normalized Nash equilibrium; we also derive some properties of the equilibrium. In Section 5, we extend the discussion to general topology networks. We study the uniqueness of equilibrium in the symmetrical framework and also under the assumption of "all positive flows". The paper concludes with conclusion in Section 6.

2. The model

We consider a network $(\mathcal{N}, \mathcal{L})$ where $\mathcal{N}$ is a finite set of nodes and $\mathcal{L} = \{1, 2, \ldots, L\}$ is a set of L links. We consider an extension of directional links (see Altman and Kameda, 2001) where a link may carry traffic in both directions, but the direction for each user is fixed. For a node $v \in \mathcal{N}$, let $\text{Out}(v, i)$ be the set of outgoing links from node v available to user i, and $\text{In}(v, i)$ be the corresponding set of links in-going to node v available to user i. We consider a set $\mathcal{I}$ of I selfish users (players) who share the network. With each user $i \in \mathcal{I}$, we associate a unique pair, $(s(i), d(i))$, of source and destination nodes. Each user i has to determine the amount of flow $r^i \in R^i := [m^i, M^i]$ to ship between $s(i)$ and $d(i)$ and how to route it in the network.

Let f_l^i denote the amount of flow that user i sends over link l, which is constrained to be nonnegative, and satisfy the flow conservation law, i.e. for each node $v \notin (s(i), d(i))$,

$$f_l^i \geq 0, \qquad \sum_{l \in \text{Out}(v,i)} f_l^i = \sum_{l \in \text{In}(v,i)} f_l^i, \qquad v \notin \big(s(i), d(i)\big), \tag{7.1}$$

and

$$r^i := \sum_{l \in \text{Out}\left(s(i),i\right)} f_l^i = \sum_{l \in \text{In}\left(d(i),i\right)} f_l^i \in [m^i, M^i]. \tag{7.2}$$

Further define $\mathbf{f}_l = \{f_l^1, \ldots, f_l^I\}$, $f_l = \sum_{i=1}^I f_l^i$, $\mathbf{f}^i = \{f_l^i\}_{l \in \mathcal{L}}$, $\mathbf{f}^{-i} = \{\mathbf{f}^1, \ldots, \mathbf{f}^{i-1}, \mathbf{f}^{i+1}, \ldots, \mathbf{f}^I\}$, $\mathbf{f} = \{\mathbf{f}_l\}_{l \in \mathcal{L}}$.

We consider a situation where extra side constraints are imposed. These may represent constraints on the quality of service, which may be user dependent. These are formulated as a set of flow restrictions of the form:

$$g_k(\mathbf{f}) \leq 0, \quad k \in \mathcal{K} \tag{7.3}$$

where $\mathcal{K}$ is a finite index set (e.g., formed by subsets of $\mathcal{I}$, $\mathcal{N}$, $\mathcal{L}$, $\mathcal{P}$, where $\mathcal{P}$ stands for the set of routes (paths)), and $g_k \colon \mathbb{R}_+^{|\mathcal{L}| \times I} \to \mathbb{R}$, $k \in \mathcal{K}$. Introduce the function $h \colon \mathbb{R}_+^{|\mathcal{L}| \times I} \to \mathbb{R}^m$ to describe the constraints (7.1)–(7.3), where m is the number of constraints. Hence admissible strategies will be limited by the requirement that $\mathbf{f}$ be selected from a set $\mathcal{R}$, where $\mathcal{R} = \{\mathbf{f}, h(\mathbf{f}) \leq 0\}$. We will say that $\mathcal{R}$ is a coupled constraint set. With $\mathbf{f}^{-i} := \{\mathbf{f}^j, j \in \mathcal{I}; j \neq i\}$ fixed, we also introduce the set $\mathcal{R}^i(\mathbf{f}^{-i}) := \{\mathbf{f}^i \colon (\mathbf{f}^i, \mathbf{f}^{-i}) \in \mathcal{R}\}$. This is the set of allowable flows for user i with all other users' flows fixed.

Example 7.1 The most immediate example of a set of flow restrictions is that of upper bounds on the end-to-end packet delay. For each $i \in \mathcal{I}$, let $w(i)$ be its corresponding O-D pair, i.e., $w(i) = \big(s(i), d(i)\big)$. Let $\mathcal{P}_{w(i)}$ be the set of routes connecting the O-D pair $w(i)$. The delay over link l can be given by $(c_l - f_l)^{-1}$ where c_l is the capacity of link l; see Remark 17. In the framework of (7.3) such constraints are described by letting $\mathcal{K} = \{w(i)/w(i) = \big(s(i), d(i)\big), i \in \mathcal{I}\}$, and

$$\sum_{l \in \mathcal{L}} \frac{\delta_{lp}}{c_l - f_l} \le D^i, \quad p \in \mathcal{P}_{w(i)}, \quad i \in \mathcal{I}, \tag{7.4}$$

where $\delta_{\ell p} = 1$ if route p uses link ℓ, and is 0 otherwise. Constraint (7.4) requires that for each O-D pair $\big(s(i), d(i)\big)$, $i \in \mathcal{I}$, the end-to-end packet delay should be no larger than D^i.

The performance objective of user i is quantified by means of a cost function $J^i(\mathbf{f})$. User i wishes to find a strategy f^i that minimizes its cost. This optimization depends on the routing decisions of the other users, described by the strategy profile $\mathbf{f}^{-i}$, since J^i is a function of the system flow configuration $\mathbf{f}$, and the constraints (7.3) are coupled.

Definition 7.1 [*Cost functions and Nash equilibrium*] *Let* $J^i(\mathbf{f})$ *be the cost for user* i *when the flows of all users are given by* $\mathbf{f} \in \mathcal{R}$. *A coupled Nash equilibrium of the routing game is a strategy profile from which no user finds it beneficial to unilaterally deviate. Accordingly, we seek a Coupled Nash Equilibrium* (CNE) $\tilde{\mathbf{f}}$, *that is an* $\tilde{\mathbf{f}} \in \mathcal{R}$ *satisfying*

$$J^i(\tilde{\mathbf{f}}) = \min_{(\mathbf{f}^i, \tilde{\mathbf{f}}^{-i}) \in \mathcal{R}} J^i(\tilde{\mathbf{f}}^{-i}, \mathbf{f}^i) \ \textit{where} \tag{7.5}$$
$$J^i(\tilde{\mathbf{f}}^{-i}, \mathbf{f}^i) := J^i(\tilde{\mathbf{f}}^1, \ldots, \tilde{\mathbf{f}}^{i-1}, \mathbf{f}^i, \tilde{\mathbf{f}}^{i+i}, \ldots \mathbf{f}^I).$$

We make the following assumptions on the cost function J^i for user i, which will be invoked throughout the paper:

G1: J^i is given as the sum of link costs $J^i_l(\mathbf{f}_l)$ minus the utility function $U^i(r^i)$: $J^i(\mathbf{f}) = \sum_{l \in \mathcal{L}} J^i_l(\mathbf{f}_l) - U^i(r^i)$.

G2: $J^i_l \colon [0, \infty]^I \to [0, \infty]$ is continuous and U^i is continuous in its argument.

G3: J^i_l is convex in f^i_l, and g_k is convex in f^i_l and U^i is concave in its argument.

G4: J^i_l is continuously differentiable in f^i_l, g_k is continuously differentiable in f^i_l, and U^i is continuously differentiable in its argument. We set $K^i_l := \partial J^i_l(\mathbf{f}_l)/\partial f^i_l$, $l \in \mathcal{L}$, and $K^i_0(r^i) := -\partial U^i(r^i)/\partial r^i$.

G5: The feasible set of (7.1)–(7.3) is non-empty. Moreover, for any player i and any strategy of the other players, the set of feasible strategies for player i contains a point that is strictly interior to every nonlinear constraint. Functions that comply with the above assumptions shall be referred to as *type*-**G** functions.

A1: Assumptions **G1**–**G5** are all satisfied, and J_l^i depends on the vector $\mathbf{f}_l$ only through user i's flow on link l and the total flow on that link. In other words, J_l^i can be written (with some abuse of notation) as $J_l^i(\mathbf{f}_l) = J_l^i(f_l^i, f_l)$.

A2: g_k is strictly increasing in each of its arguments, for each $k \in \mathcal{K}$.

A3: Viewing $K_l^i = K_l^i(f_l^i, f_l)$ now as a function of two arguments, whenever J_l^i is finite, $K_l^i(f_l^i, f_l)$, $l \in \mathcal{L}$, is increasing in each of its two arguments, and (due to **G3**) strictly increasing in the first one.

We will refer to functions that meet the conditions of these three assumptions as *type*-**A** functions.

Typically, the performance of a link l is manifested through some function $T_l(f_l)$, which measures the cost per unit of flow on the link, and depends on the link's total flow. Thus, it is of interest to consider cost functions of the following form:

B1: $J^i(\mathbf{f}) = \sum_{l \in \mathcal{L}} f_l^i T_l(f_l) - U(r^i)$.

B2: $T_l\colon [0, \infty) \to (0, \infty]$.

B3: $T_l(f_l)$ is continuously differentiable and $T_l{}'(f_l) = dT_l(f_l)/df_l$ is increasing in f_l, for all $l \in \mathcal{L}$.

We will refer to functions that meet the conditions of these three assumptions as *type*-**B** functions.

REMARK 17 Cost functions used in real networks are either related to actual pricing, or they are related to some performance measure such as expected delay. In the first case, a frequently used cost is that of linear link costs, i.e. for each user i, $J^i(\mathbf{f}) = \sum_{i=1}^{L} f_l^i T_l(f_l)$ where $T_l(f_l) = a_l f_l + b_l$ (Lutton, 2001). When the costs represent delays, they typically have the same form but with $T_l(f_l) = (c_l - f_l)^{-1} + d_l$, where the first term represents queuing delay, with c_l standing for the queue capacity, and d_l represents the propagation delay related to link l. The queuing delay is that of an M/M/1 queue operating under the FIFO regime (packets are served in the same order as they arrive, see Orda, Rom and Shimkin, 1993) or of an M/G/1 queue operating under the processor sharing regime. Other, more complicated costs can be found in Altman, El Azouzi and Vyacheslav (2002).

3. Existence of equilibria and pricing

3.1 Characterization of equilibria and normalized Nash equilibria

If assumptions **G** hold, it follows that the minimization in (7.5) is equivalent to the following Kuhn–Tucker conditions: For every $i \in \mathcal{I}$, there exists a set of Lagrange multipliers $(\lambda_u^i)_{u \in \mathcal{N}}$, $(\beta_k^i)_{k \in \mathcal{K}}$, γ^i and μ^i such that, for every link $(u, v) \in \mathcal{L}$,

$$K_{uv}^i(\mathbf{f}_{uv}) + \lambda_v^i - \lambda_u^i + \sum_{k \in \mathcal{K}} \beta_k^i \frac{\partial g_k(\mathbf{f})}{\partial f_{uv}^i} = 0 \quad \text{if} \quad f_{uv}^i > 0. \tag{7.6}$$

$$K_{uv}^i(\mathbf{f}_{uv}) + \lambda_v^i - \lambda_u^i + \sum_{k \in \mathcal{K}} \beta_k^i \frac{\partial g_k(\mathbf{f})}{\partial f_{uv}^i} \geq 0 \quad \text{if} \quad f_{uv}^i = 0. \tag{7.7}$$

$$K_0^i(r^i) - \lambda_{d(i)}^i + \lambda_{s(i)}^i + \gamma^i - \mu^i = 0. \tag{7.8}$$

$$\gamma^i(r^i - M^i) = 0, \quad \mu^i(m^i - r^i) = 0, \quad \beta_k^i g_k(\mathbf{f}) = 0. \tag{7.9}$$

$$r^i - M^i \leq 0, \quad m^i - r^i \leq 0, \quad g_k(\mathbf{f}) \leq 0, \quad f_{uv}^i \geq 0. \tag{7.10}$$

$$\mu^i \geq 0, \quad \gamma^i \geq 0, \quad \beta_k^i \geq 0. \tag{7.11}$$

3.2 Normalized Nash equilibria and pricing

We now consider a special kind of equilibrium, called *normalized (Rosen) Nash equilibrium*, defined below.

DEFINITION 7.2 *A coupled Nash equilibrium* $\mathbf{f}$ *is a normalized Nash equilibrium (see* Rosen (1965)) *if there exist a vector* $\tilde{\alpha} > \mathbf{0}$ *where* $\tilde{\alpha} = (\alpha^1, \ldots, \alpha^I)$ *and* $\mathbf{0}$ *is a vector of zeros, and some constants* $\beta_k \geq 0$, $k \in \mathcal{K}$, *such that conditions* (7.6)-(7.11) *are satisfied with*

$$\beta_k^i = \beta_k / \alpha^i, \qquad k \in \mathcal{K}, \quad i \in \mathcal{I} \tag{7.12}$$

Notice that if a user's weight α^i is greater than those of his competitors, then his corresponding Lagrange multipliers are smaller.

The normalized Nash equilibrium can be used in relation to an appealing pricing scheme in which additional congestion costs are imposed by the network. Congestion pricing will allow us to relax the original constraints $g_k(\mathbf{f}) \leq 0$; yet the resulting equilibrium will have the following three appealing properties:

1. It will be a coupled Nash equilibrium (*CNE*) for the original problem.

2. Non-zero congestion prices will only be imposed for saturated constraints, which represent congestion; in the absence of congestion, no congestion cost is imposed.

3. The most interesting feature of this pricing is that congestion costs may be chosen to be user independent. This allows us to implement them in a decentralized way without requiring per-flow information.

More precisely, assume that the utility of user i can be written as $-J^i(\mathbf{f}) - 1/a^i \sum_{k\in\mathcal{K}} C_k(\mathbf{f})$, where $C_k(\mathbf{f})$ is a user-independent cost function that all users are charged due to congestion related to the kth constraint. Let $(\beta_k^i)^*$ be Lagrange multipliers that correspond to a *CNE* induced by taking in (7.12) $\tilde{\alpha} = (a^1, \dots, a^I)$. We set $C_k(\mathbf{f}) = \beta_k^* \cdot g_k(\mathbf{f}_l)$. With this cost function we may now consider a competitive routing problem in which we ignore constraints (7.3). The equilibrium obtained is a *CNE* for the original constrained model, and the complementary slackness conditions imply that at the normalized equilibrium no user actually pays any congestion cost. Under various conditions, there is a unique Nash Equilibrium, see Orda, Rom and Shimkin (1993), to the pricing game (where constraints (7.3) are removed) and the corresponding Kuhn–Tucker conditions obviously coincide with our original ones. We conclude that a simple pricing can replace the QoS (Quality of Service) constraints and yet force users to choose a *CNE* (so the constraints still hold). Since the pricing does not depend on the user, the charging can be performed in a distributed way without any need for per flow information. The existence of such a pricing is equivalent to the existence of a Normalized Nash equilibrium.

3.3 Existence of equilibria

Under assumption **G5**, the set $\mathcal{R}$ contains a point that is strictly interior to every nonlinear constraint. This is a sufficient condition for the Kuhn–Tucker constraint qualification (Arrow, Hurwicz and Uzawa, 1961). Hence, the routing game (7.5) using the cost functions of *type*-**G** is equivalent to a convex game in the sense of Rosen (1965) and, thus the existence of a *CNE* as well as a normalized equilibrium is guaranteed (Rosen, 1965, Thm. 1) if the costs are finite for any strategy. Note that the proof of existence in Orda, Rom and Shimkin (1993) is based on Rosen (1965) with the restriction to finite costs assured using the following assumption: "For every flow configuration $\mathbf{f}$, if not all costs are finite then at least one user with infinite cost $J^i(\mathbf{f})$ can change its flow configuration to make its cost finite".

THEOREM 7.1 *Consider the cost function of* type-**G**. *Then there exists a normalized Nash equilibrium point for every specified vector* $\tilde{\alpha} > \mathbf{0}$ *(componentwise) where* $\vec{\alpha} = (\alpha^1, \alpha^2, \ldots, \alpha^I)$.

We analyze the routing problem in two phases. First, we consider a case of two nodes connected by a set of parallel links. Second, we extend some results to a general network, for the case of symmetric users and positive flows.

4. Parallel links

In this section, we consider the case of two nodes $\{1, 2\}$ connected by a set $\mathcal{L}$ of L links. Such a system of parallel links may represent a network in which resources are pre-allocated to various paths, or an internetworking in which each link models a different subnetwork. With each user i, we associate a unique pair, $(s(i), d(i))$, of source and destination nodes, where $s(i), d(i) \in \{1, 2\}$. For the parallel links we further impose a quality of service on each link. This is captured by the constraints (7.3), where $\mathcal{K} = \mathcal{L}$ and g_l depends on the flows only through total flow on link l, i.e.,

$$g_l(f_l) \leq 0. \tag{7.13}$$

Under assumptions **A**, the functions g_l are strictly increasing in f_l, and hence g_l^{-1} exists and the constraints (7.13) become $f_l \leq d_l$, $l \in \mathcal{L}$, where $d_l = g_l^{-1}(0)$ (positive real).

REMARK 18 Equation (7.13) can be interpreted as capturing the link capacity constraints i.e., the total flow at each link l can not exceed the link capacity $d_l = c_l$. In many cases, however, performance measures (such as loss probabilities or delays) are monotone in the link load, which then implies that bounds on these measures are obtained by bounding the link load as in (7.13).

In El Azouzi and Altman (2003), an example has been presented, which demonstrates that the above weak convexity conditions are not sufficient for uniqueness of a Nash equilibrium (without flow control). These indeed point at the complexity of the coupled constraint (7.13). This nonuniqueness of Nash equilibria motivates us to study normalized (*Rosen*) Nash equilibrium (defined earlier) in the parallel links topology, particularly its uniqueness and some of its characteristics.

First, we prove in the following under some hypotheses that the link utilizations and the user demands are unique at each Nash equilibrium.

THEOREM 7.2 *Let the cost function of each user be of type-*$\mathbf{A}$. *Further let* $\mathbf{f}$ *and* $\hat{\mathbf{f}}$ *be two coupled Nash equilibria, and* $(\beta_l^i, \mu^i, \gamma^i)$ *and* $(\hat{\beta}_l^i, \hat{\mu}^i, \hat{\gamma}^i)$

be the corresponding Lagrange multipliers. Assume that for each link $l \in \mathcal{L}$, $\beta_l^i \leq \hat{\beta}_l^i$, $\forall i \in \mathcal{I}$ *or* $\hat{\beta}_l^i \leq \beta_l^i$, $\forall i \in \mathcal{I}$. *Then* $f_l = \hat{f}_l$ $\forall l \in \mathcal{L}$, *and* $r^i = \hat{r}^i$, $\forall i \in \mathcal{I}$ *(i.e., the link utilizations and the user demands are the same under* $\mathbf{f}$ *and* $\hat{\mathbf{f}}$*).*

Proof. Let $\mathbf{f}$ and $\hat{\mathbf{f}}$ be two Nash equilibria. Then we have from (7.6)–(7.8):

$$\begin{cases} K_l^i(f_l^i, f_l) + K_0^i(r^i) \geq \mu^i - \gamma^i - \beta_l^i; \\ K_l^i(f_l^i, f_l) + K_0^i(r^i) = \mu^i - \gamma^i - \beta_l^i; & \text{if} \quad f_l^i > 0 \end{cases} \tag{7.14}$$

$$\begin{cases} K_l^i(\hat{f}_l^i, \hat{f}_l) + K_0^i(\hat{r}^i) \geq \hat{\mu}^i - \hat{\gamma}^i - \hat{\beta}_l^i; \\ K_l^i(\hat{f}_l^i, \hat{f}_l) + K_0^i(\hat{r}^i) = \hat{\mu}^i - \hat{\gamma}^i - \hat{\beta}_l^i; & \text{if} \quad \hat{f}_l^i > 0 \end{cases} \tag{7.15}$$

As in the proof of (El Azouzi and Altman, 2003, Thm. 3.1), we have the following relations:

(i) $\{\hat{\beta}_l^i < \beta_l^i; \hat{f}_l \geq f_l\} \Longrightarrow \hat{f}_l = f_l$ moreover if $(\mu^i - \gamma^i \leq \hat{\mu}^i - \hat{\gamma}^i$ and $\hat{r}^i \leq r^i)$ then $\hat{f}_l^i \geq f_l^i$ and the last inequality is strict if $f_l^i > 0$.

(ii) $\{\hat{\beta}_l^i > \beta_l^i; \hat{f}_l \leq f_l\} \Longrightarrow \hat{f}_l = f_l$ moreover if $(\mu^i - \gamma^i \geq \hat{\mu}^i - \hat{\gamma}^i$ and $\hat{r}^i \geq r^i)$ then $\hat{f}_l^i \leq f_l^i$ and the last inequality is strict if $\hat{f}_l^i > 0$.

(iii) $\{\hat{\mu}^i - \hat{\gamma}^i \leq \mu^i - \gamma^i; \hat{r}^i \geq r^i; \hat{\beta}_l^i \geq \beta_l^i; \hat{f}_l \geq f_l\} \Longrightarrow \hat{f}_l^i \leq f_l^i$

(vi) $\{\hat{\mu}^i - \hat{\gamma}^i \geq \mu^i - \gamma^i; \hat{r}^i \leq r^i; \hat{\beta}_l^i \leq \beta_l^i; \hat{f}_l \leq f_l\} \Longrightarrow \hat{f}_l^i \geq f_l^i$.

Let $\mathcal{L}_1 = \{l \colon \hat{f}_l > f_l\}$. Also, denote $\mathcal{I}_1 = \{i \colon \hat{r}^i \leq r^i; \hat{\mu}^i - \hat{\gamma}^i \geq \mu^i - \gamma^i\}$, $\mathcal{L}_2 = \{l \colon \hat{f}_l \leq f_l; \hat{\beta}_l^i \leq \beta_l^i\}$ and $\mathcal{L}_3 = \{l \colon \hat{f}_l \leq f_l; \hat{\beta}_l^i > \beta_l^i\}$. We observe that $\mathcal{L} = \mathcal{L}_1 \cup \mathcal{L}_2 \cup \mathcal{L}_3$. Assume that $\mathcal{L}_1$ is nonempty; then it follows by (iv) that for $i \in \mathcal{I}_1$:

$$\begin{aligned} \sum_{l \in \mathcal{L}_1} \hat{f}_l^i &= \hat{r}^i - \sum_{l \in \mathcal{L}_2} \hat{f}_l^i - \sum_{l \in \mathcal{L}_3} \hat{f}_l^i \\ &\leq r^i - \sum_{l \in \mathcal{L}_2} f_l^i - \sum_{l \in \mathcal{L}_3} \hat{f}_l^i \\ &= \sum_{l \in \mathcal{L}_1} f_l^i + \sum_{l \in \mathcal{L}_3} (f_l^i - \hat{f}_l^i). \end{aligned}$$

We now proceed to show that

$$i \notin \mathcal{I}_1 \text{ implies that } \hat{r}^i \geq r^i \quad \text{and} \quad \hat{\mu}^i - \hat{\gamma}^i \leq \mu^i - \gamma^i. \tag{7.16}$$

To this end, it suffices to show that $\hat{r}^i > r^i \rightarrow \hat{\mu}^i - \hat{\gamma}^i \leq \mu^i - \gamma^i$ and $\hat{\mu}^i - \hat{\gamma}^i < \mu^i - \gamma^i \rightarrow \hat{r}^i \geq r^i$. Indeed, if $\hat{r}^i > r^i$ then $r^i < M^i$, and $\hat{r}^i > m^i$, consequently, $\gamma^i = \hat{\mu}^i = 0$, and $-\hat{\gamma}^i \leq \mu^i$. On the other hand, if $\hat{\mu}^i - \hat{\gamma}^i < \mu^i - \gamma^i$, we consider two cases:

Case 1: $\mu^i - \gamma^i \leq 0$, then $0 \leq \hat{\mu}^i < \hat{\gamma}^i$, we deduce that $\hat{r}^i = M^i$, in particular, we obtain $\hat{r}^i \geq r^i$.

Case 2: $\mu^i - \gamma^i > 0$, it follows that $\mu^i > 0$ and $r^i = m^i$, in particular, we obtain $\hat{r}^i \geq r^i$. Therefore we conclude (7.16).

Noting that (i) implies that $\{l \in \mathcal{L}_1 / \hat{\beta}_l^i < \beta_l^i\} = \emptyset$, it follows that if $l \in \mathcal{L}_1$ and $i \notin \mathcal{I}_1$, we have from (iii) and (7.16) $\hat{f}_l^i \leq f_l^i$. It further follows that:

$$\begin{aligned}
\sum_{l\in\mathcal{L}_1} \hat{f}_l &= \sum_{l\in\mathcal{L}_1}\sum_{i\in\mathcal{I}_1} \hat{f}_l^i + \sum_{l\in\mathcal{L}_1}\sum_{i\notin\mathcal{I}_1} \hat{f}_l^i \\
&\leq \sum_{l\in\mathcal{L}_1}\sum_{i\in\mathcal{I}_1} f_l^i + \sum_{l\in\mathcal{L}_3}\sum_{i\in\mathcal{I}_1}(f_l^i - \hat{f}_l^i) + \sum_{l\in\mathcal{L}_1}\sum_{i\notin\mathcal{I}_1} f_l^i \\
&= \sum_{l\in\mathcal{L}_1} f_l + \sum_{l\in\mathcal{L}_3}\sum_{i\in\mathcal{I}_1}(f_l^i - \hat{f}_l^i) \\
&= \sum_{l\in\mathcal{L}_1} f_l + \sum_{l\in\mathcal{L}_3}(f_l - \hat{f}_l) - \sum_{l\in\mathcal{L}_3}\sum_{i\notin\mathcal{I}_1}(f_l^i - \hat{f}_l^i) \\
&\leq \sum_{l\in\mathcal{L}_1} f_l.
\end{aligned} \tag{7.17}$$

The last inequality follows from (ii), since for $l \in \mathcal{L}_3$, $f_l = \hat{f}_l$ and for $l \in \mathcal{L}_3$ and $i \notin \mathcal{I}_1$, $f_l^i \geq \hat{f}_l^i$.

Inequality (7.17) and the definition of $\mathcal{L}_1$ are contradictory, which implies that $\mathcal{L}_1$ is an empty set. By symmetry, it can be concluded that the set $\mathcal{L}_1{}' = \{l \colon \hat{f}_l < f_l\}$ is also empty. Therefore we conclude that, $\hat{f}_l = f_l, \forall l \in \mathcal{L}$.

We now proceed to show that $r^i = \hat{r}^i$, $\forall i \in \mathcal{I}$. To this end, let $\mathcal{I}_1 = \{i \colon \hat{r}^i > r^i\}$. Also, denote $\mathcal{L}_1 = \{l \colon \hat{\beta}_l^i \leq \beta_l^i\}$, $\mathcal{I}_2 = \{i \colon \hat{r}^i \leq r^i; \hat{\mu}^i - \hat{\gamma}^i \geq \mu^i - \gamma^i\}$ and $\mathcal{I}_3 = \{i \colon \hat{r}^i \leq r^i; \hat{\mu}^i - \hat{\gamma}^i < \mu^i - \gamma^i\}$. We observe that $\mathcal{I} = \mathcal{I}_1 \cup \mathcal{I}_2 \cup \mathcal{I}_3$. Assume that $\mathcal{I}_1$ is nonempty; then, it follows by (iv) that for $l \in \mathcal{L}_1$:

$$\begin{aligned}
\sum_{i\in\mathcal{I}_1} \hat{f}_l^i = \hat{f}_l - \sum_{i\in\mathcal{I}_2} \hat{f}_l^i - \sum_{i\in\mathcal{I}_3} \hat{f}_l^i &\leq f_l - \sum_{i\in\mathcal{I}_2} f_l^i - \sum_{i\in\mathcal{I}_3} \hat{f}_l^i \\
&= \sum_{i\in\mathcal{I}_1} f_l^i + \sum_{i\in\mathcal{I}_3}(f_l^i - \hat{f}_l^i).
\end{aligned}$$

Noting that from (7.16), the relation (iii) implies that $\hat{f}_l^i \leq f_l^i$ for $i \in \mathcal{I}_1$ and $l \notin \mathcal{L}_1$, it follows that:

$$\begin{aligned}
\sum_{i\in\mathcal{I}_1} \hat{r}^i &= \sum_{i\in\mathcal{I}_1}\sum_{l\in\mathcal{L}_1} \hat{f}_l^i + \sum_{i\in\mathcal{I}_1}\sum_{l\notin\mathcal{L}_1} \hat{f}_l^i \\
&\leq \sum_{i\in\mathcal{I}_1}\sum_{l\in\mathcal{L}_1} f_l^i + \sum_{i\in\mathcal{I}_3}\sum_{l\in\mathcal{L}_1} (f_l^i - \hat{f}_l^i) + \sum_{i\in\mathcal{I}_1}\sum_{l\notin\mathcal{L}_1} f_l^i \\
&= \sum_{i\in\mathcal{I}_1} r^i + \sum_{i\in\mathcal{I}_3}\sum_{l\in\mathcal{L}_1} (f_l^i - \hat{f}_l^i) \qquad (7.18) \\
&= \sum_{i\in\mathcal{I}_1} r^i + \sum_{i\in\mathcal{I}_3} (r^i - \hat{r}^i) - \sum_{i\in\mathcal{I}_3}\sum_{l\notin\mathcal{L}_1} (f_l^i - \hat{f}_l^i) \\
&\leq \sum_{i\in\mathcal{I}_1} r^i.
\end{aligned}$$

The last inequality follows from (i), since for $i \in \mathcal{I}_3$, $\hat{r}^i = r^i$ and for $i \in \mathcal{I}_3$ and $l \notin \mathcal{L}_1$, $\hat{f}_l^i \leq f_l^i$. Hence, the inequality (7.18) and the definition of $\mathcal{I}_1$ are contradictory, which implies that $\mathcal{I}_1$ is an empty set. By symmetry, it can be concluded that the set $\mathcal{I}_1{}' = \{i: \hat{r}^i < r^i\}$ is also empty. Therefore we conclude that $\hat{r}^i = r^i$, $\forall i \in \mathcal{I}$. □

4.1 Uniqueness of the normalized Nash equilibrium

A set of sufficient conditions for uniqueness of the normalized Nash equilibrium has been established by Rosen in Rosen (1965) under some strict diagonal convexity conditions. These conditions will not be satisfied in our case, and hence we need to prove uniqueness in some other way.

REMARK 19 Let $\vec{\alpha}$ and $\vec{\hat{\alpha}}$ be two positive vectors such that $\vec{\hat{\alpha}} = a\vec{\alpha}$ for some positive real a. Let $\mathcal{A}(\vec{\alpha})$ and $\mathcal{A}(\vec{\hat{\alpha}})$ be the corresponding normalized Nash equilibria sets. Then $\mathcal{A}(\vec{\alpha}) = \mathcal{A}(\vec{\hat{\alpha}})$.

The following result shows that the parallel-links network also has a unique normalized Nash equilibrium for every specified vector $\vec{\alpha} > 0$.

THEOREM 7.3 *In a network of parallel links where the cost function of each user is of type-***A***, the normalized Nash equilibrium for every specified $\vec{\alpha} > 0$ is unique.*

Proof. The hypotheses of Theorem 7.2 hold in this case, and hence we have the uniqueness for link utilizations, and each user has the same demand under all normalized Nash equilibria. Then the theorem follows directly from (El Azouzi and Altman, 2003, Thm. 4.1) (i.e., the case where the demands are fixed). □

COROLLARY 7.1 *In a network of parallel links where the cost function of each user is of type-***A***, and in the absence of the side constraints (i.e., $\beta_l = 0$, $\forall l \in \mathcal{L}$), there is a single Nash Equilibrium.*

The above result can be considered an extension of (Rosen, 1965, Thm 1) in noncooperative flow control and routing games.

4.2 Properties of the normalized Nash equilibrium

Here, we assume that the cost functions of all users are symmetrically identical, i.e., $J_l^i \equiv J_l$ and $U^i \equiv U$ for all $i \in \mathcal{I}$ and $l \in \mathcal{L}$. We also let $\eta^i = \mu^i - \gamma^i, \forall i \in \mathcal{I}$.

LEMMA 7.1 *Assume that all users have the same weight and assume that the condition $f_{\hat{l}}^i > f_{\hat{l}}^j$ holds for some link $\hat{l}$ and some users i and j. Then $f_l^i \geq f_l^j$ for all $l \in \mathcal{L}$; moreover, the inequality is strict if $f_l^j > 0$.*

Proof. From Remark 19 we shall only prove the lemma for $\alpha^i = 1$, $\forall i \in \mathcal{I}$. Choose an arbitrary link l. If $f_l^j = 0$, then the implication is trivial. Otherwise, i.e., if $f_l^j > 0$, from the Kuhn–Tucker conditions we have that

$$\eta^j = \beta_l + K_l(f_l^j, f_l) + K_0(r^j) \leq \beta_{\hat{l}} + K_{\hat{l}}(f_{\hat{l}}^j, f_{\hat{l}}) + K_0(r^j)$$

and since $f_{\hat{l}}^i > f_{\hat{l}}^j$ implies $f_{\hat{l}}^i > 0$, we have

$$\eta^i = \beta_{\hat{l}} + K_{\hat{l}}(f_{\hat{l}}^i, f_{\hat{l}}) + K_0(r^i) \leq \beta_l + K_l(f_l^i, f_l) + K_0(r^i).$$

Thus, we have

$$\begin{aligned}\beta_l + K_l(f_l^j, f_l) &\leq \beta_{\hat{l}} + K_{\hat{l}}(f_{\hat{l}}^j, f_{\hat{l}}) < \beta_{\hat{l}} + K_{\hat{l}}(f_{\hat{l}}^i, f_{\hat{l}}) \\ &\leq \beta_l + K_l(f_l^i, f_l)\end{aligned}$$

i.e., $K_l(f_l^j, f_l) < K_l(f_l^i, f_l)$, which implies $f_l^j < f_l^i$. □

PROPOSITION 7.1 *Consider the identical type-***A** *cost functions and the identical user weights. Assume that $m^i \geq m^j$ and $M^i \geq M^j$. Then $r^i \geq r^j$ and $f_l^i \geq f_l^j$ for all links $l \in \mathcal{L}$. If $m^i = m^j$ and $M^i = M^j$, then $f_l^i = f_l^j$ for all $l \in \mathcal{L}$.*

Proof. Note that $r^i \geq r^j$ holds trivially if $r^j = m^j$. Otherwise, if $r^j > m^j$, by contradiction assume that $r^i < r^j$; then, $\eta^i \geq \eta^j$. Since $r^j > r^i$, then there must be at least one link $\hat{l}$ for which $f^i_{\hat{l}} < f^j_{\hat{l}}$. From the Kuhn–Tucker conditions, we have that

$$\eta^j - \beta_{\hat{l}} = K_{\hat{l}}(f^j_{\hat{l}}, f_{\hat{l}}) + K_0(r^j) > K_{\hat{l}}(f^i_{\hat{l}}, f_{\hat{l}}) + K_0(r^i) \geq \eta^i - \beta_{\hat{l}}.$$

We then have a contradiction since $\eta^i \geq \eta^j$.
Now we show that $f^i_l \geq f^j_l$ for all links $l \in \mathcal{L}$. Assume that to the contrary $f^i_{\hat{l}} < f^j_{\hat{l}}$ for some $\hat{l}$. Then, by Lemma 7.1 we have $f^i_l \leq f^j_l$ on all other links, which upon summation yields $r^i < r^j$, which contradict $r^i \geq r^j$. □

The next proposition shows that, for identical type-**A** cost functions, and identical interval demands (i.e.,$m^i = m$ and $M^i = M$, $\forall i \in \mathcal{I}$), there is a monotonicity among users in their demands, i.e., a user with a higher weight has a higher demand.

PROPOSITION 7.2 *Assume that all users have the same type-***A** *cost function. Then for some vector $\vec{\alpha} = (\alpha^1, \alpha^2, \ldots, \alpha^I) > 0$, we have:*

- $\alpha^i > \alpha^j \Longrightarrow r^i \geq r^j$, $\forall i, j \in \mathcal{I}$.
- $\alpha^i = \alpha^j \Longrightarrow r^i = r^j$, $\forall i, j \in \mathcal{I}$ *and* $f^i_l = f^j_l$ $\forall l \in \mathcal{L}$.

Proof. Note that $r^i \geq r^j$ holds trivially if $r^j = m^j$. Otherwise, if $r^j > m^j$, by contradiction assume that $r^i < r^j$; then, $\eta^i \geq \eta^j$. Since $r^j > r^i$, then there must be at least one link $\hat{l}$ for which $f^i_{\hat{l}} < f^j_{\hat{l}}$. From the Kuhn–Tucker conditions, we have that

$$\eta^j - \beta_{\hat{l}} = K_{\hat{l}}(f^j_{\hat{l}}, f_{\hat{l}}) + K_0(r^j) > K_{\hat{l}}(f^i_{\hat{l}}, f_{\hat{l}}) + K_0(r^i) \geq \eta^i - \beta_{\hat{l}}$$

We then have a contradiction since $\eta^i \geq \eta^j$.
Now we show that $f^i_l \geq f^j_l$ for all links $l \in \mathcal{L}$. Assume that to the contrary $f^i_{\hat{l}} < f^j_{\hat{l}}$ for some $\hat{l}$. Then, by Lemma 7.1 we have $f^i_l \leq f^j_l$ on all other links, which upon summation yields $r^i < r^j$, which contradict $r^i \geq r^j$. Similarly we obtain the second point. □

4.3 Application: Virtual path allocation with level QoS

An interesting application of our model (parallel links) is to virtual path bandwidth allocation with level QoS. We consider a set of users that share a resource of total capacity (bandwidth) of B units. The resource may stand for more than a single link if different reservation rules apply at each link or if users prefer one link over the other (e.g., due to their qualities); then such a system cannot be treated as a single link and will thus be investigated in the more general context (e.g., parallel links).

The system of parallel links may represent various distinguishable resources that directly interconnect a source to a destination. Each user reserves an amount of capacity on a link or several links.

Let B_l^i be the amount of capacity reserved on link l by user i, which is constrained to be nonnegative and not exceed the capacity C_l. Further define $\mathbf{B}_l := (B_l^1, B_l^2, \ldots, B_l^I)$, $\mathbf{B}^i := (B_1^i, B_2^i, \ldots, B_L^i)$, $B_l := \sum_{i \in \mathcal{I}} B_l^i$ and $B^i := \sum_{l \in \mathcal{L}} C_l^i$. The cost function for user i, denoted by J^i, is of the following form:

$$J^i(\mathbf{B}) = \sum_l F_l^i(B_l^i, B_l) + G^i(B^i). \tag{7.19}$$

Here F_l^i accounts for the cost of reserving capacity for a user on link l, as perceived by that user. Note that, according to this formulation, a user may pay different prices for reserving capacity, but distributing it differently among the links. Moreover, the F_l^i functions can reflect user preferences among the link, for example, if some link l has some undesirable property, such as high propagation delay, a user i which is sensitive to end-to-end could add to the respective F_l^i function a large additive constant or multiply it by a large constant factor, etc. The function G^i accounts for the effect the amount of reserved capacity has on the performance of that user. Moreover, the G^i takes as its argument the total capacity B^i reserved by user i. Indeed, a user can assign an incoming call to any link(or combination of links) on which the amount of capacity, which has been reserved by the user but was not used yet, can accommodate that call. This means that the loss process of a user depends only on the total amount of capacity reserved by that user and not on the precise distribution of that capacity among the links.

This virtual path network is transparent to the users in which the users calculate the routes and capacities of virtual paths in the network such that the following requirements are met:

1. *Capacity constraints*: The sum of **VP** capacities on each link does not exceed its capacity, i.e.,

$$B_l = \sum_{i \in \mathcal{I}} B_l^i \leq c_l, \quad \forall l \in \mathcal{L}. \tag{7.20}$$

2. *Constraints of quality of service*: Let the blocking constraint faced by user i, which enforces QoS at the call level, be denoted by δ^i.

Further let the blocking probability be denoted by P_i which is the percentage of call attempts of user i that are denied service due to the unavailability of the resources. We must always have that

$$P_i \leq \delta^i. \tag{7.21}$$

The blocking probability P_i is a function of two variables, the total of capacity allowed by user i (B^i) and the total arrival rate of user i, which we assume to be fixed. Moreover, this function is decreasing with respect to B^i. Then (7.21) is equivalent to $m^i \leq B^i$, $\forall i \in \mathcal{I}$, where $m^i = P_i^{-1}(\delta^i)$.

The following result establishes the uniqueness of normalized Nash equilibrium for the **VP** allocation game.

COROLLARY 7.2 *Consider the cost functions of type-***A***. Then the normalized Nash equilibrium of* **VP** *allocation game is unique.*

Proof. Follows directly from Theorem 7.3. □

REMARK 20 In the absence of the capacity constraint (7.20), if we use property $F5$ in Lazar, Orda and Pendarakis (1997), then Corollary 7.1 implies that the Nash equilibrium that corresponds to the **VP** allocation game is unique.

5. General topology

In this section, we study an extension to a general network. We assume that all users have the same source and destination (s, d). For the general topology, we further impose a quality of service constraint on each path. The goals are formulated as a set of flow restrictions of the form:

$$\sum_{l \in p} \hat{g}_l(f_l) \leq d_p, \quad p \in \mathcal{P} \tag{7.22}$$

where $\mathcal{P}$ is the set of paths connecting the O-D pair (s, d) and $\hat{g}_l \colon \mathbb{R}_+ \to \mathbb{R}$, $l \in \mathcal{L}$. Constraints (7.22) require that for each path connected the source s and destination d, the end-to-end packet cost (delay) should be no larger than d_p.

5.1 The Symmetric Case

LEMMA 7.2 *Consider the identical type-***A** *cost functions and identical intervals for demand (i.e., $m^i = m$ and $M^i = M, \forall i \in \mathcal{I}$). Let a vector $\vec{\alpha}$ be picked such that $\vec{\alpha} = (\alpha^1, \alpha^2, \ldots, \alpha^I) > 0$. Then for $i, j \in \mathcal{I}$ such that $\alpha^i = \alpha^j$, $f_l^i = f_l^j$ for all $l \in \mathcal{L}$. Moreover, if α^i, $i \in \mathcal{I}$, are the same, then $f_l^i = f_l/I$ $\forall i, l$.*

Proof. We first show that $r^i = r^j$ for $i, j \in \mathcal{I}$ such that $\alpha^i = \alpha^j$. Suppose that, to the contrary, $r^i \neq r^j$, and without loss of generality assume that $r^i > r^j$.

Now we construct a directed network $(\mathcal{N}', \mathcal{L}')$, where $\mathcal{N}' = \mathcal{N}$ and the set of links $\mathcal{L}'$ is constructed as follows:

1. For each link $l = (u, v) \in \mathcal{L}$, such that $f_l^i \geq f_l^j$, we have a link $l' = (u, v) \in \mathcal{L}'$; to such a link l' we assign a (flow) value $z_{l'} = f_l^i - f_l^j$.
2. For each link $l = (u, v)$, such that $f_l^i < f_l^j$, we have a link $l' = (v, u) \in \mathcal{L}'$; to such a link we assign a (flow) value $z_{l'} = f_l^j - f_l^i$.

It is easy to verify that the value $z_{l'}$ constitutes a nonnegative, directed flow in the network. Since $r^i > r^j$, $z_{l'}$ must carry some flow (the amount of $r^i - r^j$) from the source s to the destination d. This implies that there exists a path p^* from s to d, such that $z_{l'} > 0$, for all $l' \in p^*$.

Consider now a link $l' = (u, v) \in p^*$. Since $z_{l'} > 0$ either $f_{uv}^i > f_{uv}^j$ or $f_{vu}^j > f_{vu}^i$. In the case where $f_{uv}^i > f_{uv}^j$, we have:

$$\begin{aligned}\alpha^i(\lambda_u^i - \lambda_v^i) &= \alpha^i K_{uv}(f_{uv}^i, f_{uv}) + \beta_{uv} g'(f_{uv}) \\ &> \alpha^j K_{uv}(f_{uv}^j, f_{uv}) + \beta_{uv} g'(f_{uv}) \\ &\geq \alpha^j(\lambda_u^j - \lambda_v^j).\end{aligned} \tag{7.23}$$

Thus

$$\lambda_u^i - \lambda_v^i > \lambda_u^j - \lambda_v^j. \tag{7.24}$$

If $f_{vu}^j > f_{vu}^i$, we have by symmetry that $\alpha^j(\lambda_v^j - \lambda_u^j) > \alpha^i(\lambda_v^i - \lambda_u^i)$; thus we obtain (7.24).

Define more precisely the path p^* by $p^* = (s, u_1, u_2, \ldots, u_{n^*}, d)$, where u_k, $k = 1, 2, \ldots, n^*$, is the kth node after the source s on the path p^* and n^* is the number of nodes between the source s and the destination d. Hence, from (7.24) we have:

$$\lambda_s^i - \lambda_s^j > \lambda_{u_1}^i - \lambda_{u_1}^j > \cdots > \lambda_{u_{n^*}}^i - \lambda_{u_{n^*}}^j > \lambda_d^i - \lambda_d^j. \tag{7.25}$$

On the other hand, we have $m \leq r^j < r^i \leq M$, it follows that $\mu^i = \gamma^j = 0$, and from (7.8) we have:

$$K_0(r^i) - \lambda_d^i + \lambda_s^i \leq K_0(r^j) - \lambda_d^j + \lambda_s^j.$$

Since K_0 is strictly increasing and $r^i > r^j$, we have from the last inequality that $-\lambda_d^i + \lambda_s^i < -\lambda_d^j + \lambda_s^j$, which contradict (7.25).

We now proceed to show that $f_l^i = f_l^j$, for all $l \in \mathcal{L}$. Suppose that, to the contrary, $f_{l_0}^i > f_{l_0}^j$ for some $l_0 \in \mathcal{L}$. Since $r^i = r^j$, by using the same procedure as previously we can show that there exists a cycle $\mathbf{S} = (u_0, u_1, u_2, \ldots, u_{n^*}, u_0)$, such that $z_{u_{k-1}u_k} > 0$ and $z_{u_{n^*}u_0} > 0$, where u_k, $k = 1, 2, \ldots, n^*$, is the kth node after the node u_0 on the cycle $\mathbf{S}$ and n^* is the number of nodes in the cycle. Similarly we have:

$$\lambda^i_{u_0} - \lambda^j_{u_0} > \lambda^i_{u_1} - \lambda^j_{u_1} > \cdots > \lambda^i_{u_{n^*}} - \lambda^j_{u_{n^*}} > \lambda^i_{u_0} - \lambda^j_{u_0}$$

which is a contradiction. □

THEOREM 7.4 *Consider the identical type-$\mathbf{A}$ cost functions and assume that all users have the same interval for demand (i.e., $m^i = m$ and $M^i = M$, $\forall i \in \mathcal{I}$) and the same weight (i.e., $\alpha^i = \alpha$, $\forall i \in \mathcal{I}$). Such a network with symmetrical users has a unique normalized Nash equilibrium for every $\alpha > 0$. Moreover $f^i_l = f_l/I$, $\forall i, l$, where $\mathbf{f}$ is the unique normalized Nash equilibrium.*

Proof. In view of Remark 19, it suffices to establish the statement of the theorem for $\alpha = 1$ only. Suppose that, to the contrary, there are two normalized equilibria, $\tilde{\mathbf{f}}$ and $\hat{\mathbf{f}}$. The first step is to establish that $\hat{f}_l = \tilde{f}_l$ $\forall l \in \mathcal{L}$. From the Kuhn–Tucker conditions (7.6) and (7.7), for $\tilde{\mathbf{f}}$ and $\hat{\mathbf{f}}$ we have:

$$\begin{aligned}
K_{uv}(\hat{f}^i_{uv}, \hat{f}_{uv}) + \hat{\beta}_{uv} g'_{uv}(\hat{f}_{uv}) + \hat{\lambda}^i_v &\geq \hat{\lambda}^i_u; \\
K_{uv}(\hat{f}^i_{uv}, \hat{f}_{uv}) + \hat{\beta}_{uv} g'_{uv}(\hat{f}_{uv}) + \hat{\lambda}^i_v &= \hat{\lambda}^i_u \quad \text{if} \quad \hat{f}^i_{uv} > 0 \\
K_{uv}(\tilde{f}^i_{uv}, \tilde{f}_{uv}) + \tilde{\beta}_{uv} g'_{uv}(\tilde{f}_{uv}) + \tilde{\lambda}^i_v &\geq \tilde{\lambda}^i_u; \\
K_{uv}(\tilde{f}^i_{uv}, \tilde{f}_{uv}) + \tilde{\beta}_{uv} g'_{uv}(\tilde{f}_{uv}) + \tilde{\lambda}^i_v &= \tilde{\lambda}^i_u \quad \text{if} \quad \tilde{f}^i_{uv} > 0
\end{aligned} \tag{7.26}$$

where $\tilde{\beta}_l = \sum_p \delta_{lp}\tilde{\beta}_p$ and $\hat{\beta}_l = \sum_p \delta_{lp}\hat{\beta}_p$. From Lemma 7.2, we have that, for all $i \in \mathcal{I}$ and $l \in \mathcal{L}$, $\hat{f}^i_l = \hat{f}_l/I$ and $\tilde{f}^i_l = \frac{\tilde{f}_l}{I}$. Thus (7.26) become:

$$\begin{aligned}
K_{uv}\left(\frac{\hat{f}_{uv}}{I}, \hat{f}_{uv}\right) + \hat{\beta}_{uv} g'_{uv}(\hat{f}_{uv}) + \hat{\lambda}^i_v &\geq \hat{\lambda}^i_u; \\
K_{uv}\left(\frac{\hat{f}_{uv}}{I}, \hat{f}_{uv}\right) + \hat{\beta}_{uv} g'_{uv}(\hat{f}_{uv}) + \hat{\lambda}^i_v &= \hat{\lambda}^i_u \quad \text{if} \quad \hat{f}_{uv} > 0 \\
K_{uv}\left(\frac{\tilde{f}_{uv}}{I}, \tilde{f}_{uv}\right) + \tilde{\beta}_{uv} g'_{uv}(\tilde{f}_{uv}) + \tilde{\lambda}^i_v &\geq \tilde{\lambda}^i_u; \\
K_{uv}\left(\frac{\tilde{f}_{uv}}{I}, \tilde{f}_{uv}\right) + \tilde{\beta}_{uv} g'_{uv}(\tilde{f}_{uv}) + \tilde{\lambda}^i_v &= \tilde{\lambda}^i_u \quad \text{if} \quad \tilde{f}_{uv} > 0.
\end{aligned}$$

We define the function G_l for $l \in \mathcal{L}$ by $G_l(f_l) = K_l(f_l/I, f_l)$. Summing each of these equations over i, we get:

$$\begin{aligned} G_{uv}(\hat{f}_{uv}) + \hat{\beta}_{uv} g'_{uv}(\hat{f}_{uv}) + \hat{\lambda}_v &\geq \hat{\lambda}_u; \\ G_{uv}(\hat{f}_{uv}) + \hat{\beta}_{uv} g'_{uv}(\hat{f}_{uv}) + \hat{\lambda}_v &= \hat{\lambda}_u \quad \text{if} \quad \hat{f}_{uv} > 0 \\ G_{uv}(\tilde{f}_{uv}) + \tilde{\beta}_{uv} g'_{uv}(\tilde{f}_{uv}) + \tilde{\lambda}_v &\geq \tilde{\lambda}_u; \\ G_{uv}(\tilde{f}_{uv}) + \tilde{\beta}_{uv} g'_{uv}(\tilde{f}_{uv}) + \tilde{\lambda}_v &= \tilde{\lambda}_u \quad \text{if} \quad \tilde{f}_{uv} > 0 \end{aligned} \tag{7.27}$$

where $\tilde{\lambda}_u = 1/I \sum_i \tilde{\lambda}^i_u$ and $\hat{\lambda}_u = 1/I \sum_i \lambda^i_u$. Note that these equations are very similar to the Kuhn–Tucker conditions for a single-user optimization problem of link flow, with respect to a modified (convex) link cost function with derivative G_l and the following constraints.

$$f_l \geq 0, \quad \sum_{l \in \text{Out}(v)} f_l = \sum_{l \in \text{In}(v)} f_l, \qquad v \notin (s, d),$$

$$\sum_{l \in p} g_l(f_l) \leq d_p, \quad p \in \mathcal{P}.$$

$$r = \sum_{l \in \text{Out}(s)} f_l = \sum_{l \in \text{In}(d)} f_l \in [Im, IM].$$

Since the link cost function $\int G_l$ is convex (see Assumption **A**), then the uniqueness of their solution is actually a consequence of standard convex programming results. It therefore follows that $\hat{f}_l = \tilde{f}_l$ for all $l \in \mathcal{L}$, and this implies by Lemma 7.2 that $\hat{f}^i_l = \tilde{f}^i_l$ for every l, i. Uniqueness of the normalized Nash equilibrium then readily follows. □

COROLLARY 7.3 *Consider the identical type-***A** *cost functions, and assume that all users have the same interval demands (i.e., $m^i = m$ and $M^i = M, \forall i \in \mathcal{I}$). Then in the absence of the QoS constraints* (7.22)*, the Nash equilibrium is unique.*

5.2 Positive flows

In this subsection, we suppose that all users use the type-**B** cost functions. The following result establishes the uniqueness of equilibrium among those that satisfy the *"all-positive flow"* assumption: whenever a player sends a positive amount of flow to some link, then all other players also do so.

THEOREM 7.5 *Consider cost functions of type-***B***, and let $\tilde{\mathbf{f}}$ and $\hat{\mathbf{f}}$ be two Nash equilibria. Assume that there exists a set $\mathcal{L}_1$ of links, $\mathcal{L}_1 \subset \mathcal{L}$, such that $\{\tilde{f}^i_l > 0$ and $\hat{f}^i_l > 0, i \in \mathcal{I}\}$ for $l \in \mathcal{L}_1$, and $\{\tilde{f}^i_l = \hat{f}^i_l = 0, i \in \mathcal{I}\}$ for $l \notin \mathcal{L}_1$. Then, $\tilde{f}_l = \hat{f}_l$, $\forall l \in \mathcal{L}$.*

Proof. By using the same procedure as in the proof of Theorem 7.4, with the assumption of positive flows, we can show that the Kuhn Tucker conditions for Nash equilibria $\hat{\mathbf{f}}$ and $\tilde{\mathbf{f}}$ implies the following conditions:

$$G_{uv}(\hat{f}_{uv}) + \hat{\beta}_{uv}g'_{uv}(\hat{f}_{uv}) + \hat{\lambda}_v \geq \hat{\lambda}_u;$$
$$G_{uv}(\hat{f}_{uv}) + \hat{\beta}_{uv}g'_{uv}(\hat{f}_{uv}) + \hat{\lambda}_v = \hat{\lambda}_u \quad \text{if} \quad \hat{f}_{uv} > 0$$
$$G_{uv}(\tilde{f}_{uv}) + \tilde{\beta}_{uv}g'_{uv}(\tilde{f}_{uv}) + \tilde{\lambda}_v \geq \tilde{\lambda}_u;$$
$$G_{uv}(\tilde{f}_{uv}) + \tilde{\beta}_{uv}g'_{uv}(\tilde{f}_{uv}) + \tilde{\lambda}_v = \tilde{\lambda}_u \quad \text{if} \quad \tilde{f}_{uv} > 0$$

where $\lambda_u = \sum_i \lambda^i_u$, $\beta_{uv} = \sum_i \beta^i_{uv}$, and $G_{uv}(f_{uv}) = f_{uv}T'_{uv}(f_{uv}) + I.T_{uv}(f_{uv})$. Now proceeding as in the proof of Theorem 7.4 (starting from equation (7.27)) it can be inferred that $\hat{f}_l = \tilde{f}_l$, $l \in \mathcal{L}$. □

The next, final result shows in particular that under the *"all-positive flow"* assumption, there exists at most one normalized Nash equilibrium.

THEOREM 7.6 *Consider cost functions of type-***B***. For some vector $\vec{\alpha} > 0$, let $\tilde{\mathbf{f}}$ and $\hat{\mathbf{f}}$ be two normalized Nash equilibria. Assume that there exists a set $\mathcal{L}_1$ of links, $\mathcal{L}_1 \subset \mathcal{L}$, such that $\{\tilde{f}^i_l > 0$ and $\hat{f}^i_l > 0, i \in \mathcal{I}\}$ for $l \in \mathcal{L}_1$, and $\{\tilde{f}^i_l = \hat{f}^i_l = 0, i \in \mathcal{I}\}$ for $l \notin \mathcal{L}_1$. Then $\tilde{\mathbf{f}} = \hat{\mathbf{f}}$.*

Proof. Assume that for some $\vec{\alpha}$ we have two normalized Nash equilibrium points $\hat{\mathbf{f}}$ and $\tilde{\mathbf{f}}$. Then we have from (7.6) and (7.7):

$$K^i_{uv}(\tilde{f}^i_{uv}, \tilde{f}_{uv}) + \frac{\tilde{\beta}_{uv}}{\alpha^i}g'_{uv}(\tilde{f}_{uv}) = \tilde{\lambda}^i_u - \tilde{\lambda}^i_v \quad \text{if} \quad \tilde{f}^i_{uv} > 0,$$
$$K^i_{uv}(\hat{f}^i_{uv}, \hat{f}_{uv}) + \frac{\hat{\beta}_{uv}}{\alpha^i}g'_{uv}(\hat{f}_{uv}) = \hat{\lambda}^i_u - \hat{\lambda}^i_v \quad \text{if} \quad \hat{f}^i_{uv} > 0.$$

Assume that, to the contrary, there exists a pair $(l_0, i) \in \mathcal{L} \times \mathcal{I}$ such that $\hat{f}^i_{l_0} \neq \tilde{f}^i_{l_0}$, and without loss of generality assume that $\hat{f}^i_{l_0} < \tilde{f}^i_{l_0}$. In the sequel, we consider two cases:

Case 1: I is even. Since $\hat{f}^i_{l_0} < \tilde{f}^i_{l_0}$ and $\tilde{f}_l = \hat{f}_l$, $l \in \mathcal{L}$, then, it is easy to show that there exist two disjoint sets $\mathcal{I}_1$ and $\mathcal{I}_2$ such that $\mathcal{I}_1 \cup \mathcal{I}_2 = \mathcal{I}$, $|\mathcal{I}_1| = |\mathcal{I}_2| = I/2$ and

$$\tilde{f}_{l_0,1} := \sum_{i \in \mathcal{I}_1} \tilde{f}^i_{l_0} > \hat{f}_{l_0,1} := \sum_{i \in \mathcal{I}_1} \hat{f}^i_{l_0,1}, \quad \text{and}$$
$$\tilde{f}_{l_0,2} := \sum_{i \in \mathcal{I}_2} \tilde{f}^i_{l_0} < \hat{f}_{l_0,2} := \sum_{i \in \mathcal{I}_2} \hat{f}^i_{l_0,2}.$$

Now, we construct a directed network $(\mathcal{N}', \mathcal{L}')$, where $\mathcal{N}' = \mathcal{N}$ and the set of links $\mathcal{L}'$ is constructed as follows:

1. For each link $l = (u, v) \in \mathcal{L}$, such that $\tilde{f}_{l,1} \geq \hat{f}_{l,1}$, we have a link $l' = (u, v) \in \mathcal{L}'$; to such a link l' we assign a (flow) value $z_{l'} = \tilde{f}_{l,1} - \hat{f}_{l,1}$.

2. For each link $l = (u, v) \in \mathcal{L}$, such that $\tilde{f}_{l,1} < \hat{f}_{l,1}$, we have a link $l' = (v, u) \in \mathcal{L}'$; to such a link we assign a (flow) value $z_{l'} = \hat{f}_{l,1} - \tilde{f}_{l,1}$.

It is easy to verify that the value $z_{l'}$ constitutes a nonnegative, directed flow in the network. Let $\hat{r}_1 = \sum_{i \in \mathcal{I}_1} \hat{r}^i$ and $\tilde{r}_1 = \sum_{i \in \mathcal{I}_1} \tilde{r}^i$. If $\hat{r}_1 = \tilde{r}_1$ and since $\tilde{f}_{l_0,1} > \hat{f}_{l_0,1}$, then there exists a cycle $\mathbf{D}$; such that $z_{l'} > 0$, $\forall l \in \mathbf{D}$, else (i.e., $\hat{r}_1 \neq \tilde{r}_1$) $z_{l'}$ must carry some flow (the amount of $|\hat{r}_1 - \tilde{r}_1|$) from the source s to the destination d, which implies that there exists a path p^* from s to d, such that $z_{l'} > 0$ for all $l' \in p^*$. Let $\mathbf{S}$ be a set that represents $\mathbf{D}$ if $\hat{r}_1 = \tilde{r}_1$ and p^* otherwise.

Consider now a link $l' = (u, v) \in \mathbf{S}$. Since $z_{l'} > 0$, either $\tilde{f}_{uv,1} > \hat{f}_{uv,1}$ or $\hat{f}_{vu,1} > \tilde{f}_{vu,1}$. Under the assumption of positive flows, we can show that the Kuhn–Tucker conditions for Nash equilibria $\tilde{\mathbf{f}}$ and $\hat{\mathbf{f}}$ implies the following conditions:

$$\tilde{\lambda}_{u,1} - \tilde{\lambda}_{v,1} = \tilde{f}_{uv,1} T'(f_{uv}) + \frac{I}{2} T_{uv}(f_{uv}) + \delta_1 \tilde{\beta}_{uv} g'_{uv}(f_{uv}) \quad \text{if} \quad f_{uv} > 0,$$

$$\hat{\lambda}_{u,1} - \hat{\lambda}_{v,1} = \hat{f}_{uv,1} T'(f_{uv}) + \frac{I}{2} T_{uv}(f_{uv}) + \delta_1 \hat{\beta}_{uv} g'_{uv}(f_{uv}) \quad \text{if} \quad f_{uv} > 0$$

where $f_{uv} = \tilde{f}_{uv} = \hat{f}_{uv}$ (see Theorem 7.5),

$$\lambda_{u,1} = \sum_{i \in \mathcal{I}_1} \lambda_u^i \quad \text{and} \quad \delta_1 = \sum_{i \in \mathcal{I}_1} \frac{1}{\alpha^i}.$$

In the case where $\tilde{f}_{uv,1} > \hat{f}_{uv,1}$, we have:

$$\begin{aligned} \tilde{\lambda}_{u,1} - \tilde{\lambda}_{v,1} &= \tilde{f}_{uv,1} T'(f_{uv}) + \frac{I}{2} T_{uv}(f_{uv}) + \delta_1 \tilde{\beta}_{uv} g'_{uv}(f_{uv}) \\ &> \hat{f}_{uv,1} T'(f_{uv}) + \frac{I}{2} T_{uv}(f_{uv}) + \delta_1 \tilde{\beta}_{uv} g'_{uv}(f_{uv}) \\ &= \hat{\lambda}_{u,1} - \hat{\lambda}_{v,1} + \delta_1 \frac{I}{2} g'_{uv}(f_{uv})(\tilde{\beta}_{uv} - \hat{\beta}_{uv}). \end{aligned} \tag{7.28}$$

Furthermore, since $\tilde{f}_{uv,2} < \hat{f}_{uv,2}$, similarly we have:

$$\hat{\lambda}_{u,2} - \hat{\lambda}_{v,2} > \tilde{\lambda}_{u,2} - \tilde{\lambda}_{v,2} + \delta_2 g'_{uv}(f_{uv})(\hat{\beta}_{uv} - \tilde{\beta}_{uv}) \tag{7.29}$$

where $\lambda_{u,2} = \sum_{i \in \mathcal{I}_2} \lambda_u^i$ and $\delta_2 = \sum_{i \in \mathcal{I}_2} 1/\alpha^i$.

If $\hat{f}_{vu,1} > \tilde{f}_{vu,1}$, we have by symmetry

$$\hat{\lambda}_{v,1} - \hat{\lambda}_{u,1} > \tilde{\lambda}_{v,1} - \tilde{\lambda}_{u,1} + \delta_1 g'_{vu}(f_{vu})(\hat{\beta}_{vu} - \tilde{\beta}_{vu}). \tag{7.30}$$

Furthermore, since $\tilde{f}_{vu,2} > \hat{f}_{vu,2}$, we have:

$$\hat{\lambda}_{u,2} - \hat{\lambda}_{v,2} > \tilde{\lambda}_{u,2} - \tilde{\lambda}_{v,2} + \delta_2 g'_{vu}(f_{vu})(\tilde{\beta}_{vu} - \hat{\beta}_{vu}). \tag{7.31}$$

Summing each of the inequalities (7.28) over $uv \in \{\mathbf{S} \cap \mathcal{L}\}$ and each of the inequalities (7.30) over $uv \in \{\mathbf{S} - \mathcal{L}\}$, we obtain:

$$\sum_{uv \in \{\mathbf{S}-\mathcal{L}\}} g'_{vu}(f_{vu})(\tilde{\beta}_{vu} - \hat{\beta}_{vu}) > \sum_{uv \in \{\mathbf{S}\cap\mathcal{L}\}} g'_{uv}(f_{uv})(\tilde{\beta}_{uv} - \hat{\beta}_{uv}) \tag{7.32}$$

and by summing each of the inequalities (7.29) or (7.31), we obtain

$$\sum_{uv \in \{\mathbf{S}-\mathcal{L}\}} g'_{vu}(f_{vu})(\hat{\beta}_{vu} - \tilde{\beta}_{vu}) > \sum_{uv \in \{\mathbf{S}\cap\mathcal{L}\}} g'_{uv}(f_{uv})(\hat{\beta}_{uv} - \tilde{\beta}_{uv}) \tag{7.33}$$

which leads to a contradiction between inequalities (7.32) and (7.33).

Case 2: I is odd. Since $\hat{f}^i_{l_0} < \tilde{f}^i_{l_0}$ and $\tilde{f}_l = \hat{f}_l, l \in \mathcal{L}$, then it is easy to show that there exist a user i_0 and two disjoint sets $\mathcal{I}_1$ and $\mathcal{I}_2$ such that $\mathcal{I}_1 \cup \mathcal{I}_2 \cup \{i_0\} = \mathcal{I}$, $|\mathcal{I}_1| = |\mathcal{I}_2| = (I-1)/2$ and

$$\tilde{f}_{l_0,1} := \sum_{i \in \mathcal{I}_1} \tilde{f}^i_{l_0} + \frac{\tilde{f}^{i_0}_{l_0}}{2} > \hat{f}_{l_0,1} := \sum_{i \in \mathcal{I}_1} \hat{f}^i_{l_0,1} + \frac{\hat{f}^{i_0}_{l_0}}{2}, \quad \text{and}$$

$$\tilde{f}_{l_0,2} := \sum_{i \in \mathcal{I}_2} \tilde{f}^i_{l_0} + \frac{\tilde{f}^{i_0}_{l_0}}{2} < \hat{f}_{l_0,1} := \sum_{i \in \mathcal{I}_2} \hat{f}^i_{l_0,2} + \frac{\hat{f}^{i_0}_{l_0}}{2}.$$

If we continue with the same procedure as in the case where I is even, we obtain analogous results. □

COROLLARY 7.4 *Consider cost functions of type-***B***. Then, in the absence of the QoS constraints* (7.22) *and under the "all-positive flow" assumption, the Nash equilibrium is unique.*

6. Conclusion

We have established in this paper conditions for existence and uniqueness (in some weak sense) of Nash equilibria in combined noncooperative flow control and routing where users are faced with a multi-criteria optimization problem. The multiobjective problem faced by a user has been formulated as the minimization of one criterion subject to constraints on others. We saw that the standard notion of Nash equilibrium was often too general to yield a single equilibrium which raised the question of how one could further select a candidate among possibly many Nash equilibria. We then introduced such a candidate, known as the normalized Nash equilibrium (due to Rosen) which has been shown in El Azouzi and Altman (2003) to possess nice properties in relation to decentralized pricing schemes. We established its existence and uniqueness. Establishing uniqueness of equilibria is a first important step in engineering noncooperative networks: when a single equilibrium exists, the network designer or administrator can tune the network's parameters so as to induce an equilibrium with desired properties. This will be the subject of future work.

Acknowledgments

The work of the first and third author was supported by a research contract with France Telecom R&D 001B001. And for the fifth author the work was supported in part by the NSF Grants CCR 00-85917 ITR and ANI 03-12976 ITR.

References

Altman, E. and Kameda, H. (2001). Equilibria for Multiclass Routing Problems in Multi-Agent Networks. Submitted in 2001; `http://www-sop.inria.fr/mistral/personnel/Eitan.Altman/ntkgame.html`.

Altman, E., Başar, T., and Srikant, R. (2002). Nash equilibria for combined flow control and routing in networks: Asymptotic behavior for a large number of users. *IEEE Transactions on Automatic Control*, 47(6):917–930; *Proceedings of the 38th IEEE Conference on Decision and Control*, pp. 4002–4007, Phoenix, Arizona, USA, 1999.

Altman, E., El Azouzi, R., and Vyacheslav, V. (2002). Non-cooperative routing in loss networks. *Performance Evaluation*, 49(1-4):257–272.

Arrow, K., Hurwicz, L., and Uzawa, H. (1961). Constraint qualifications in maximization problems. *Naval Research Logistics Quarterly*, 8:175–191.

Aubin, J.P. (1980). *Mathematical Methods of Game and Economic Theory*. Elsevier, Amsterdam.

Ching, W.K. (1999). A note on the convergence of asynchronous greedy algorithm with relaxation in a multiclass queueing environment. *IEEE Communications Letters*, 3(2):34–36.

El Azouzi, R. and Altman, E. (2003). Constrained traffic equilibrium in routing. *IEEE Transaction on Automatic Control*,48(9):1656 1660.

Gibbons, R. (1992). *Game Theory of Applied Economists.* Princeton University Press.

Haurie, A. and Marcotte, P. (1985). On the relationship between Nash-Cournot and wardrop equilibria. *Networks*, 15:295–308.

Lazar, A., Orda, A., and Pendarakis, D.E. (1997). Virtual path bandwidth allocation in multiuser networks. *IEEE/ACM Transaction on Networking*, 5(6):861–871.

Lutton, J.L. (2001). *On Link Costs in Networks.* Private communication, France Telecom.

Nikaido, H. and Isoda, K. (1955). Note on cooperative convex games. *Pacific Journal of Mathematics*, 5(1):807–815.

Orda, A., Rom, R., and Shimkin, N. (1993). Competitive routing in multi-user communication networks. *IEEE/ACM Transactions on Networks*, 1:510–520.

Patriksson, M. (1994). *The Traffic Assignment Problem: Models and Methods.* VSP BV, Utrecht, The Netherlands.

Rhee, S.H. and Konstantopoulos, T. (1998). Optimal flow control and capacity allocation in multi-service networks. *37th IEEE Conf. Decision and Control*, Tampa, Florida.

Rhee, S.H. and Konstantopoulos, T. (1999). Decentralized optimal flow control with constrained source rates. *IEEE Communications Letters*, 3(6):188–200.

Rosen, J. (1965). Existence and uniqueness of equilibrium points for concave n-person games. *Econometrica*, 33:520–534.

The ATM Forum Technical Committee. (1999). *Traffic Management Specification*, Version 4.1, AF-TM-0121.000.

Chapter 8

GENERALIZED UPLIFTS IN POOL-BASED ELECTRICITY MARKETS

François Bouffard
Francisco D. Galiana

Abstract The electricity pool is a fundamental coordination mechanism for scheduling short-term forward electricity markets. It is fundamental theoretically in the sense that its main goal is the best use of society's resources subject to basic reliability criteria. However, the pool scheduling and its associated marginal pricing procedure may not be able to unambiguously coordinate its self-interested participant generators. In other words, the centrally-determined operation and marginal price schedules may not lead to a competitive equilibrium. Several out-of-market mechanisms have been proposed to bridge the objectives of the pool with those of the market players. Among them are uplifts whose current use still does not induce true competitive equilibria. In this chapter, generalized uplifts are proposed as an alternative to traditional uplifts. An innovative mechanism to compute the generalized uplifts is developed. A simple illustrative example is presented.

1. Introduction

The electricity pool (Stoft, 2002) is one of the most fundamental coordination scheme used for short-term forward electricity markets. It is fundamental, at least theoretically, in the sense that its main goal is the best use of society's resources while respecting engineering constraints minimizing the risks of costly blackouts, that is it maximizes social welfare. However, the marginal pricing method generally associated with the electricity pool is not always able to coordinate the self-interested, profit-seeking participant generators to the centrally-determined pool operational schedule (Motto et al., 2001; Madrigal and Quintana, 2001). In other words, the pool operation schedule may be different from the one obtained independently by the generators when they maximize their profit from the given marginal price schedule; namely dual coordination, as defined by White and Simmons (1977), fails. Thus, in an electricity pool market, a competitive equilibrium may fail to exist. This competitive failure stems from the non-convex nature of the operating technology of most electricity generators (Motto and Galiana, 2002; Galiana, Motto and Bouffard, 2003)). The non-convexities in those production technologies are caused generally by sunk costs attributed to startup processes and fixed charges, minimum output limitations, and increasing returns to scale (Arroyo and Conejo, 2000).

Several out-of-market mechanisms have been proposed to bridge the social welfare objectives of the pool with those of the profit-seeking generators. Among them are uplifts, whose current use still cannot induce competitive equilibria (O'Neill et al., 2002; Hogan and Ring, 2003; Galiana, Motto and Bouffard, 2003; Bouffard, 2003). In this chapter, generalized uplifts are proposed as an alternative to traditional uplifts (Bouffard and Galiana, 2003; Galiana, Motto and Bouffard, 2003). We show that they guarantee the existence of a competitive equilibrium. In addition, we illustrate that those uplifts possess the flexibility required to execute wealth transfers among the market players in the most equitable fashion possible (Bouffard and Galiana, 2003).

In previous works (Motto and Galiana, 2002; Galiana, Motto and Bouffard, 2003; Bouffard, 2003), the minimum generalized uplift problem was formulated. The solution of the minimum uplift problem has to a) steer all the profit-maximizing market players to the centralized pool solution; and, b) do this with as little uplifts as possible, in other words with the least out-of-market manipulations. However, so far, the methods developed to solve the minimum generalized uplift problem were confined to the special case where time couplings are absent. Thus, the impossibility to account for those essential practical cases reduced significantly the appeal of generalized uplifts to resolve market disequilibrium. This is because most, if not all, short-term forward electricity market scheduling algorithms span over several hours.

The difficulty with coordinating the time-coupled markets comes from the fact that it is generally impossible to formulate *explicit* necessary and sufficient profit optimality conditions for each of the market players. In this monograph, we propose a way to circumvent this difficulty through a mechanism similar to a cutting-plane algorithm. This algorithm iteratively updates the feasible set of prices and uplifts so as to impose indirectly necessary profit optimality conditions. We show that after enough iterations, this feasible set converges to an equivalent set of profit optimality conditions sufficient to coordinate all the profit-seeking generators to *willingly* follow the pool operational schedule.

The monograph is organized as follows. First, in Section 2 we define the notation we use throughout this chapter. Section 3 reviews briefly the electric pool market model, the notions of profit sub-optimality and of market equilibrium. Next, in Section 4, we state the minimum generalized uplift problem. Section 5 is devoted to our novel solution approach for the minimum uplift problem. Section 6 presents a numerical example. Lastly, we conclude in Section 7.

2. Notation

Indices.

- i Generator indices running from 1 to I;
- k Iteration indices running from 0 to K;
- t Time period indices running from 1 to T.

Sets.

- $\mathcal{G}(\cdot)$ Standardized operating set of a generator;
- Φ_i Set of maximum profit violations of generator i;
- $\Omega(\cdot)$ Feasible set for the electricity prices and the uplift parameters.

Parameters.

- d_t Power demand at time t;
- π_{it} Offering parameters of generator i at time t;
- π_i $[\pi_{it}]$, $t = 1, \ldots, T$;
- ε Profit-optimality violation threshold.

Variables.

- u_{it} Commitment status (1 if generator i is on at time t, 0 otherwise);
- g_{it} Real power output of generator i at time t;
- ξ_{it} $[u_{it}, g_{it}]$;
- ξ_i $[\xi_{it}]$, $t = 1, \ldots, T$;
- ξ $[\xi_i]$, $i = 1, \ldots, I$;
- Δa_{it} Uplift parameter associated with power production;
- $\Delta c_{it}^{\text{on}}$ Uplift parameter associated with the *on* state;
- $\Delta c_{it}^{\text{off}}$ Uplift parameter associated with the off state;
- $\Delta\pi_{it}$ $[\Delta a_{it}, \Delta c_{it}^{\text{on}}, \Delta c_{it}^{\text{off}}]$;
- $\Delta\pi_i$ $[\Delta\pi_{it}]$, $t = 1, \ldots, T$;
- $\Delta\pi$ $[\Delta\pi_i]$, $i = 1, \ldots, I$;
- λ_t Electricity price at time t;
- λ $[\lambda_t]$, $t = 1, \ldots, T$.

Functions.

$\xi_i(\cdot,\cdot,\cdot)$	Profit-maximizing schedule vector function of generator i;
$c_g(\cdot,\cdot)$	Standardized generator cost function;
$\mathrm{pr}_g(\cdot,\cdot,\cdot,\cdot)$	Standardized generator profit function;
$U_g(\cdot,\cdot)$	Standardized generator uplift function.

3. Electricity pool scheduling and profit sub-optimality

The goal of the electricity pool is the maximization of the net social welfare obtained through the production and use of electricity. For expositional purposes, we consider a pool where consumers draw constant benefits from their use of electricity and have an inelastic demand function; hence, the electricity pool problem will boil down to a minimization of the production costs of the generators subject to simple supply-demand constraints and some sets of technological constraints. Here the generator production cost and the technological data is submitted in the form of a standardized offering structure π_i which gives shape to a standard cost function, $c_g(\cdot,\cdot)$, and a standard operational set, $\mathcal{G}(\cdot)$.[1] We stress that these offering structures do not need to necessarily reflect truthfully the production characteristics of the participating generators; in fact, strategic behavior on the part of the generators is *expected* and *normal* in an open marketplace. The *abuse* of such market manipulation is what needs to be monitored and punished by the appropriate authorities; however, this aspect is outside the scope of this monograph. We state the Electricity Pool Scheduling Problem (EPSP) as the following mathematical program

$$F = \min_{\xi} \sum_{t=1}^{T} \sum_{i=1}^{I} c_g(\xi_{it}, \pi_{it}) \tag{8.1}$$

subject to,

$$\sum_{i=1}^{I} g_{it} - d_t = 0; \quad t = 1, \ldots, T, \tag{8.2}$$

$$\xi_i \in \mathcal{G}(\pi_i); \quad i = 1, \ldots, I. \tag{8.3}$$

This optimization problem minimizes in (8.1) the *offered* operational cost over the participating generators and over the scheduling horizon.

[1] This standardized set could contain, for example, provisions for minimum and maximum production limits, ramping limits, minimum up and down time restrictions, etc.

We point out to the unfamiliar reader that electricity is not storable; consequently, the equality constraints (8.2) require that the production and demand sum up to zero for each of the time periods of the scheduling horizon. Lastly, in (8.3), the operating variables of each of the generators are constrained to their feasible operating sets. The descriptions of the specifics of the cost functions and of the feasible operating sets are outside the scope of this monograph; we refer the interested reader to Arroyo and Conejo (2000) for further details. One point, however, that we must point out is that the feasible sets of the generators are non-convex because of the binary commitment variables u_{it}, and that their offered costs may show increasing returns to scale.

Assuming that the EPSP, (8.1)–(8.3), has a non-empty feasible set, we shall denote its optimal solution by ξ^*.[2] Moreover, in electricity pools it is common practice to derive from the Lagrange multipliers associated with real power balance constraints, (8.2), the marginal prices of electricity (Stoft, 2002). This said differently, the marginal price at time t, denoted by λ_t^*, is then used to compensate generator i, during period t (in the amount $\lambda_t^* g_{it}^*$), and is used to charge consumers for electricity used up during period t (in the amount $\lambda_t^* d_t$).

We calculate the offer-based profits of a generator i under the EPSP as the difference between its revenues and its standardized cost function

$$\mathrm{pr}_g(\lambda, \xi_i, \pi_i, 0) = \sum_{t=1}^{T} \big(\lambda_t g_{it} - c_g(\xi_{it}, \pi_{it})\big). \tag{8.4}$$

Thus, at the optimum of the EPSP, the profits of generator i are given by (8.4), evaluated for ξ_i^* and λ^*, that is $\mathrm{pr}_g(\lambda^*, \xi_i^*, \pi_i, 0)$. Given the offering structure it submitted, π_i, a generator i cannot be dissatisfied with the profits it obtains at the prices λ^* if

$$\mathrm{pr}_g(\lambda^*, \xi_i^*, \pi_i, 0) = \mathrm{pr}_g\big(\lambda^*, \xi_i(\lambda^*, \pi_i, 0), \pi_i, 0\big), \tag{8.5}$$

where

$$\mathrm{pr}_g\big(\lambda^*, \xi_i(\lambda^*, \pi_i, 0), \pi_i, 0\big) = \max_{\xi_i \in \mathcal{G}(\pi_i)} \mathrm{pr}_g(\lambda^*, \xi_i, \pi_i, 0), \tag{8.6}$$

and

$$\xi_i(\lambda^*, \pi_i, 0) = \arg \max_{\xi_i \in \mathcal{G}(\pi_i)} \mathrm{pr}_g(\lambda^*, \xi_i, \pi_i, 0). \tag{8.7}$$

[2] We have to be aware, however, that this optimal solution may not be unique. If this is the case, we shall assume that there exists a tie-breaking rule to select one solution over the others.

In other words, the operational schedule for generator i, at the prices found by the EPSP, is as profitable as the schedule that it would obtain if it were to optimize its profits given the prices λ^* and its offering structure π_i.

In the event that (8.5) is satisfied for *all* the generators $i = 1, \ldots, I$, then we say that the pool market has reached a competitive solution because all the generators maximize their profits while the consumers' demand is fulfilled at the lowest offered cost to society. The converse situation whereby one or more generators have profits which are less than the maximum is termed *profit sub-optimality.* Profit sub-optimality is not an uncommon situation in an electricity pool (Johnson, Oren and Svoboda, 1996; Stoft, 2002). Generally, cases of profit sub-optimality fall under two categories. Cases in which generators do not break even fall in the first category, that is those cases for which $\text{pr}_g(\lambda^*, \xi_i^*, \pi_i, 0) < 0$. The second category groups the other cases for which $0 \leq \text{pr}_g(\lambda^*, \xi_i^*, \pi_i, 0) < \text{pr}_g\big(\lambda^*, \xi_i(\lambda^*, \pi_i, 0), \pi_i, 0\big)$. From an equity point-of-view, forcing generators to operate at less than maximum profit is certainly unfair given that generally generators cannot withdraw from the pool at will. In the long run, repeated manifestations of profit sub-optimality may drive generators out of business and may also delay some of the necessary capacity investments.

4. Resolving profit sub-optimality

To handle cases of profit sub-optimality in pool-based electricity markets, the use of out-of-market financial mechanisms has been proposed and used Madrigal and Quintana (2001); Motto and Galiana (2002); O'Neill et al. (2002); Hogan and Ring (2003); Galiana, Motto and Bouffard (2003)). Some of these mechanisms use lump-sum monetary transfers made between subsets of consumers and generators while other types of mechanisms use non-linear pricing principles. Further design differences between the approaches include the number of degrees of freedom used, the degree of participation of the consumers in the removal of profit sub-optimality and whether or not there can exist cross-subsidies between the generators.

All of the mechanisms, except those proposed by Motto and Galiana (2002) and by Galiana, Motto and Bouffard (2003), only treat the instances of profit sub-optimality wherein generators do not break even. In other words, most authors leave out the instances of generators earning non-negative but less than maximum profits. The argument put forward in favor of ignoring these profit losses is twofold. First, ignoring these profit losses reduces the amount of money which needs to be raised from the consumers and the generators already earning maximum profits. Second, not considering these losses reduces the complexity of the out-of-market resolution procedure. We believe, however, that leaving out these profit losses is first unfair; in addition, this may lead capacity under-investments because of the associated higher profit risks.

4.1 Generalized uplifts

The groundwork of Galiana, Motto and Bouffard (2003) proposes to treat all profit sub-optimality cases without discrimination. In this approach, not earning the maximum nonnegative profit is as economically incongruous as not breaking even. In this vain, the authors extended the concept of uplift, which, so far, had been used only to offset the financial losses of the generators not breaking even. For that reason, they suggest the concept of the generalized uplift, which applies to all the generators in the market whether they were scheduled on or off. Moreover, they propose to share more evenly the social cost of profit sub-optimality between the consumers and the generators. The sharing of these costs is justified by the incentives it should give to the generators not to try to artificially cause profit sub-optimality by their offering strategies.

The generalized uplift, $U_g(\cdot,\cdot)$, is a standardized mathematical function that applies to a single generator i during a time step t; it depends on a) the scheduling state, ξ_{it}, and b) a set of parameters, $\Delta\pi_{it}$, in the following way

$$U_g(\xi_{it}, \Delta\pi_{it}) = \Delta c_{it}^{on} u_{it} + \Delta c_{it}^{\text{off}}(1 - u_{it}) + \Delta a_{it} g_{it}. \tag{8.8}$$

The market operator uses the generalized uplift to modify (augment) the profit function of each of the generators

$$\text{pr}_g(\lambda, \xi_i, \pi_i, \Delta\pi_i) = \sum_{t=1}^{T} \big(\lambda_t g_{it} - c_g(\xi_{it}, \pi_{it}) + U_g(\xi_{it}, \Delta\pi_{it})\big). \tag{8.9}$$

Thus, by appropriately adjusting the price of electricity,[3] λ_t, and the set of uplift parameters for each of the generators, the market operator can ensure that operating at the EPSP solution ξ^* (or, in fact, at any feasible operating solution) is profit optimum. In other words, the market operator has the power to steer the market to a competitive solution. Mathematically, the market operator must find electricity prices, λ, and uplift parameters, $\Delta\pi$, which will ensure that

[3] Electricity price increases are the means through which the consumers contribute to the removal of profit sub-optimality.

$$\xi_i^* = \xi_i(\lambda, \pi_i, \Delta\pi_i) = \arg\max_{\xi_i \in \mathcal{G}(\pi_i)} \mathrm{pr}_g(\lambda, \xi_i, \pi_i, \Delta\pi_i) \qquad i = 1, \ldots, I. \quad (8.10)$$

In plain words, this means that the marginal prices and the uplift parameters must make sure that operating at the EPSP solution is the best way all the generators can operate.

With its three parameters per time step, the generalized uplift provides more that enough degrees of freedom to the market operator to bring all the generators in the market to a competitive solution. In addition, it is possible with the adjustable parameters $\Delta c_{it}^{\mathrm{on}}$ and $\Delta c_{it}^{\mathrm{off}}$ to affect the profit of a generator during any time step whether it is to be on or off. Moreover, when the generator is online, the parameter Δa_{it} permits the enforcement of any desired output power level.

Concretely, the generalized uplift is to act as a threat or a reward or to a given generator. A threat is an uplift without a tangible monetary value; it materializes only if the unit receiving it does not comply with the orders of the operator. It is applied to keep generators from changing their scheduling state. On the other hand, we call a reward an uplift that applies when a unit remains in a given operating state; rewards have a real monetary value. A given generalized uplift can act both as a reward and as a threat. We identify four rules which the threats and rewards should logically obey:

1. If a unit is scheduled on and were to turn off, then its uplift function is non-positive. This uplift is a threat simulating a potential loss if the generator were not to comply with the schedule.

2. If a unit is scheduled off and were to turn on, then its uplift function is non-positive. This uplift is also a threat simulating a potential loss for non-compliance with the scheduling instructions.

3. If a unit is on and stays on, then its uplift function can be positive or negative. This means that the unit may receive a positive reward to write-off a financial loss or it may provide a subsidy via a negative reward.

4. If a unit is off and stays off, then its uplift function is non-negative. In other words, when a unit is off it could receive a compensation for any forgone profits it could make by turning on.

We already mentioned that the generalized uplift as defined in (8.9) provides more degrees of freedom to the market operator than needed to guarantee that the EPSP solution ξ^* is a competitive equilibrium.[4] As a result, there exists infinitely many combinations of electricity prices and uplift parameters which allow the market operator to attain a competitive solution. The above four rules restrict the sign of the uplift functions, but they are not sufficient to uniquely determine the uplift parameters. Therefore, to eliminate this indefiniteness Galiana, Motto and Bouffard (2003) proposed to solve an optimization problem over the space of uplift parameters, $\Delta\pi$, and electricity prices, λ, the *minimum generalized uplift problem*.

4.2 The minimum generalized uplift problem

For a given EPSP solution, ξ^*, the basic minimum generalized uplift problem is formulated

$$F = \min_{\lambda, \Delta\pi} \sum_{t=1}^{T} \sum_{i=1}^{I} \|\Delta\pi_{it}\| \tag{8.11}$$

subject to,

$$\sum_{t=1}^{T} \sum_{i=1}^{I} U_g(\xi_{it}^*, \Delta\pi_{it}) = 0, \tag{8.12}$$

$$(\lambda, \Delta\pi_i) \in \Omega(\xi_i^*); \quad i = 1, \ldots, I. \tag{8.13}$$

The goal of the minimum generalized uplift problem is to minimize some weighted norm of the uplift parameters, (8.11). The first constraint of the problem (8.12) enforces the well accepted principle that the market operator should be revenue neutral with respect to the administration of the uplifts.[5] The sets (8.13) constrain the electricity prices and the uplift parameters for each generator. Mathematically, the sets $\Omega(\xi_i^*)$ are defined as

$$\Omega(\xi_i^*) = \{(\lambda, \Delta\pi_i) \in \mathbb{R}^T \times \mathbb{R}^{3T} : \\ \mathrm{pr}_g(\lambda, \xi_i, \pi_i, \Delta\pi_i) \leq \mathrm{pr}_g(\lambda, \xi_i^*, \pi_i, \Delta\pi_i) - \varepsilon; \ \forall \xi_i \in \mathcal{G}(\pi_i)\}, \tag{8.14}$$

where ε is a small positive number. Constraining the prices and the uplift parameters to be in the I sets $\Omega(\xi_i^*)$ makes sure that all the profit

[4] The total number of degrees of freedom is $3IT + T$. The market operator will use up $2IT$ degrees of freedom to meet profit optimality of the generators. Thus, there will remain $IT+T$ unaccounted degrees of freedom.

[5] Other, more stringent, revenue neutrality conditions could be imposed. For instance, one could impose revenue neutrality of the market operator in each time period of the scheduling.

maximum conditions in (8.10) will hold. Here the parameter $\varepsilon > 0$ could well be omitted; however, it becomes necessary in the event that the market operator encounters more than one profit-maximizing schedule for a given generator. In other words, ε guarantees that the EPSP solution ξ^* will always be the schedule producing the highest profit for each of the generators. Furthermore, other constraints could be added to the basic set in (8.12) and (8.13). These may be bounds on the prices, bounds on the profits, and rules for sharing profit sub-optimality "costs" between the generators and the consumers (Galiana, Motto and Bouffard, 2003; Bouffard and Galiana, 2003).

The sets $\Omega(\xi_i^*)$ are non-convex, and they are impossible to express explicitly in terms of the prices and the uplift parameters because of the infinity of feasible operating schedules $\xi_i \in \mathcal{G}(\pi_i)$. Notwithstanding this difficulty, for a single-time period electricity market, Motto and Galiana (2002); Galiana, Motto and Bouffard (2003) derived an easily solvable formulation of the minimum generalized uplift problem. In those, the non-convex constraints sets $\Omega(\xi_i^*)$ were transformed to equivalent linear constraints on the prices and the uplift parameters which are necessary and sufficient conditions for profit optimality of each generator. In the single-period case, there are only two profit optimality conditions per generator. The resulting formulations were shown to be convex programming problems (Motto and Galiana, 2002). However, attempts at deriving sets of equivalent profit optimality conditions for large-scale multi-period cases are impractical because the count of profit optimality conditions increases exponentially with the number of time periods in the scheduling horizon. We note also that this count increases polynomially with the number of generators.

5. An iterative solution method

Deriving the equivalent profit-optimality constraints is counter- productive because of dimensionality reasons for multi-period problems. In addition, deriving the complete set of equivalent profit optimality conditions may not be sound intuitively because we might expect that at the optimum of the minimization problem, (8.11)–(8.13), only a few of the equivalent constraints are going to be active. This reason thus motivates the investigation of alternate problem formulations where the market operator solves the problem subject to only those constraints delimiting the sets $\Omega(\xi_i^*)$ most likely to be active at the optimum solution. However, an efficient procedure to find this "most likely" constraint subset may prove to be difficult to discover, and most importantly it may lack robustness.

The proposed solution method which we outline next avoids the pitfalls of dimensionality and uncertain robustness.

To solve the minimum generalized uplift problem we propose to use an iterative optimization method based on the sequential generation of cutting planes to partially delimit the sets $\Omega(\xi_i^*)$. The work of Kelley (1960) on cutting plane algorithms for linear programming provides the main impetus for implementing this solution approach. The following subsections are devoted respectively to the description of the proposed solution algorithm and the proof of its optimality and convergence.

5.1 Solution algorithm

1. **Initialization:** Solve the EPSP, (8.1)–(8.3), to obtain the minimum cost schedule ξ^* and the corresponding marginal electricity price schedule, λ^*.

   Set the iteration count to zero, $k = 0$.
   Set the trial price schedule to the marginal prices found by the EPSP, $\lambda^{[0]} = \lambda^*$.
   Set the trial uplift parameters to zero, $\Delta\pi^{[0]} = 0$.
   Set the the set of maximum profit violations to the entire space of electricity prices and uplift parameters, $\Phi_i^{[0]} = \mathbb{R}^T \times \mathbb{R}^{3T}$, for $i = 1, \ldots, I$.

2. **Maximum profit calculation:** For all the generators $i = 1, \ldots, I$, calculate their profit-maximizing schedule for the current trial prices and uplift parameters, $(\lambda^{[k]}, \Delta\pi_i^{[k]})$. In other words, for $i = 1, \ldots, I$ evaluate

$$\xi_i^{[k+1]} = \arg \max_{\xi_i \in \mathcal{G}(\pi_i)} \text{pr}_g(\lambda^{[k]}, \xi_i, \pi_i, \Delta\pi_i^{[k]})$$

3. **Detection of maximum profit violations:** For all the generators $i = 1, \ldots, I$, if the schedule $\xi_i^{[k+1]}$ obtained in Step 2 for the prices $\lambda^{[k]}$ and the uplift parameters $\Delta\pi_i^{[k]}$ is different from the EPSP schedule ξ_i^*, then we add a new profit optimality constraint cut to the set, Φ_i. That is, for $i = 1, \ldots, I$ if

$$\xi_i^{[k+1]} \neq \xi_i^*$$

   then

$$\Phi_i^{[k+1]} = \Phi_i^{[k]} \cap \{(\lambda, \Delta\pi_i) \in \mathbb{R}^T \times \mathbb{R}^{3T} : \\ \text{pr}_g(\lambda, \xi_i^{[k+1]}, \pi_i, \Delta\pi_i) \leq \text{pr}_g(\lambda, \xi_i^*, \pi_i, \Delta\pi_i) - \varepsilon\},$$

else

$$\Phi_i^{[k+1]} = \Phi_i^{[k]}.$$

4. **Termination or computation of a new trial solution:** If, for all generators $i = 1, \ldots, I$, the constraint violation set has not been augmented in Step 3, that is,

$$\Phi_i^{[k+1]} = \Phi_i^{[k]}.$$

then stop. Else, obtain the next trial pair $(\lambda^{[k+1]}, \Delta\pi^{[k+1]})$ through the solution of

$$(\lambda^{[k+1]}, \Delta\pi^{[k+1]}) = \arg\min_{\lambda, \Delta\pi} \sum_{t=1}^{T} \sum_{i=1}^{I} \|\Delta\pi_{it}\|$$

subject to,

$$\sum_{t=1}^{T} \sum_{i=1}^{I} U_g(\xi_{it}^*, \Delta\pi_{it}) = 0,$$

$$(\lambda, \Delta\pi_i) \in \Phi_i^{[k+1]}; \quad i = 1, \ldots, I.$$

Then increment the iteration counter, $k \leftarrow k + 1$, and return to Step 2.

5.2 Some remarks

Before going any further, we complement the description of the steps of the proposed iterative solution algorithm.

During the initialization step, we solve the electricity pool scheduling problem. We recall that in practice this problem is a mixed-integer problem of large dimensions. Its outcome is the operational schedule, ξ^*, which uses society's resources optimally and ensures the desired reliability criteria. In other words, it is the schedule that maximizes social welfare.

Together, Steps 2 and 3 determine whether a given pair $(\lambda^{[k]}, \Delta\pi^{[k]})$ is appropriate to steer all the generators to the desired maximum social welfare solution found in Step 1. Step 2 involves the solution of a single-generator profit-driven unit commitment for each generator. Next, in Step 3 the profit-driven schedules found in Step 2 are compared with the maximum social welfare schedule found in Step 1. If the two schedules match then there is no profit sub-optimality. Otherwise, this means that there exists generators which are not earning the maximum profit for the

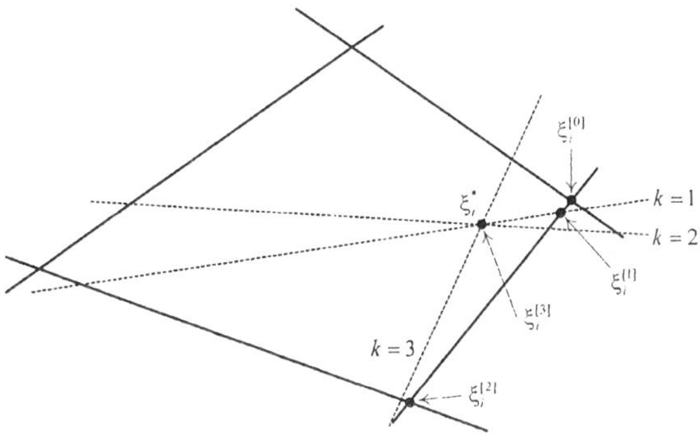

Figure 8.1. Successive generation of profit-violation cutting planes change the profit-maximizing behavior of the generator through the generalized uplifts. The solid lines represent the original feasible space of the generator. The dotted lines represent virtual constraints on the generator's scheduling space created by the uplifts. Here the algorithm converges after solving the minimum generalized uplift problem three times.

given prices and uplifts.[6] This also means that the trial pairs considered at the moment are outside of the maximum profit constraint sets $\Omega(\xi_i^*)$ for some $i = 1, \ldots, I$. That is, these pairs must be discarded when the next trial pairs are computed. The exclusion is done by cutting them off through the generation of the appropriate cutting planes. Hence, the sets Φ_i always include $\Omega(\xi_i^*); i = 1, \ldots, I$, that is $\Phi_i \supseteq \Omega(\xi_i^*); i = 1, \ldots, I$. Lastly, it is worth mentioning that these two steps, because they involve calculations pertaining to individual generators only, could well be paralleled to enhance the speed of the algorithm.

In Step 4 the algorithm stops if there is no more profit sub-optimality detected in Step 3. Otherwise, the minimum generalized uplift problem is solved with the new constrains that reject the previous trial price-uplift pairs. Once the new trial pair is found, it is sent to Step 2 to check whether it can steer all the generators to the maximum social welfare schedule. A comment must be made about the parameter $\varepsilon > 0$ which appeared first in (8.14). The sole usefulness of this parameter is to ensure that the profits obtained for ξ^* will always be the largest attainable by the generators. Figure 8.1 shows how the algorithm modifies the profit-

[6]One could argue that it should be computationally faster to compare profit levels rather than the operation schedules. However, this may fail if some generators have non-unique optimal profit-driven operational schedules.

maximizing behavior of a generator. The effect of ε is seen clearly in this picture as all the previous profit-driven schedules are rejected in a strict manner at each iteration.

In the previous works of Motto and Galiana (2002); Galiana, Motto and Bouffard (2003) on generalized uplifts, it was proved that the optimal solution of the minimum generalized uplift problem had the added property of inducing a strongly dualizable "uplifted EPSP".[7] By "uplifted EPSP", we mean a variant of the EPSP resulting from adding (8.12) to the objective function (8.1) (with ξ_{it} left free). In the works cite above, the uplifts ensure that the strong duality theorem holds for the EPSP when it is augmented with the optimal uplifts. The current algorithm does not go as far, because it does not ensure that the optimal uplifts produce Lagrange multipliers of the uplifted EPSP, λ^*, which are equal to the price schedules found by the iterative algorithm. Nonetheless, the generalized uplift concept does not necessarily require strong duality of the uplifted EPSP. The optimal results of the iterative algorithm are consistent with the notions of equilibrium used by several authors (Madrigal and Quintana, 2001; Hogan and Ring, 2003, for instance), and in other ad hoc uplifting schemes. Obviously, the results of the algorithm would gain in elegance if strong duality was to be guaranteed. This is subject of ongoing investigation.

5.3 Optimality and convergence proofs

Optimality: We assume that the algorithm has converged after K iterations. The optimality of the solution is established from the fact that the feasible set of the partially-specified minimum uplift problem solved at iteration K contained the feasible set of the original problem namely, $\bigcap_i \Phi_i^{[K]} \supseteq \bigcap_i \Omega(\xi_i^*)$. Hence, the optimum of the partially-specified problem has to be less than or equal to that obtained for the original problem. This completes the proof.

Convergence: The proof of convergence is established from the following lemma.

LEMMA 8.1 *After k iterations, the upper bound on the number of remaining iterations before convergence, $N^{[k]}$, is given by*

$$N^{[k]} = \frac{1}{\varepsilon} \max_{i=1,\ldots,I} \{\mathrm{pr}_g(\lambda^{[k]}, \xi_i^{[k+1]}, \pi_i, \Delta\pi_i^{[k]}) - \mathrm{pr}_g(\lambda^{[k]}, \xi_i^*, \pi_i, \Delta\pi_i^{[k]})\}. \quad (8.15)$$

[7] This means that the "uplifted EPSP" has no duality gap.

Proof. At iteration k after Step 2, the magnitude of the profit-optimality violation of generator i is calculated

$$\Delta\mathrm{pr}_{gi}^{[k]} = \mathrm{pr}_g(\lambda^{[k]}, \xi_i^{[k+1]}, \pi_i, \Delta\pi_i^{[k]}) - \mathrm{pr}_g(\lambda^{[k]}, \xi_i^*, \pi_i, \Delta\pi_i^{[k]}); i = 1, \ldots, I$$

In Step 3, if $\Delta\mathrm{pr}_{gi}^{[k]} \geq \varepsilon$ then it implies that $\xi_i^{[k+1]} \neq \xi_i^*$ meaning that there exists a schedule which is weakly more profitable than the maximum social welfare schedule. Therefore, the constraint violation set of i gets modified so that this undesired schedule will always be rejected in the next iterations through the generation of a new cutting plane,

$$\Phi_i^{[k+1]} = \Phi_i^{[k]} \cap \{(\lambda, \Delta\pi_i) \in \mathbb{R}^T \times \mathbb{R}^{3T} : \\ \mathrm{pr}_g(\lambda, \xi_i^{[k+1]}, \pi_i, \Delta\pi_i) \leq \mathrm{pr}_g(\lambda, \xi_i^*, \pi_i, \Delta\pi_i) - \varepsilon\}.$$

After solving the partial minimum uplift problem with the new constraint violation sets of all the other generators, the lower bound on the profit-optimality violation of generator i will be once more $\Delta\mathrm{pr}_{gi}^{[k]} \geq \varepsilon$ if it has not converged. We conclude, by induction, that the lower bound on the convergence rate is ε per iteration. Hence, the upper bound on the number of iterations left before convergence for generator i in iteration k is $\Delta\mathrm{pr}_{gi}^{[k]}/\varepsilon$. Now if we consider the system as a whole, the upper bound on the number of remaining iterations before convergence is the maximum over all the generators.

COROLLARY 8.1 *Since $\varepsilon > 0$ and the profit-optimality violations are finite, the upper bound on the remaining number of iterations $N^{[k]}$ is always finite. As a result, the algorithm must converge in a finite number of iterations.*

This completes the proof. □

6. Illustrative example

We now show an example of the proposed algorithm at work on a small four-generator pool market spreading over four consecutive time periods of one hour each. The generators' standard running cost offering follows the quadratic model used by Galiana, Motto and Bouffard (2003) augmented with corresponding startup costs, that is

$$c_g(\xi_{it}, \pi_i) = c_{0i}u_it + a_ig_{it} + \tfrac{1}{2}b_ig_{it}^2 + S_i(u_{it}). \tag{8.16}$$

In (8.16), the offer parameters (c_{0i}, a_i, b_i) are embedded in the standardized offering structure π_i. The offering structure also contains the information on the startup cost of the generator, $S_i(u_{it})$. Lastly, in this example, the generators' feasible operating sets $\mathcal{G}(\cdot)$ contain limitations on a) the generators' maximum output (the minimum outputs are equal to zero); and, on b) the generators' up and down ramping capabilities. The generation offering data and unit commitment initial conditions are found in Table 8.1.

Table 8.1. Generation offering data and initial conditions.

i	1	2	3	4
c_{0i} (\$/h)	500	200	200	300
a_i (\$/MWh)	20	27	28	26
b_i (\$/MW^{2}h)	0.02	0.05	0.05	0.01
S_i (\$/h)	100	150	200	200
Maximum output (MW)	100	100	100	100
Ramp limit (MW/h)	75	50	50	90
$\xi_{i0} = [u_{i0}, g_{i0}]$	[1,100]	[0,0]	[0,0]	[1,95]

Table 8.2. Power demand.

t (h)	1	2	3	4
d_t (MW)	210	287	367	161

For the demand pattern given in Table 8.2, the maximum social welfare operation and associated marginal price schedules are found by solving the EPSP. The operation schedule is shown Figure 8.2, and the price schedule is found in the first row of Table 8.3.

Table 8.3. Price schedules found by the EPSP and by the solution of the minimum generalized uplift problem.

t (h)	1	2	3	4
λ_t^* (\$/MWh)	27.5000	29.6500	39.6200	21.6800
λ_t (\$/MWh)	27.4692	29.7467	39.5467	25.2970

If the generators were to self-schedule by maximizing their profit from the price schedule in the first row of Table 8.3, i.e., according to the EPSP price schedule, then they would operate as seen in Figure 8.3. By inspection of Figure 8.2 and Figure 8.3, the EPSP price schedule does not direct all the generators to operate according to the EPSP operational

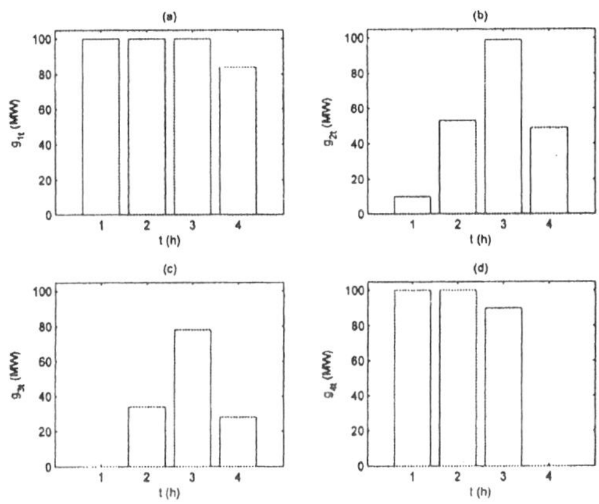

Figure 8.2. Operational schedule found by the EPSP. a) Generator 1; b) Generator 2; c) Generator 3; and, d) Generator 4.

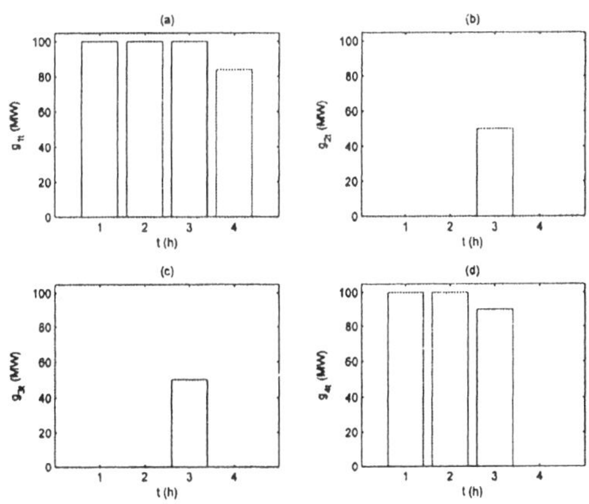

Figure 8.3. Profit-driven self-scheduling from the EPSP price schedule. a) Generator 1; b) Generator 2; c) Generator 3; and, d) Generator 4.

schedule. In other words, we are witnessing a situation of profit sub-optimality. In the first and second rows of Table 8.4, the profit figures are given for these two cases. Clearly here, generators 2 and 3 have

Table 8.4. Profit levels for the generators.

i	1	2	3	4
$\text{pr}_g(\lambda^*, \xi_i^*, \pi_i, 0)$ ($)	1451.40	-188.35	-211.60	701.75
$\text{pr}_g(\lambda^*, \xi_i(\lambda^*, \pi_i, 0), \pi_i, 0)$ ($)	1451.40	219.75	119.75	701.75
$\text{pr}_g(\lambda, \xi_i^*, \pi_i, \Delta\pi_i)$ ($)	1449.82	62.62	116.16	701.74

grounds to claim that the EPSP scheduling is unfair because it forces them to operate at a loss while forgoing a positive profit.

To remove the profit sub-optimality incongruity, we solve the minimum generalized uplift problem using the iterative algorithm proposed before. Moreover in doing so, we add two extra constraint sets to the basic set found in (8.12) and (8.13). The first constraint set requests that the profits of the generators with the uplifts should be less than or equal to the profits obtained via self-scheduling given the EPSP price schedule

$$\text{pr}_g(\lambda, \xi_i^*, \pi_i, \Delta\pi_i) \leq \text{pr}_g(\lambda^*, \xi_i(\lambda^*, \pi_i, 0), \pi_i, 0) - \varepsilon; \quad i = 1, \ldots, I. \quad (8.17)$$

The second one limits the use of the uplift parameter $\Delta c_{it}^{\text{off}}$

$$\Delta c_{it}^{\text{off}} \geq 0; \quad i = 1, \ldots, I; \; t = 1, \ldots, T. \quad (8.18)$$

The goal of these limits is to eliminate the possibility that the market operator imposes financial losses on the generators. In other words, it is not possible for the market operator to use the threat of more losses on a generator if chooses to remain offline when it loses money under the EPSP schedule.

With the following uplift-parameter weighted-norm as the objective function,

$$\sum_{t=1}^{T} \sum_{i=1}^{I} \{|\Delta c_{it}^{on}| + \Delta c_{it}^{\text{off}} + |\Delta a_{it} g_{it}^*|\} \quad (8.19)$$

the iterative algorithm converges in 15 iterations with $\varepsilon = 0.01$.[8] The results respective to the various generators are found in Tables 8.5—8.8. The second row of Table 8.3 shows the updated price schedule. Finally, the third row of Table 8.4 shows the profit levels of the generators with the uplifts.

[8]The profits associated with the maximum welfare schedule have to be higher at least by one penny with respect to the other possible profit levels. Faster convergence could be achieved with a larger ε, but the resulting uplifts and price schedule may become unnecessarily large.

Table 8.5. Minimum generalized uplift problem solution: generator 1.

t (h)	1	2	3	4
Δc_{1t}^{on} ($/h)	0	0	0	0
$\Delta c_{1t}^{\text{off}}$ ($/h)	0	0	0	0
Δa_{1t} ($/MWh)	0	0	0	-3.6270
$U_g(\xi_{1t}^*, \Delta\pi_{1t})$ ($/h)	0	0	0	-304.67

Table 8.6. Minimum generalized uplift problem solution: generator 2.

t (h)	1	2	3	4
Δc_{2t}^{on} ($/h)	197.67	129.33	0	59.15
$\Delta c_{2t}^{\text{off}}$ ($/h)	0	0	0	0
Δa_{2t} ($/MWh)	0	-0.1082	-3.0767	0
$U_g(\xi_{2t}^*, \Delta\pi_{2t})$ ($/h)	197.67	123.59	-304.60	59.15

Table 8.7. Minimum generalized uplift problem solution: generator 3.

t (h)	1	2	3	4
Δc_{3t}^{on} ($/h)	0	168.67	0	158.75
$\Delta c_{3t}^{\text{off}}$ ($/h)	0	0	0	0
Δa_{3t} ($/MWh)	0	0	0	-3.5203
$U_g(\xi_{3t}^*, \Delta\pi_{3t})$ ($/h)	0	168.67	0	60.18

With the uplifts, all the generators earn the maximum possible profit when they operate at the EPSP operational schedule, ξ^*. Moreover, by comparing the price schedules in Table 8.3, we see that the consumers end up paying more for the electricity (now $32,892.71 versus $32,315.57 under the EPSP prices, an increase of $577.14 or 1.75%). From the profits in Table 8.4 we see that the generators together gave away $162.31 (6.51%) from their profit optimum achievable with the EPSP price schedule. The sum of these consumer and generator contributions to the removal of profit sub-optimality is equal to the total profit losses ($739.45) that the generators would have faced if the EPSP price and operation schedules had been implemented. We should note here that a sharing of the initial EPSP profit losses between the consumers and the generators could have been pre-specified and included as a constraint in the minimum uplift problem. This was done by Galiana,

Table 8.8. Minimum generalized uplift problem solution: generator 4.

t (h)	1	2	3	4
Δc_{4t}^{on} (\$/h)	0	0	0	0
$\Delta c_{4t}^{\text{off}}$ (\$/h)	0	0	0	0
Δa_{4t} (\$/MWh)	0	0	0	2.9377
$U_g(\xi_{4t}^*, \Delta\pi_{4t})$ (\$/h)	0	0	0	0

Motto and Bouffard (2003) with a 50-50 split between generators and consumers. Further pro-rata profit shedding rules between generators were also defined investigated by Galiana, Motto and Bouffard (2003) and by Bouffard and Galiana (2003).

7. Conclusion

In this monograph, we briefly surveyed the intricacies of generator profit sub-optimality in pool-based electricity markets. We gave a detailed description of the generalized uplift as a means to eliminate profit sub-optimality. The generalized uplifts are used by the pool market operator to properly steer the generators to the preferred maximum social welfare operating schedule through economic control rather than by a set of ad hoc scheduling enforcement measures. We saw that the combined means of the generalized uplifts and the price schedules provide the market operator with more degrees of freedom than necessary to properly control the generators. Hence, in order to fully specify the uplifts and the corresponding price schedule, the use of an optimization problem is proposed. This, optimization problem, denoted as the minimum generalized uplift problem, was formulated. It is a complex non-convex and highly combinatorial problem which is hard to solve especially when the market has intertemporal couplings. To overcome these difficulties, we proposed a novel procedure based on an iterative updating of the feasible space of the minimum generalized uplift problem. We showed that this procedure must converge in a finite number of iterations and that it must yield an optimum solution. Finally, we showed how the proposed solution procedure works with a simple example.

Future research opportunities in the area are many. Among them we distinguish a) the large-scale and parallel solution of the minimum generalized uplift problem; b) the use of trade-offs between uplifts and non-maximum social welfare schedules; c) the imposition of strong duality conditions for the uplifted EPSP; d) the consideration of the electricity transmission grid and nodal pricing; e) the assessment of market manipulation strategies via uplifts; and, f) the investigation of consumer uplifts in pool markets with demand-side bidding.

Acknowledgments

The authors wish to acknowledge the financial support received from the Natural Sciences and Engineering Research Council of Canada, and from le Fond québécois de la recherche sur la nature et les technologies. The thoughtful comments of Dr. Alexis L. Motto were also greatly appreciated.

References

Arroyo, J.M. and Conejo, A.J. (2000). Optimal response of a thermal unit to an electricity spot market. *IEEE Transactions on Power Systems*, 15(3):1098–1104.

Bouffard, F. and Galiana, F.D. (2003). Equitable rules for generalized uplifts in electricity markets. *INFORMS General Meeting*, Atlanta, GA, Oct.

Bouffard, F. (2003). Minimum uplifts in electricity pools: From market distortion to market equilibrium. *Technical report*, Power Engineering Research Laboratory, McGill University, Canada.

Galiana, F.D., Motto, A.L., and Bouffard, F. (2003). Reconciling social welfare, agent profits and consumer payments in electricity pools. *IEEE Transactions on Power Systems*, 18(2):452–459.

Hogan, W.W. and Ring, B.J. (2003). On Minimum-uplift pricing for electricity markets. *Working paper*, Harvard University.

Johnson, R.B., Oren, S.S., and Svoboda, A.J. (1996). Equity and efficiency of unit commitment in competitive electricity markets. *Working paper*, University of California Energy Institute.

Kelley, J.E. Jr. (1960). The cutting-plane method for solving convex programs. *Journal of the Society for Industrial and Applied Mathematics*, 8(4):703–712.

Madrigal, M. and Quintana, V.H. (2001). Existence and determination of competitive equilibrium in unit commitment power pool auctions: Price setting and scheduling alternatives. *IEEE Transactions on Power Systems*, 16(3):380–388.

Motto, A.L. and Galiana, F.D. (2001). Equilibrium of auction markets with unit commitment: The need for augmented pricing. *IEEE Transactions on Power Systems*, 17(2):798–805.

Motto, A.L., Galiana, F.D., Conejo, A.J., and Huneault, M. (2001). Decentralized nodal-price self-dispatch and unit commitment. In: B.F. Hobbs, M.H. Rothkopf, R.P. O'Neill, and H.-p. Chao (eds.), *The Next Generation of Electric Power Unit Commitment Models*, pages 271–292, Kluwer.

O'Neill, R.P, Sotkiewicz, P.M., Hobbs, B.F., Rothkopf, M.H., and Steward, W.R. Jr. (2002). Efficient market-clearing prices in markets with non-convexities. *Working paper*, Rutgers University.

Stoft, S. (2002). *Power System Economics: Designing Markets for Electricity.* John Wiley & Sons.

White, G.W.T. and Simmons, M.D. (1977). Analysis of complex systems. *Philosophical Transactions of the Royal Society of London, Series A*, 287(1346):405–423.

Chapter 9

NASH EQUILIBRIA FOR LARGE-POPULATION LINEAR STOCHASTIC SYSTEMS OF WEAKLY COUPLED AGENTS

Minyi Huang
Roland P. Malhamé
Peter E. Caines

Abstract We consider dynamic games in large population conditions where the agents evolve according to non-uniform dynamics and are weakly coupled via their dynamics and the individual costs. A state aggregation technique is developed to obtain a set of decentralized control laws for the individuals which possesses an ε-Nash equilibrium property. An attraction property of the mass behaviour is established. The methodology and the results contained in this paper reveal novel behavioural properties of the relationship of any given individual with respect to the mass of individuals in large-scale noncooperative systems of weakly coupled agents.

1. Introduction

The control and optimization of large-scale complex systems is evidently of importance due to their ubiquitous appearance in engineering, industrial, social and economic settings. These systems are usually characterized by features such as high dimensionality and uncertainty, and the system evolution is associated with complex interactions among its constituent parts or sub-systems.

In the past decades considerable research effort has been devoted to a significant variety of large-scale systems, and a range of techniques has been developed for their analysis and optimization, including model reduction, aggregation, and hierarchical optimization, etc.

To date, although considerable progress has been made in different directions concerning the optimization of large-scale dynamical systems, general theoretical principles and methodologies are still lacking; this may well be an inherent problem in this domain given the great diversity in the nature of the systems under consideration and their associated optimization problems. So far, most work on optimization of large-scale dynamical systems is based upon centralized performance measures. However, in many social, economic, and engineering models, the individuals or agents involved have conflicting objectives and it is more appropriate to consider optimization based upon individual payoffs or costs. This gives rise to noncooperative game theoretic approaches partly based upon the vast corpus of relevant work within economics and the social sciences. In particular, game theoretic methods have been used in the engineering context in the study of wireless and wired networks optimization, as in Altman, Basar and Srikant (2002); Dziong and Mason (1996).

Game theoretic approaches are intended to capture the individual interest seeking nature of agents in many social, economic and manmade systems; however, in a large-scale dynamic model this approach results in an analytic complexity which is in general prohibitively high, and correspondingly leads to few implementable results on dynamic optimization. We note that the so-called evolutionary games which have been used to treat large population dynamic models at reduced complexity (see Fudenberg and Levine, 1998) are useful mainly for analyzing the asymptotic behaviour of the overall system, and do not lead to a satisfactory framework for the dynamic quantitative optimization of individual performance since the revision of agents' strategies is specified a priori via heuristic rules.

In this paper, we investigate the optimization of large-scale linear control systems wherein many agents (also to be called players) are each coupled with others via the individual dynamics and the costs in a particular form. We view this to be the characteristic property of a class of situations which we term (*distributed*) control problems with weak coupling. The study of such large-scale weakly coupled systems is motivated by a variety of scenarios, for instance, dynamic economic models involving agents linked via a market, and power control in mobile wireless communications. In the latter case, different users have independent power control mechanisms and statistically independent communication channels, but they interact with each other via mutual interference as reflected by the resulting signal-to-interference ratio (SIR) performance indices (cf. Huang, Caines and Malhamé (2003, 2004b)). Indeed, the model studied in this paper is also related to the research on swarming, flocking, behaviour of human crowds, and formation control of autonomous mobile agents, where each agent has its individual dynamics in which an average effect by all others or the surrounding agents acts as a nominal driving term. For relevant literature, see, e.g., Helbing, Farkas and Vicsek (2000); Tanner, Jadbabaie and Pappas (2003); Liu and Passino (2004); Low (2000). Also, see the large-scale electric load model in Malhamé and Chong (1985).

In the literature, within the optimal control context weakly interconnected systems was studied by Bensoussan (1988). Dynamic LQG games were considered by Papavassilopoulos (1982); Petrovic and Gajic (1988) proposed an iterative computing procedure with small coupling coefficients for two players assuming existence of a solution. In a two player noncooperative nonlinear dynamic game setting, the Nash equilibria was analyzed in Srikant and Basar (1991) where the coefficients for the coupling terms in the dynamics and costs are restricted to be sufficiently small. In contrast to existing work, our concentration is on games with large populations. We analyze the ε-Nash equilibrium properties for a control law by which each individual optimizes using *local information* its cost function depending upon the state of the individual agent and the average effect of all agents taken together, hereon referred to as "the mass". In preceding work (see Huang, Caines and Malhamé, 2003) we considered the LQG game for a population of uniform agents and introduced a state aggregation procedure for the design of decentralized control with an ε-Nash equilibrium property. In the non-uniform case studied in Huang, Caines and Malhamé (2004a) a given agent only has exact information on its own dynamics, and the information concerning other agents is available in a statistical sense as described by a randomized parametrization for agents' dynamics across the population. Building upon our previous results, in this paper we consider the more general model where the aggregated population effect is incorporated into the individual dynamics. Due to the particular structure of the individual dynamics and costs, the mass formed by all agents impacts any given agent as a nearly deterministic quantity. In response to any known mass influence, a given individual will select its localized control strategy to minimize its own cost. In a practical situation the mass influence cannot be assumed known *a priori*. It turns out, however, that this does not present any difficulty for applying the individual-mass interplay methodology as described below.

In the noncooperative game setup studied here, an overall rationality assumption for the population, to be characterized further down, implies the potential of achieving a stable predictable mass behaviour in the following sense: if some deterministic mass behaviour were to be given, rationality would require that each agent synthesize its individual cost based optimal response as a *tracking* action. Thus the mass trajectory corresponding to rational behaviour would guide the agents to collectively generate the trajectory which, individually, they were assumed to be reacting to in the first place. Indeed, if a mass trajectory with the above fixed point property existed, if it were unique, and furthermore, if each individual had enough information to compute it, then rational agents who were assuming all other agents to be rational would anticipate their collective state of agreement and select a control policy consistent with that state. Thus, in the context of this paper, we make the following rationality assumption: Each agent is rational in the sense that it both

(i) optimizes its own cost function, and

(ii) assumes that all other agents are being simultaneously rational when evaluating their competitive behaviour. This justifies and motivates the search for mass trajectories with the fixed point property; in fact the resulting situation is seen to be that of a Nash equilibrium holding between any agent and the mass of the other agents.

The central results of this paper consists of the precise characterization of

1. the Nash equilibrium associated with the individual cost functions depending on both the individual and mass behaviour,

2. the consistency (fixed point property) of the mass trajectory under the Nash equilibrium inducing individual feedback controls, and

3. the global attraction property of the mass behaviour in function spaces of policy iterations with respect to such individual optimizing behaviour. This equilibrium then has the rationality and optimality interpretations but we underline that these hypotheses are not employed in the mathematical derivation of the results.

The framework presented in this paper is particularly suitable for optimization of large-scale systems where individuals seek to optimize for their own reward and where it is effectively impossible to achieve global optimality through close coordination between all agents. In this context, the methodology of noncooperative games and state aggregation (particularly stochastic aggregation as presented in Malhamé and Chong (1985)) developed in this paper provides a feasible approach for building simple (decentralized) optimization rules which under appropriate conditions lead to stable population behaviour. Our methodology could potentially provide effective methods for analyzing complex systems arising in socio-economic and engineering areas; see, e.g., Huang, Malhamé and Caines (2004); Baccelli, Hong and Liu (2001).

It is worthwhile noting that the large population limit formulation presented in this paper is relevant to economic problems concerning (mainly static) models with a large number or a continuum of agents; see e.g. Green (1984). However, instead of directly assigning a prior measure in a continuum space for labelling an infinite number of agents, we induce a probability distribution on a parameter space in a natural way via empirical statistics; this approach avoids certain measurability difficulties arising in the direct introduction of dynamics labelled by a continuum (see Judd, 1985). Furthermore, based upon the resulting induced measure, we develop *state aggregation* for the underlying *dynamic* models, and our approach differs from the well-known aggregation techniques initiated by Simon and Ando (1961) based upon time-scales which leads to a form of hierarchical optimization (Sethi and Zhang, 1994; Phillips and Kokotovic, 1981).

The paper is organized as follows. In Section 2 we introduce the dynamic model. Section 3 gives preliminary results on linear tracking. Section 4 contains the individual and mass behaviour analysis via a state aggregation procedure. In Section 5 we establish the ε-Nash equilibrium property of the decentralized individual control laws. Section 6 concludes the paper.

2. The weakly coupled systems

We consider an n dimensional linear stochastic system where the evolution of each state component is described by

$$dz_i = (a_i z_i + b_i u_i)\,dt + \alpha z^{(n)}\,dt + \sigma_i dw_i, \quad 1 \le i \le n,\ t \ge 0, \tag{9.1}$$

where $\{w_i, 1 \le i \le n\}$ denotes n independent standard scalar Wiener processes and $z^{(n)} = 1/n \sum_{i=1}^{n} z_i$, $\alpha \in \mathbb{R}$. Hence, $z^{(n)}$ may be looked at as a nominal driving term imposed by the population. The initial states $z_i(0)$ are mutually independent and are also independent of $\{w_i, 1 \le i \le n\}$. In addition, $E|z_i(0)|^2 < \infty$ and $b_i \neq 0$. Each state component shall be referred to as the state of the corresponding individual (also to be called an agent or a player).

In this paper we investigate the behaviour of the agents when they interact with each other through specific coupling terms appearing in their cost functions; this is displayed in the following set of *individual* cost functions which shall be used henceforth in the analysis:

$$J_i(u_i, v_i) \triangleq E \int_0^\infty e^{-\rho t}[(z_i - v_i)^2 + ru_i^2]\, dt. \tag{9.2}$$

For simplicity of analysis we assume in this paper that

$$b_i = b > 0, \quad 1 \le i \le n.$$

In particular we assume the cost-coupling to be of the following form:

$$v_i = \Phi(z^{(n)}) = \Phi\left(\frac{1}{n}\sum_{k=1}^{n} z_k\right),$$

where Φ is a continuous function on $\mathbb{R}$, and we study the large-scale system behaviour in the dynamic noncooperative game framework. Evidently the linking term v_i gives a measure of the average effect of the mass formed by all agents in this type of group tracking problem. Here we assume ρ, $r > 0$ and unless otherwise stated, throughout the paper z_i is described by the dynamics (9.1).

3. The preliminary linear tracking problem

With the particular set of individual costs (9.2), the first key step in our analysis is to construct a certain deterministic approximation of the aggregate impact of the mass on a given player. In the tracking analysis, we begin by replacing the average driving term $z^{(n)}$ in (9.1) by a deterministic function f. This suggests we introduce the auxiliary dynamics

$$d\hat{z}_i = a_i \hat{z}_i\, dt + bu_i\, dt + \alpha f\, dt + \sigma_i dw_i, \tag{9.3}$$

where f is bounded and continuous on $[0, \infty)$. For distinction, the state variable $\hat{z}_i$ is used in (9.3), and all other terms are specified in a similar manner as in (9.1).

For large n, we intend to approximate the term $v_i \triangleq \Phi(1/n \sum_{k=1}^n z_k)$ in Section 2 by a *deterministic* continuous function z^* defined on $[0, \infty)$. Here we choose z^* in a more general setting without relating it to the function f introduced above. For a given z^*, we construct the *individual cost* associated with (9.3) as follows:

$$J_i(u_i, z^*) = E \int_0^\infty e^{-\rho t}\{[\hat{z}_i - z^*]^2 + ru_i^2\}\, dt. \tag{9.4}$$

We shall consider the tracking problem with bounded z^*. For minimization of J_i, the admissible control set is taken as $\mathcal{U}_i \triangleq \{u_i \mid u_i$ adapted to $\sigma(\hat{z}_i(0), w_i(s), s \le t)$, and $E\int_0^\infty e^{-\rho t}(\hat{z}_i^2 + u_i^2)\,dt < \infty\}$. Define

$$C_b[0,\infty) \triangleq \{x \in C[0,\infty), |x|_\infty < \infty\},$$

where $|x|_\infty = \sup_{t\ge 0}|x(t)|$, for $x \in C[0,\infty)$. Under the norm $|\cdot|_\infty$, $C_b[0,\infty)$ is a Banach space; see Yosida (1980).

Let Π_i be the positive solution to the algebraic Riccati equation

$$\rho\Pi_i = 2a_i\Pi_i - \frac{b^2}{r}\Pi_i^2 + 1. \tag{9.5}$$

It is easy to verify that $-a_i + b^2\Pi_i/r + \rho/2 > 0$. Denote

$$\beta_1 = -a_i + \frac{b^2}{r}\Pi_i, \quad \beta_2 = -a_i + \frac{b^2}{r}\Pi_i + \rho. \tag{9.6}$$

Clearly, $\beta_2 > \rho/2$. The proofs of Propositions 9.1 and 9.2 below may be obtained following an algebraic approach as in (Bensoussan, 1992, pp. 21–25).

PROPOSITION 9.1 *Assume*

(i) $E|\hat{z}_i(0)|^2 < \infty$ *and* f, $z^* \in C_b[0,\infty)$;

(ii) $\Pi_i > 0$ *is the solution to* (9.5) *and* $\beta_1 = -a_i + b^2/r\Pi_i > 0$; *and*

(iii) $s_i \in C_b[0,\infty)$ *is determined by the differential equation*

$$\rho s_i = \frac{ds_i}{dt} + a_i s_i - \frac{b^2}{r}\Pi_i s_i + \alpha\Pi_i f - z^*. \tag{9.7}$$

Then the control law

$$\hat{u}_i = -\frac{b}{r}(\Pi_i\hat{z}_i + s_i) \tag{9.8}$$

minimizes $J_i(u_i, z^*)$, *for all* $u_i \in \mathcal{U}_i$.

PROPOSITION 9.2 *Suppose assumptions* (i)–(iii) *in Proposition 9.1 hold and* $q \in C_b[0,\infty)$ *satisfies*

$$\rho q = \frac{dq}{dt} - \frac{b^2}{r}s_i^2 + (z^*)^2 + 2\alpha f s_i + \sigma_i^2\Pi_i. \tag{9.9}$$

Then the cost for the control law (9.8) *is given by* $J_i(\hat{u}_i, z^*) = \Pi_i E\hat{z}_i^2(0) + 2s(0)E\hat{z}_i(0) + q(0)$.

REMARK In Proposition 9.1, assumption (i) insures that J_i has a finite minimum attained at some $u_i \in \mathcal{U}_i$. Assumption (ii) means that the resulting closed-loop system has a stable pole.

REMARK s_i in Proposition 9.1 may be uniquely determined only utilizing its boundedness, and it is unnecessary to specify the initial condition for (9.7) separately. Similarly, after $s_i \in C_b[0, \infty)$ is obtained, q in Proposition 9.2 can be uniquely determined from its boundedness.

PROPOSITION 9.3 *Under the assumptions of Proposition* 9.1, *there exists a unique initial condition* $s_i(0) \in \mathbb{R}$ *such that the associated solution* s_i *to* (9.7) *is bounded, i.e.,* $s_i \in C_b[0, \infty)$. *And moreover, for the obtained* $s_i \in C_b[0, \infty)$, *there is a unique initial condition* $q(0) \in \mathbb{R}$ *for* (9.9) *such that the solution* $q \in C_b[0, \infty)$.

Proof. Consider (9.7) for an initial condition $s_i(0)$ which leads to

$$s_i(t) = s_i(0)e^{\beta_2 t} + e^{\beta_2 t} \int_0^t e^{-\beta_2 \tau}[z^*(\tau) - \alpha \Pi_i f(\tau)]\, d\tau.$$

Since $\beta_2 > 0$ always holds, the integral $\int_0^\infty e^{-\beta_2 \tau}[z^*(\tau) - \alpha \Pi_i f(\tau)]\, d\tau$ exists and is finite. We take initial condition $s_i(0) = -\int_0^\infty e^{-\beta_2 \tau}[z^*(\tau) - \alpha \Pi_i f(\tau)]\, d\tau$ which yields

$$s_i(t) = e^{\beta_2 t} \int_t^\infty e^{-\beta_2 \tau}[\alpha \Pi_i f(\tau) - z^*(\tau)]\, d\tau \in C_b[0, \infty),$$

and it is easily verified that any initial condition other than $s_i(0)$ yields an unbounded solution. Similarly, a unique initial condition $q(0)$ in (9.9) may be determined to give $q \in C_b[0, \infty)$. □

4. Competitive behaviour and continuum mass behaviour

In the weakly coupled situation with individual costs, each agent is assumed to be rational in the sense that it both optimizes its own cost and its strategy is based upon the assumption that the other agents are rational. In other words each agent believes (i.e., has as a hypothesis in the derivation of its strategy) the other agents are optimizers.

Due to the specific structure of the dynamics and cost, under the rationality assumption it is possible to approximate the driving term $z^{(n)}$ and the linking term $v_i = \Phi(z^{(n)})$ by a purely deterministic process f and $z^* = \Phi(f)$, respectively, and as a result, if a deterministic tracking is employed by the i-th agent, its optimality loss will be negligible in large population conditions. Hence, all agents would tend to adopt such a tracking based control strategy if an approximating f and the associated $z^* = \Phi(f)$ were to be given.

However, we stress that the rationality notion is only used to construct the aggregation procedure, and the main theorems in the paper will be based solely upon their mathematical assumptions.

4.1 State aggregation via large population limit

Assume $f \in C_b[0,\infty)$ is given for approximation of $z^{(n)}$, and $s_i \in C_b[0,\infty)$ is a solution to (9.7) computed with $z^* = \Phi(f)$. For the i-th agent, after applying the optimal tracking based control law (9.8), the closed-loop equation is approximated by

$$dz_i = \left(a_i - \frac{b^2}{r}\Pi_i\right) z_i\, dt - \frac{b^2}{r} s_i\, dt + \alpha f\, dt + \sigma_i dw_i, \tag{9.10}$$

where f replaces $z^{(n)}$ in (9.1). Taking expectation on both sides of (9.10) yields

$$\frac{d\bar{z}_i}{dt} = \left(a_i - \frac{b^2}{r}\Pi_i\right)\bar{z}_i - \frac{b^2}{r}s_i + \alpha f, \tag{9.11}$$

where $\bar{z}_i(t) = Ez_i(t)$ and the initial condition is $\bar{z}_i\big|_{t=0} = Ez_i(0)$.

We further define the population average of means (simply called population mean) as $\bar{z}^{(n)} \triangleq 1/n \sum_{i=1}^n \bar{z}_i$. Note that in the case all agents have i.i.d. dynamics the evolution of $\bar{z}^{(n)}$ is simply expressed using the dynamics of any $\bar{z}_i$ combined with the initial condition $\bar{z}^{(n)}\big|_{t=0}$.

So far, the individual reaction is determined in a straightforward manner if a mass effect f is given *a priori*. Here one naturally comes up with the important questions: How is the deterministic process f chosen to approximate the overall influence of all players on the given player? In what way does it capture the dynamic behaviour of the collection of many individuals? Since we wish to have $f \approx 1/n \sum_{k=1}^n z_k$, for large n it is plausible to express

$$f = \bar{z}^{(n)}, \quad z^*(t) = \Phi\big(\bar{z}^{(n)}(t)\big). \tag{9.12}$$

As n increases, accuracy of the approximations given in (9.12) is expected to improve. After introducing such an equality relation, a dynamic interaction is built up between the individual and the mass: by averaging over the individual mean trajectories, the pair f and z^* is constructed, in response to which the individuals, in turn, optimize their own objectives. Notice that by taking $f = \bar{z}^{(n)}$, the resulting dynamics (9.11) for $\bar{z}_i$ associated with (9.1) are exact, as long as u_i takes the form (9.8).

Our analysis below will be based upon the observation that the large population limit may be employed to determine the effect of the mass of the population on any given individual, and that the population limit is characterized by an empirical distribution, which is assumed to exist. Specifically, our interest is in the case when a_i, $i \geq 1$, is "adequately randomized" in the sense that the population exhibits certain statistical properties. In this context, the association of the value a_i, $i \geq 1$, and the specific index i plays no essential role, and the more important fact is the frequency of occurrence of a_i on different segments in the range space of the sequence $\{a_i, i \geq 1\}$. Within this setup, we assume that the sequence $\{a_i, i \geq 1\}$, has an empirical distribution function $F(a)$.

For the sequence $\{a_i, i \geq 1\}$, we define the empirical distribution associated with the first n agents

$$F_n(x) = \frac{\sum_{i=1}^{n} 1_{(a_i < x)}}{n}, \quad x \in \mathbb{R}.$$

We introduce the following assumption:

(H1) There exists a distribution function F on $\mathbb{R}$ such that $F_n \to F$ weakly as $n \to \infty$, i.e., $\lim_{n\to\infty} F_n(x) = F(x)$ if F is continuous at $x \in \mathbb{R}$.

(H1') There exists a distribution function F on $\mathbb{R}$ such that $F_n \to F$ uniformly as $n \to \infty$, i.e., $\lim_{n\to\infty} \sup_{x\in\mathbb{R}} |F_n(x) - F(x)| = 0$.

REMARK It is obvious that **(H1')** implies **(H1)**. Notice that if the sequence $a_1^\infty \stackrel{\Delta}{=} \{a_i, i \geq 1\}$ is sufficiently "randomized" such that a_1^∞ is generated by independent observations on the same underlying distribution function F, then with probability one **(H1')** holds by Glivenko-Cantelli theorem; see Chow and Teicher (1997).

For the Riccati equation (9.5), when the coefficient a is used in place of a_i, we denote the corresponding solution by Π_a. Accordingly, we express $\beta_1(a)$ and $\beta_2(a)$ when a and Π_a are substituted into (9.6). Straightforward calculation gives

$$\Pi_a = \left(\frac{b^2}{r}\right)^{-1} \left[a - \frac{\rho}{2} + \sqrt{\left(a - \frac{\rho}{2}\right)^2 + \frac{b^2}{r}} \right],$$

$$\beta_1(a) = -\frac{\rho}{2} + \sqrt{\left(a - \frac{\rho}{2}\right)^2 + \frac{b^2}{r}}, \tag{9.13}$$

$$\beta_2(a) = \frac{\rho}{2} + \sqrt{\left(a - \frac{\rho}{2}\right)^2 + \frac{b^2}{r}}. \tag{9.14}$$

EXAMPLE 9.1 For the set of parameters: $a = 1$, $b = 1$, $\alpha = 1$, $\sigma = 0.3$, $\rho = 0.5$, $r = 0.1$, we have $\Pi_a = 0.4$, $\beta_1(a) = 3$, $\beta_2(a) = 3.5$.

To simplify the aggregation procedure we assume zero mean for initial conditions of all agents, i.e., $Ez_i(0) = 0$, $i \geq 1$. The above analysis suggests we consider the large population limit and introduce the equation system:

$$\rho s_a = \frac{ds_a}{dt} + as_a - \frac{b^2}{r}\Pi_a s_a + \alpha \Pi_a \bar{z} - z^*, \tag{9.15}$$

$$\frac{d\bar{z}_a}{dt} = \left(a - \frac{b^2}{r}\Pi_a\right)\bar{z}_a - \frac{b^2}{r}s_a + \alpha\bar{z}, \tag{9.16}$$

$$\bar{z} = \int_{\mathcal{A}} \bar{z}_a dF(a), \tag{9.17}$$

$$z^* = \Phi(\bar{z}). \tag{9.18}$$

In the above, each individual differential equation is indexed by the parameter a. For the same reasons as noted in Proposition 9.3, here it is unnecessary to specify the initial condition for s_a derived from optimal tracking, which shall be determined later in an inherent manner. Equation (9.16) with $\bar{z}_a\big|_{t=0} = 0$ is based upon (9.11). Hence $\bar{z}_a$ is regarded as the expectation given the parameter a in the individual dynamics. Also, in contrast to the arithmetic average for computing $\bar{z}^{(n)}$ appearing in (9.12), (9.17) is derived by use of the empirical distribution function $F(a)$ for the sequence of parameters $a_i \in \mathcal{A}$, $i \geq 1$, with the range space $\mathcal{A}$. Notice that, had the dynamics of (9.1) been nonlinear the calculation of the mean $\bar{z}_a(t)$ dynamics would have involved an integration with respect to the density generated by an associated Fokker-Planck equation as in Malhamé and Chong (1985). Equation (9.17) describing the stochastic aggregation over parameter space would however remain in the same form as in the linear case.

With a little abuse of terminology, we shall conveniently refer to either $\bar{z}$, or in some cases $\Phi(\bar{z})$, as the mass trajectory.

REMARK In the more general case with non-zero $Ez_i(0)$, we may introduce a joint empirical distribution $F_{a,z}$ for the two dimensional sequence $\{(a_i, Ez_i(0)), i \geq 1\}$. Then the function in (9.16) is to be labelled by both the dynamic parameter a and an associated initial condition, and furthermore, the integration in (9.17) is to be computed with respect to $F_{a,z}$. In this paper we only consider the zero initial mean case in order to avoid notational complication.

We introduce the assumptions:

(H2) The function Φ is Lipschitz continuous on $\mathbb{R}$ with a Lipschitz constant $\gamma > 0$, i.e., $|\Phi(y_1) - \Phi(y_2)| \leq \gamma |y_1 - y_2|$ for all $y_1, y_2 \in \mathbb{R}$.

(H3) $\beta_1(a) > 0$ for all $a \in \mathcal{A}$, and

$$\int_{\mathcal{A}} [|\alpha|/\beta_1(a) + b^2(\gamma + |\alpha|\Pi_a)/r\beta_1(a)\beta_2(a)]\, dF(a) < 1,$$

where $\beta_1(a)$, $\beta_2(a)$ are defined by (9.13)–(9.14), $\mathcal{A}$ is a measurable subset of $\mathbb{R}$ and contains all a_i, $i \geq 1$, and $F(a)$ is the empirical distribution function for $\{a_i, i \geq 1\}$, which is assumed to exist. The constant $\gamma > 0$ is specified in (**H2**).

(**H4**) All agents have mutually independent initial conditions of zero mean, i.e. $Ez_i(0) = 0$, $i \geq 1$. In addition, $\sup_{i\geq 1}[\sigma_i^2 + Ez_i^2(0)] < \infty$.

We state a sufficient condition to insure $\beta_1(a) > 0$ for $a \in \mathbb{R}$. The proof is trivial and is omitted.

PROPOSITION 9.4 *If $b^2 > r\rho^2/4$, then $\beta_1(a) > 0$ for all $a \in \mathbb{R}$.*

REMARK Under (**H3**), we have $-\beta_2(a) < -\beta_1(a) < 0$ where $-\beta_1(a)$ is the stable pole of the closed-loop system for the agent with parameter a. $\beta_1(a)$ measures the stability margin. To avoid triviality for the linking term in the cost, we assume $\gamma > 0$ for Φ in (**H2**).

The following procedure is used to illustrate the interaction between the individual and the mass. First, given $\bar{z} \in C_b[0, \infty)$, Proposition 9.3 implies that (9.15) has the bounded solution

$$s_a(t) = e^{\beta_2(a)t} \int_t^\infty e^{-\beta_2(a)\tau} \big[\alpha\Pi_a \bar{z}(\tau) - \Phi(\bar{z}(\tau))\big]\, d\tau \triangleq \mathcal{T}_1 \bar{z}. \qquad (9.19)$$

Then under (**H4**), equations (9.16) and (9.17) correspond to the equations below:

$$\bar{z}_a(t) = \int_0^t e^{-\beta_1(a)(t-s)}$$

$$\times \left[\alpha\bar{z}(s) + \frac{b^2}{r} e^{\beta_2(a)s} \int_s^\infty e^{-\beta_2(a)\tau} \big[\Phi(\bar{z}(\tau)) - \alpha\Pi_a \bar{z}(\tau)\big]\, d\tau\right] ds, \quad (9.20)$$

$$\bar{z}(t) = \int_{\mathcal{A}} \int_0^t e^{-\beta_1(a)(t-s)} \left[\alpha\bar{z}(s) + \frac{b^2}{r} e^{\beta_2(a)s} \times \left[\int_s^\infty e^{-\beta_2(a)\tau} \big[\Phi(\bar{z}(\tau)) - \alpha\Pi_a \bar{z}(\tau)\big] d\tau\right] ds\right] dF(a)$$

$$\triangleq (\mathcal{T}\bar{z})(t). \qquad (9.21)$$

Here (9.20) indicates what would be the individual mean trajectory resulting from the optimal tracking of a given mass trajectory. Appendix A contains the proof of the following lemma which establishes that $\mathcal{T}$ defined above is a map from $C_b[0, \infty)$ to itself.

LEMMA 9.1 *Under* (**H2**)–(**H3**), *we have* $\mathcal{T}x \in C_b[0,\infty)$, *for any* $x \in C_b[0,\infty)$.

THEOREM 9.1 *Under* (**H2**)–(**H3**), *the map* $\mathcal{T}\colon C_b[0,\infty) \to C_b[0,\infty)$ *has a unique fixed point which is uniformly Lipschitz continuous on* $[0,\infty)$.

Proof. By Lemma 9.1, $\mathcal{T}$ is a map from the Banach space $C_b[0,\infty)$ to itself. For any $x, y \in C_b[0,\infty)$ we have

$$
\begin{aligned}
&|(\mathcal{T}x - \mathcal{T}y)(t)| \\
&\quad \le |x-y|_\infty \int_{\mathcal{A}} \int_0^t |\alpha| e^{-\beta_1(a)(t-s)}\, ds\, dF(a) \\
&\quad + \frac{b^2|x-y|_\infty}{r} \int_{\mathcal{A}} \int_0^t \int_s^\infty [\gamma + |\alpha|\Pi_a] e^{-\beta_1(a)(t-s)} e^{-\beta_2(a)(\tau-s)}\, d\tau\, ds\, dF(a) \\
&\quad \le |x-y|_\infty \int_{\mathcal{A}} \frac{|\alpha|}{\beta_1(a)}\, dF(a) \\
&\qquad + \frac{b^2|x-y|_\infty}{r} \int_{\mathcal{A}} \int_0^t \frac{\gamma + |\alpha|\Pi_a}{\beta_2(a)} e^{-\beta_1(a)(t-s)}\, ds\, dF(a) \\
&\quad \le |x-y|_\infty \int_{\mathcal{A}} \frac{|\alpha|}{\beta_1(a)}\, dF(a) + \frac{b^2|x-y|_\infty}{r} \int_{\mathcal{A}} \frac{\gamma + |\alpha|\Pi_a}{\beta_1(a)\beta_2(a)}\, dF(a) \\
&\quad = |x-y|_\infty \left[\int_{\mathcal{A}} \frac{|\alpha|}{\beta_1(a)}\, dF(a) + \frac{b^2}{r} \int_{\mathcal{A}} \frac{\gamma + |\alpha|\Pi_a}{\beta_1(a)\beta_2(a)}\, dF(a) \right].
\end{aligned}
$$

Then from (**H3**) it follows that $\mathcal{T}$ is a contraction and therefore has a unique fixed point $\bar{z} \in C_b[0,\infty)$.

From the proof of Lemma 9.1 we see that the fixed point $\bar{z} \in C_b[0,\infty)$ is uniformly Lipschitz continuous on $[0,\infty)$ since for any given $x \in C_b[0,\infty)$, $\mathcal{T}x$ is uniformly Lipschitz continuous on $[0,\infty)$ by (A.1). □

THEOREM 9.2 *Under* (**H2**)–(**H4**), *the equation system* (9.15)–(9.18) *admits a unique bounded solution.*

Proof. By Theorem 9.1, we obtain a unique $\bar{z} \in C_b[0,\infty)$ solving $\bar{z} = \mathcal{T}\bar{z}$. Let z^* be computed by (9.18). Then $\bar{z}$ together with z^* leads to a unique bounded solution to (9.15) by Proposition 9.3, and subsequently a unique bounded solution to (9.16). The solution $\bar{z}$ to (9.17) is just equivalently given by (9.21). Uniqueness of the bounded solution to (9.15)–(9.18) is obvious by the unique determination of $\bar{z}$ and hence of $z^* = \Phi(\bar{z})$. □

4.2 The virtual agent, policy iteration and attraction to mass behaviour

We proceed to investigate certain asymptotic properties on the interaction between the individual and the mass, and the formulation shall be interpreted in the large population limit (i.e., an infinite population) context. Corresponding to a large population (deterministic) mass effect $\bar{z}$, let the dynamics for the individual be given as

$$dz_i = a_i z_i \, dt + bu_i \, dt + \alpha \bar{z} \, dt + \sigma_i \, dw_i.$$

At this stage, however, we do not relate $\bar{z}$ to the fixed point equation (9.21). Assume each agent is assigned a cost according to (9.4) with $z^* = \Phi(\bar{z})$, i.e.,

$$J_i\big(u_i, \Phi(\bar{z})\big) = E \int_0^\infty e^{-\rho t}\{[z_i - \Phi(\bar{z})]^2 + ru_i^2\} \, ds, \quad i \geq 1. \qquad (9.22)$$

We now introduce a so-called *virtual agent* to represent the mass effect and use $\bar{z} \in C_b[0, \infty)$ to describe the behaviour of the virtual agent. Here the virtual agent acts as a passive player in the sense that $\bar{z}$ appears as an exogenous function of time and $\Phi(\bar{z})$ is to be tracked by the agents.

Then after each selection of the set of individual control laws, a new $\bar{z}$ will be induced as specified below; subsequently, the individual shall consider its optimal policy (over the whole time horizon) to respond to this updated $\bar{z}$. Thus, the interplay between a given individual and the virtual agent representing the mass may be described as a sequence of virtual plays which may be employed by the individual as a calculation device to eventually learn the mass behaviour. In the following policy iteration analysis in function spaces, we take the virtual agent as a *passive leader* and the individual agents as *active followers*.

It is of interest to note that the virtual play described in this section has a resemblance in spirit to the so-called tâtonnement in economic theory which was first proposed by Walras in 1874 and formalized in a modern version in terms of ordinary differential equations by Samuelson in 1947 (for relevant literature, the reader is referred to (Mas-Colell, Whinston and Green, 1995, pp. 620–626) and references therein). Specifically, in price tâtonnement, given an initial non-equilibrium price, the economic agents will each dynamically adjust its price in a trial and error process where the ensemble of all excess demands is assumed to be announced to all agents by a certain central planner. Such a process is continuously carried out in fictional time (i.e., with infinitesimal duration of iterations) and is highly informative in illuminating behavioural properties of the (Walrasian) equilibrium price. When the process converges to an equilibrium, it is termed as possessing tâtonnement stability.

In contrast, our virtual play here takes a more abstract form since the interaction of agents is specified in policy spaces for feedback controls, and the agents update their strategies via an optimal tracking action in response to an envisaged population effect at each step, which differs from the qualitative adjustment of agents in tâtonnement.

Now, we describe the iterative update of an agent's policy from its *policy space.* For a fixed iteration number $k \geq 0$, suppose that there is *a priori* $\bar{z}^{(k)} \in C_b[0,\infty)$. Then by Proposition 9.1 the optimal control for the i-th agent using the cost (9.22) with respect to $\bar{z} = \bar{z}^{(k)}$ is given as

$$u_i^{(k+1)} = -\frac{b}{r}(\Pi_i z_i + s_i^{(k+1)})$$

where $s_i^{(k+1)} \in C_b[0,\infty)$ is given by

$$\rho s_i^{(k+1)} = \frac{ds_i^{(k+1)}}{dt} + a_i s_i^{(k+1)} - \frac{b^2}{r}\Pi_i s_i^{(k+1)} + \alpha \Pi_i \bar{z}^{(k)} - \Phi(\bar{z}^{(k)}). \quad (9.23)$$

By Proposition 9.3, the unique solution $s_i^{(k+1)} \in C_b[0,\infty)$ to (9.23) may be represented by the map

$$s_i^{(k+1)} = e^{\beta_2(a_i)t} \int_t^{\infty} e^{-\beta_2(a_i)\tau} \left[\alpha \Pi_i \bar{z}^{(k)}(\tau) - \Phi\big(\bar{z}^{(k)}(\tau)\big)\right] d\tau. \quad (9.24)$$

Subsequently, the control laws $\{u_i^{(k+1)}, i \geq 1\}$ produce a mass trajectory

$$\bar{z}^{(k+1)} = \int_{\mathcal{A}} \bar{z}_a^{(k+1)}\, dF(a),$$

where

$$\frac{d\bar{z}_a^{(k+1)}}{dt} = -\beta_1(a)\bar{z}_a^{(k+1)} - \frac{b^2}{r}s_a^{(k+1)} + \alpha \bar{z}^{(k)}, \quad (9.25)$$

with initial condition $\bar{z}_a^{(k+1)}|_{t=0} = 0$ by (**H4**). Notice that (9.25) is indexed by the parameter $a \in \mathcal{A}$ instead of all i's. Then the virtual agent's state (as a function) $\bar{z}$ corresponding to $u_i^{(k+1)}$ is updated as $\bar{z}^{(k+1)}$. From the above and using the operator introduced in (9.21), we get the recursion for $\bar{z}^{(k)}$ as

$$\bar{z}^{(k+1)} = \mathcal{T}\bar{z}^{(k)},$$

where $\bar{z}^{(k+1)}|_{t=0} = 0$ for all k.

By the iterative adjustments of the individual strategies in response to the virtual agent, we induce the mass behaviour by a sequence of functions $\bar{z}^{(k)} = \mathcal{T}\bar{z}^{(k-1)} = \mathcal{T}^k \bar{z}^{(0)}$. The next proposition establishes that as the population grows, a statistical mass equilibrium exists and it is globally attracting.

PROPOSITION 9.5 *Under* (**H2**)–(**H4**), $\lim_{k\to\infty} \bar{z}^{(k)} = \bar{z}$ *for any* $\bar{z}^{(0)} \in C_b[0,\infty)$*, where* $\bar{z}$ *is determined by* (9.15)–(9.18).

Proof. This follows as a corollary to Theorem 9.1. □

4.3 Explicit solution with uniform agents

In the case of a system of uniform agents (i.e., $a_i \equiv a$) with a linear function Φ, a solution to the state aggregation equation system may be explicitly calculated. However, the distribution function $F(a)$ degenerates to point mass and (9.17) is no longer required. Since $\bar{z}$ coincides with $\bar{z}_a$, we simply specify it by (9.16) which is the dynamics of the latter. We consider the case $\Phi(z) = \hat{\gamma} z + \eta$. The equation system (9.15)–(9.18) specializes to

$$\rho s_a = \frac{ds_a}{dt} + a s_a - \frac{b^2}{r}\Pi_a s_a + \alpha \Pi_a \bar{z} - z^*, \tag{9.26}$$

$$\frac{d\bar{z}}{dt} = (a - \frac{b^2}{r}\Pi_a)\bar{z} + \alpha\bar{z} - \frac{b^2}{r}s_a, \tag{9.27}$$

$$z^* = \Phi(\bar{z}) = \hat{\gamma}(\bar{z} + \eta). \tag{9.28}$$

Here we shall compute a solution with a general initial condition $\bar{z}(0)$ for (9.27), which is not necessarily zero. Setting the derivatives to zero, we write a set of steady state equations as follows

$$\begin{cases} \beta_2(a)s_a(\infty) - \alpha\Pi_a\bar{z}(\infty) + z^*(\infty) = 0 \\ -\frac{b^2}{r}s_a(\infty) + \big(\alpha - \beta_1(a)\big)\bar{z}(\infty) = 0 \\ \hat{\gamma}\bar{z}(\infty) - z^*(\infty) = -\hat{\gamma}\eta. \end{cases} \tag{9.29}$$

It can be verified that under (**H3**) we have $\Theta \triangleq \beta_2(a)\big(\beta_1(a) - \alpha\big) + b^2/r(\alpha\Pi_a - \hat{\gamma}) > 0$, and therefore (9.29) is nonsingular and has a unique solution $\big(s_a(\infty), \bar{z}(\infty), z^*(\infty)\big)$. Denote

$$\begin{aligned} \lambda_1 &= \frac{\rho + \alpha - \sqrt{(\rho+\alpha)^2 + 4\Theta}}{2} < 0, \\ \lambda_2 &= \frac{\rho + \alpha + \sqrt{(\rho+\alpha)^2 + 4\Theta}}{2} > 0. \end{aligned} \tag{9.30}$$

Using the same method as in Huang, Caines, and Malhamé (2004c), an explicit solution for the equation system (9.26)–(9.27) may be computed.

PROPOSITION 9.6 *Under* **(H2)**–**(H3)**, *the unique bounded solution* $(\bar{z}, s_a)$ *in* (9.26)–(9.27) *is given by*

$$\bar{z}(t) = \bar{z}(\infty) + \big(\bar{z}(0) - \bar{z}(\infty)\big)e^{\lambda_1 t},$$
$$s_a(t) = s_a(\infty) + \frac{\hat{\gamma} - \alpha\Pi_a}{\beta_2 - \lambda_1}\big(\bar{z}(\infty) - \bar{z}(0)\big)e^{\lambda_1 t},$$

where $\lambda_1 < 0$ *is given by* (9.30) *and* $\beta_1 = -a + b^2/r\Pi_a$, $\beta_2 = -a + b^2/r\Pi_a + \rho$.

Notice that in the case **(H4)** is imposed, we need to set $\bar{z}(0) = 0$ in Proposition 9.6.

5. The cecentralized ε-Nash equilibrium

We continue to consider the system of n agents and rewrite the dynamics in Section 2 as follows:

$$dz_i = (a_i z_i + bu_i)\,dt + \alpha z^{(n)}\,dt + \sigma_i\,dw_i, \quad 1 \le i \le n,\ t \ge 0. \qquad (9.31)$$

The individual costs of the agents are given by (9.2) with the linking term $v_i = \Phi(1/n\sum_{k=1}^n z_k)$. To indicate the dependence of the cost J_i on u_i and the set of controls of all other agents, we write it as $J_i(u_i, u_{-i})$ where u_{-i} denotes the row $(u_1, \dots, u_n)$ with u_i deleted, so that

$$J_i(u_i, u_{-i}) \triangleq E\int_0^\infty e^{-\rho t}\left\{\left[z_i - \Phi\left(\frac{1}{n}\sum_{k=1}^n z_k\right)\right]^2 + ru_i^2\right\}dt. \qquad (9.32)$$

The new notation for the cost should be easily distinguished from $J(u_i, v_i)$, $J(u_i, z^*)$, etc., which have been introduced earlier. We postpone the specification of the admissible control set for each agent until when we introduce the notion of ε-Nash equilibria in Section 5.2. We use u_i^0 to denote the optimal tracking based control law,

$$u_i^0 = -\frac{b}{r}(\Pi_i z_i + s_i), \qquad (9.33)$$

where s_i is derived from (9.15)–(9.18) by matching a_i to a, and s_i implicitly depends on $\bar{z}$ therein. We also use u_{-i}^0 to denote $(u_1^0, \dots, u_n^0)$ with u_i^0 deleted. It should be emphasized that in the following asymptotic analysis the control law u_i^0 for the i-th agent among a population of n agents is constructed using the limit empirical distribution $F(a)$. This gives a conceptually simpler determination of the individual control law without explicitly using the population size.

Recall that in Section 4, no boundedness requirement is imposed on $\mathcal{A}$. For the performance analysis of this section, we need to restrict $\{a_i, i \geq 1\}$ to be bounded. We introduce the assumption:

(H5) The set $\mathcal{A}$ in (**H3**) is the union of a finite number of disjoint compact intervals and $\hat{\varepsilon} > 0$ is a constant such that $\beta_1(a) \geq \hat{\varepsilon}$ for all $a \in \mathcal{A}$.

Notice that under the positivity assumption of $\beta_1(a)$ in (**H3**), the compactness of $\mathcal{A}$ and continuity of $\beta_1(a)$ ensure that $\hat{\varepsilon}$ specified above always exists.

Concerning notation in this section, we make the important convention as follows. $\bar{z}_a$, given by (9.16), denotes the individual mean computed in the large population limit context, and $\bar{z}_a^{(n)}$ stands for the mean of the agents with $a_i = a$ in a population of n agents and it is computed using the n dimensional closed-loop dynamics associated with the control laws u_i^0, $1 \leq i \leq n$. Also, each of s_a, $\bar{z}$ and z^* is computed using (9.15)–(9.18) based upon the large population limit.

5.1 Stability guarantees for closed-loop systems

In order to analyze the closed-loop behaviour when the control law u_i^0 is applied by the i-th agent, we define the diagonal matrix

$$B_n = \begin{pmatrix} -\beta_1(a_1) & & \\ & \ddots & \\ & & -\beta_1(a_n) \end{pmatrix}$$

and denote by $1_{n\times n}$ the $n \times n$ matrix with each entry being one, i.e., $1_{n\times n} = e(n) \times e^T(n)$, where $e^T(n) = [1, \ldots, 1]$. Let

$$\bar{B}_n = B_n + \frac{\alpha}{n} 1_{n\times n}.$$

Hence $\bar{B}_n$ is a real symmetric matrix which has n real eigenvalues. Our further equilibrium analysis relies on the closed-loop stability when the control laws u_i^0, $1 \leq i \leq n$, are applied. We introduce the following property related to closed-loop stability.

(P1) There exist $\mu^* < 0$ and integer $N_0 > 0$ such that for all $n \geq N_0$,

$$\bar{B}_n \leq \mu^* I_n$$

where I_n is the $n \times n$ identity matrix.

In the case $\alpha \leq 0$ and $\inf_{i\geq 1} \beta_1(a_i) = \beta^* > 0$, it is easy to verify (**P1**). We give a sufficient condition to validate (**P1**) for the case $\alpha > 0$.

PROPOSITION 9.7 *Assume*

(i) $\alpha > 0$,

(ii) $\beta_1(a_i) \geq \beta^* > 0$ *for all* $i \geq 1$, *and*

(iii) *there exists* $N_0 > 0$ *such that*

$$\sup_{n \geq N_0} \frac{1}{n} \sum_{i=1}^{n} \frac{\alpha}{\beta_1(a_i)} < 1.$$

Then **(P1)** *holds for all* $n \geq N_0$.

Proof. See Appendix B. □

COROLLARY 9.1 *Assumptions* **(H1)**–**(H3)** *and* **(H5)** *imply* **(P1)**.

Proof. It suffices to verify condition (iii) in Proposition 9.7 for the case $\alpha > 0$. Under the assumptions in the corollary, we may use a weak convergence argument to show

$$\lim_{n\to\infty} \frac{1}{n} \sum_{i=1}^{n} \frac{\alpha}{\beta_1(a_i)} = \lim_{n\to\infty} \int_{\mathcal{A}} \frac{\alpha}{\beta_1(a)} dF_n(a) = \int_{\mathcal{A}} \frac{\alpha}{\beta_1(a)} dF(a) < 1,$$

where the inequality is implied by **(H3)**. Hence there exists N_0 such that condition (iii) holds. □

LEMMA 9.2 *Assuming* **(H1)**–**(H5)**, *we have the estimate*

$$\sup_{t\geq 0} E\left[\sum_{i=1}^{n}(z_i - Ez_i)(t)\right]^2 = O(n),$$

where the set of states z_i, $1 \leq i \leq n$, *corresponds to the control laws* u_i^0, $1 \leq i \leq n$, *given by* (9.33).

Proof. Consider the system of n agents using the control law u_i^0. Here associated with u_i^0, both s_a and z^* are computed via (9.15)–(9.18) based on the large population limit and are independent of n. By use of the closed-loop dynamics, we express each $z_i(t)$ in terms of the initial condition $z_i(0)$, $z^{(n)}$, s_{a_i} and the Wiener process w_i. It is easy to verify

$$\sum_{i=1}^{n}[z_i(t) - Ez_i(t)] = \sum_{i=1}^{n} e^{-\beta_1(a_i)t} z_i(0) + \sum_{i=1}^{n} \int_0^t e^{-\beta_1(a_i)(t-\tau)} \sigma_i \, dw_i$$
$$+ \frac{1}{n} \sum_{i=1}^{n} \int_0^t \alpha e^{-\beta_1(a_i)(t-\tau)} \sum_{k=1}^{n} [z_k(\tau) - Ez_k(\tau)] \, d\tau.$$

By (**H3**), we can take a sufficiently small but fixed $\varepsilon_0 > 0$ such that $(1+\varepsilon_0)[\int_{\mathcal{A}} |\alpha|/\beta_1(a)\,dF(a)]^2 < 1-\varepsilon_0$. Denote $\xi(t) = \sum_{i=1}^n [z_i(t) - Ez_i(t)]$, and $\Delta(t) = \int_{\mathcal{A}} \int_0^t |\alpha| e^{-\beta_1(a)(t-\tau)}\,d\tau\,dF_n(a)$. By use of the inequality $(y_1 + y_2 + y_3)^2 \le (1+\varepsilon_0)y_1^2 + 2(1+1/\varepsilon_0)(y_2^2 + y_3^2)$, we obtain

$$
\begin{aligned}
E\xi^2(t) \le \left(2 + \frac{2}{\varepsilon_0}\right) E &\left[\left(\sum_{i=1}^n e^{-\beta_1(a_i)t} z_i(0) \right)^2 \right. \\
&\left. + \left(\sum_{i=1}^n \int_0^t e^{-\beta_1(a_i)(t-\tau)} \sigma_i dw_i \right)^2 \right] \\
&+ (1+\varepsilon_0) E \left[\int_0^t \frac{1}{n} \sum_{i=1}^n \alpha e^{-\beta_1(a_i)(t-\tau)} \xi(\tau)\,d\tau \right]^2 \\
\le nC + (1+\varepsilon_0)\Delta^2(t) E &\left[\int_{\mathcal{A}} \int_0^t \Delta^{-1}(t) |\alpha| e^{-\beta_1(a)(t-\tau)} \xi(\tau)\,d\tau\,dF_n(a) \right]^2 \\
\le nC + (1+\varepsilon_0)\Delta^2(t) E &\int_{\mathcal{A}} \int_0^t \Delta^{-1}(t) |\alpha| e^{-\beta_1(a)(t-\tau)} \xi^2(\tau) d\tau dF_n(a) \quad (9.34) \\
\le nC + \sup_{0\le\tau\le t} E\xi^2(\tau)(1+\varepsilon_0) &\left[\int_{\mathcal{A}} \frac{|\alpha|}{\beta_1(a)}\,dF_n \right]^2, \quad (9.35)
\end{aligned}
$$

where $C > 0$ is a constant independent of n and t. Here (9.34) follows from Jensen's inequality since for the fixed $t > 0$, $\Delta^{-1}(t)|\alpha| \times e^{-\beta_1(a)(t-\tau)}\,d\tau dF_n(a)$ induces a measure on the product space $[0,t] \times \mathcal{A}$ with a total measure of one. Employing the weak convergence of F_n we can show that there exists $N_1 > 0$ such that for all $n \ge N_1$,

$$
(1+\varepsilon_0) \left| \left[\int_{\mathcal{A}} \frac{|\alpha|}{\beta_1(a)}\,dF_n \right]^2 - \left[\int_{\mathcal{A}} \frac{|\alpha|}{\beta_1(a)}\,dF \right]^2 \right| \le \frac{\varepsilon_0}{2}.
$$

Hence for all $n \ge N_1$ and any fixed $T > 0$, from (9.35) we have

$$
E\xi^2(t) \le nC + \left(1 - \frac{\varepsilon_0}{2}\right) \sup_{0\le\tau\le T} E\xi^2(\tau), \quad 0 \le t \le T,
$$

which yields

$$
\sup_{0\le t\le T} E\xi^2(t) \le nC + \left(1 - \frac{\varepsilon_0}{2}\right) \sup_{0\le t\le T} E\xi^2(t).
$$

And therefore,

$$
\sup_{0\le t\le T} E\xi^2(t) \le \frac{2nC}{\varepsilon_0}.
$$

Since $T > 0$ is arbitrary and C is independent of T, the lemma follows. □

In fact, given the control law u_i^0, we can refine the proof of Lemma 9.2 to show for sufficiently large $N_1 > 0$, $\sup_{n \geq N_1} \sup_{t \geq 0, 1 \leq k \leq n} E z_k^2(t) < \infty$. Thus the tracking based control law u_i^0 (depending on the fixed point aggregation procedure in Section 4) is stabilizing for n-agent systems for n sufficiently large.

5.2 The asymptotic equilibrium analysis

Within the context of a population of n agents, for any $1 \leq k \leq n$, the k-th agent's admissible control set $\mathcal{U}_k$ consists of all feedback controls u_k adapted to the σ-algebra $\sigma\big(z_i(\tau), \tau \leq t, 1 \leq i \leq n\big)$ $\big($i.e., $u_k(t)$ is a function of $\big(t, z_1(t), \ldots, z_n(t)\big)\big)$ such that a unique strong solution to the closed-loop system of the n agents exists on $[0, \infty)$. In this setup we give the definition.

Definition 9.1 *A set of controls $u_k \in \mathcal{U}_k$, $1 \leq k \leq n$, for n players is called an ε-Nash equilibrium with respect to the costs J_k, $1 \leq k \leq n$, if there exists $\varepsilon \geq 0$ such that for any fixed $1 \leq i \leq n$, we have*

$$J_i(u_i, u_{-i}) \leq J_i(u_i', u_{-i}) + \varepsilon, \tag{9.36}$$

when any alternative control $u_i' \in \mathcal{U}_i$ is applied by the i-th player.

If $\varepsilon = 0$ in (9.36), then Definition 9.1 specializes to the usual Nash equilibrium (Aubin, 1998).

Remark The admissible control set $\mathcal{U}_k$ is not decentralized since the k-th agent has perfect information on other agents' states. In effect, such admissible control sets lead to a stronger qualification of the ε-Nash equilibrium property for the decentralized control analyzed in this section.

Given the distribution function F, $\bar{z} \in C_b[0, \infty)$ and the associated $z^* = \Phi(\bar{z})$, from (9.15)–(9.16) it is seen that both s_a and $\bar{z}_a$ may be explicitly expressed as a function of $a \in \mathcal{A}$. Notice that $\bar{z}$ is determined from (9.21) and is independent of a. We introduce the following property:

(P2) Let $\mathcal{A}$ be the set specified in (**H5**).

(i) $\sup_{a \in \mathcal{A}} |\bar{z}_a|_\infty < \infty$, and

(ii) $\lim_{a' \to a} \sup_t |\bar{z}_a(t) - \bar{z}_{a'}(t)| = 0$ with a vanishing rate depending only on $|a - a'|$, for $a, a' \in \mathcal{A}$.

The proposition below gives a sufficient condition to insure (**P2**).

PROPOSITION 9.8 *Assume* (**H1**)–(**H5**) *holds. Then* $\bar{z}_a(t)$ *has the property* (P2).

Proof. See Appendix B. □

We note that for validating property (ii) in (**P2**) in Proposition 9.8, the set $\mathcal{A}$ is not required to be compact in the proof. Now we define

$$\varepsilon_n(t) = \left| \int_{\mathcal{A}} \bar{z}_a(t)\, dF_n(a) - \int_{\mathcal{A}} \bar{z}_a(t)\, dF(a) \right|, \quad t \geq 0, \tag{9.37}$$

$$\varepsilon_n'(t) = \left| \int_{\mathcal{A}} \bar{z}_a(t)\, dF_n(a) - \int_{\mathcal{A}} \bar{z}_a^{(n)}(t)\, dF_n(a) \right|, \quad t \geq 0. \tag{9.38}$$

As mentioned earlier, here $\bar{z}_a$ is determined using the large population limit and $\bar{z}_a^{(n)}$ denotes the mean of agents with $a_i = a$ in a system of n agents taking the control laws u_i^0, $1 \leq i \leq n$.

LEMMA 9.3 *Under* (**H1**)–(**H5**), *we have*

$$\lim_{n\to\infty} \bar{\varepsilon}_n \triangleq \lim_{n\to\infty} \sup_{t\geq 0} \varepsilon_n(t) = 0,$$

where $\varepsilon_n(t)$ *is defined by* (9.37).

Proof. Letting $I_C = [-C, C]$ for $C > 0$, we have

$$\begin{aligned} \varepsilon_n(t) = \Bigg| &\int_{\mathcal{A}\cap I_C} \bar{z}_a(t)\, dF_n(a) + \int_{\mathcal{A}\cap(\mathbb{R}\setminus I_C)} \bar{z}_a(t)\, dF_n(a) \\ &- \int_{\mathcal{A}\cap I_C} \bar{z}_a(t)\, dF(a) - \int_{\mathcal{A}\cap(\mathbb{R}\setminus I_C)} \bar{z}_a(t)\, dF(a) \Bigg| \\ \triangleq\ & |I_n^{(1)} + I_n^{(2)} - I^{(1)} - I^{(2)}|. \end{aligned}$$

Now for any fixed $\varepsilon > 0$, there exists a sufficiently large constant $C > 0$ such that F is continuous at $a = \pm C$ and $I_C \supset \mathcal{A}$ which leads to

$$|I_n^{(2)}| + |I^{(2)}| = 0,$$

since $\int_{I_C} dF_n(a) = \int_{I_C} dF(a) = 1$. We write

$$\begin{aligned} |I_n^{(1)} - I^{(1)}| &= \left| \int_{\mathcal{A}\cap I_C} \bar{z}_a(t)\, dF_n(a) - \int_{\mathcal{A}\cap I_C} \bar{z}_a(t)\, dF(a) \right| \\ &= \left| \int_{I_C} \bar{z}_a'(t)\, dF_n(a) - \int_{I_C} \bar{z}_a'(t)\, dF(a) \right|, \end{aligned}$$

where we make the convention that the domain of $\bar{z}_a(t)$ (as a function of a), if necessary, is extended from $\mathcal{A}$ to $\mathbb{R}$ (hence covering I_C) such that properties (i) and (ii) in (**P2**) still hold after $\mathcal{A}$ is replaced by $\mathbb{R}$. We denote the resulting function by $\bar{z}'_a(t)$ which is identical to $\bar{z}_a(t)$ on $\mathcal{A}$. For instance, in the case $\mathcal{A} = [c_1, c_2]$ with $c_2 < \infty$, we may simply set $\bar{z}'_a(t) = \bar{z}_{c_2}(t)$ when $a > c_2$. Such an extension can deal with the general case when $\mathcal{A}$ consists of a finite number of disjoint bounded and closed subintervals.

Next we combine the equicontinuity of $\bar{z}'_a(t)$ in $a \in I_C$ w.r.t. $t \in [0, \infty)$ insured by (**P2**) and the above extension procedure, continuity of F at $a = \pm C$, and the standard subinterval dividing technique for the proof of Helly-Bray theorem (see Chow and Teicher, 1997, pp. 274–275), to conclude that there exists a sufficiently large n_0 such that for all $n \geq n_0$,

$$|I_n^{(1)} - I^{(1)}| = \left| \int_{I_C} \bar{z}'_a(t)\, dF_n(a) - \int_{I_C} \bar{z}'_a(t)\, dF(a) \right| \leq \frac{\varepsilon}{2},$$

for the arbitrary but fixed ε, and consequently $\lim_{n\to\infty} \sup_{t\geq 0} \varepsilon_n(t) = 0$. This completes the proof. □

In the proof of Lemma 9.3, in order to preserve properties (i) and (ii) in (**P2**), we extend $\bar{z}_a(t)$ to $a \notin \mathcal{A}$ in a specific manner and avoid directly using (9.15)–(9.18) to calculate $\bar{z}_a(t)$, $a \notin \mathcal{A}$, even if the equation system may give a well defined $\bar{z}_a(t)$ for some $a \notin \mathcal{A}$. To show the merit of such an extension, we consider a simple scenario as follows. Suppose $\mathcal{A} = [0, 1]$ and F has discontinuities at $a = 0, 1$. Then by choosing $I_C = [-1, 2] \supset \mathcal{A}$ and using the obtained function $\bar{z}'_a$ (which is continuous on $[-1, 2]$), we can insure the applicability of the technique of the Helly-Bray theorem which requires the limit distribution function F to be continuous at the endpoints of the interval of integration of a continuous function. Notice that in this case F is continuous at $a = -1, 2$.

PROPOSITION 9.9 *Suppose* (**H1**)–(**H5**) *hold. Then we have*

$$\lim_{n\to\infty} \bar{\varepsilon}'_n \triangleq \lim_{n\to\infty} \sup_{t\geq 0} \varepsilon'_n(t) = 0,$$

where $\varepsilon'_n(t)$ *is defined by* (9.38).

Proof. First, using the closed-loop dynamics for $\bar{z}_a^{(n)}$ and $\bar{z}_a$ we get the relation

$$\frac{d(\bar{z}_a^{(n)} - \bar{z}_a)}{dt} = -\beta_1(a)(\bar{z}_a^{(n)} - \bar{z}_a) + \alpha(\bar{z}^{(n)} - \bar{z})$$

with initial condition $(\bar{z}_a^{(n)} - \bar{z}_a)|_{t=0} = 0$. $\bar{z}_a^{(n)}$ is the mean of agents with $a_i = a$, $1 \leq i \leq n$, and $\bar{z}^{(n)} = 1/n \sum_{i=1}^n \bar{z}_{a_i}$. Notice that the assumptions here implies (**P1**) for sufficiently large n (see Corollary 9.1) which further insures $|\bar{z}_a^{(n)}|_\infty < \infty$. Here $|x|_\infty = \sup_{t\geq 0}|x(t)|$ for $x \in C_b[0,\infty)$. Hence it follows that

$$|\bar{z}_a^{(n)} - \bar{z}_a|_\infty \leq \frac{|\alpha|}{\beta_1(a)}|\bar{z}^{(n)} - \bar{z}|_\infty < \infty \tag{9.39}$$

which further leads to

$$\left|\frac{1}{n}\sum_{i=1}^n \bar{z}_{a_i}^{(n)} - \bar{z} + \bar{z} - \frac{1}{n}\sum_{i=1}^n \bar{z}_{a_i}\right|_\infty \leq |\bar{z}^{(n)} - \bar{z}|_\infty \frac{1}{n}\sum_{i=1}^n \frac{|\alpha|}{\beta_1(a_i)}.$$

Hence we have

$$\begin{aligned}|\bar{z}^{(n)} - \bar{z}|_\infty \leq |\bar{z}^{(n)} - \bar{z}|_\infty &\int_{\mathcal{A}} \frac{|\alpha|}{\beta_1(a)}\, dF_n(a) \\ &+ \left|\int_{\mathcal{A}} \bar{z}_a\, dF(a) - \int_{\mathcal{A}} \bar{z}_a\, dF_n(a)\right|_\infty .\end{aligned} \tag{9.40}$$

On the other hand, for sufficiently large n_0, we may use the continuity and boundedness of $1/\beta_1(a)$ together with the weak convergence of F_n to get $\sup_{n\geq n_0}\int_{\mathcal{A}} |\alpha|/\beta_1(a)\, dF_n(a) < 1$ resulting from (**H3**), and moreover, by the uniform boundedness and equicontinuity of $\bar{z}_a$ as shown by Proposition 9.8, we can prove $|\int_{\mathcal{A}} \bar{z}_a\, dF(a) - \int_{\mathcal{A}} \bar{z}_a\, dF_n(a)|_\infty = o(1)$ by Lemma 9.3. Hence by (9.40) we conclude that $|\bar{z}^{(n)} - \bar{z}|_\infty = o(1)$ as $n \to \infty$. Finally the proof follows from $\bar{\varepsilon}'_n \leq |\bar{z}^{(n)} - \bar{z}|_\infty \int_{\mathcal{A}} |\alpha|/\beta_1(a)\, dF_n(a) = o(1)$. □

LEMMA 9.4 *Under* (**H1**)–(**H5**), *for* $z^* \in C_b[0,\infty)$ *determined by* (9.15)–(9.18), *we have*

$$E\int_0^\infty e^{-\rho t}\left[z^* - \Phi\left(\frac{1}{n}\sum_{k=1}^n z_k\right)\right]^2_{(u_i^0, u_{-i}^0)} dt$$
$$= O\left(\gamma^2(\bar{\varepsilon}_n + \bar{\varepsilon}'_n)^2 + \frac{\gamma^2}{n}\right), \tag{9.41}$$

where $\bar{\varepsilon}_n$ *and* $\bar{\varepsilon}'_n$ *are given in* (9.37)-(9.38), *and the set of states* z_k, $1 \leq k \leq n$, *is associated with* u_k^0 *given by* (9.33).

Proof. In the proof we shall omit the control u_k^0 associated with z_k in various places. Obviously we have

$$\frac{1}{n}\sum_{k=1}^n Ez_k = \int_{\mathcal{A}} \bar{z}_a^{(n)}\, dF_n(a).$$

Setting

$$\Psi \triangleq \left| z^* - \Phi\left(\frac{1}{n}\sum_{k=1}^{n} z_k\right)\right| \le \gamma \left| \int_{a\in\mathcal{A}} \bar{z}_a \, dF(a) - \int_{a\in\mathcal{A}} \bar{z}_a \, dF_n(a) + \int_{a\in\mathcal{A}} \bar{z}_a \, dF_n(a) - \frac{1}{n}\sum_{k=1}^{n} z_k \right|$$

where the inequality follows from (**H2**), we obtain

$$\begin{aligned} E\Psi^2(t) &\le \gamma^2 E\Bigg[\left(\int_{a\in\mathcal{A}} \bar{z}_a \, dF(a) - \int_{a\in\mathcal{A}} \bar{z}_a \, dF_n(a)\right) \\ &\quad + \left(\int_{\mathcal{A}} \bar{z}_a \, dF_n(a) - \frac{1}{n}\sum_{k=1}^{n} Ez_k\right) + \left(\frac{1}{n}\sum_{k=1}^{n} Ez_k - \frac{1}{n}\sum_{k=1}^{n} z_k\right)\Bigg]^2 \\ &\le 2\gamma^2(\bar{\varepsilon}_n + \bar{\varepsilon}_n')^2 + 2\gamma^2 E\left[\frac{1}{n}\sum_{k=1}^{n}(z_k - Ez_k)\right]^2 \\ &\le 2\gamma^2(\bar{\varepsilon}_n + \bar{\varepsilon}_n')^2 + O\left(\frac{\gamma^2}{n}\right), \end{aligned}$$

where the upper bound given as the last term holds uniformly w.r.t. $t \ge 0$ by virtue of Lemma 9.2, and therefore (9.41) follows. □

LEMMA 9.5 *Letting $0 < T < \infty$, for any Lebesgue measurable function $x\colon [0,T] \to \mathbb{R}$ such that $\int_0^T x_t^2 \, dt < \infty$, we have*

$$\int_0^T e^{-\rho t}\left[x_t^2 - \delta \int_0^t x_s^2 ds\right] dt \ge 0 \tag{9.42}$$

for any $\delta \le \rho$.

Proof. Since both x_t^2 and $\int_0^t x_s^2 \, ds$ (as functions of t) have finite integral on $[0,T]$, we may split the integrand in (9.42) to compute

$$\begin{aligned} &\int_0^T e^{-\rho t} x_t^2 \, dt - \delta \int_0^T \int_0^t e^{-\rho t} x_s^2 \, ds \, dt \\ &\qquad = \int_0^T e^{-\rho t} x_t^2 \, dt - \delta \int_0^T \int_s^T e^{-\rho t} x_s^2 \, dt \, ds \\ &\qquad \ge \int_0^T e^{-\rho t} x_t^2 \, dt - \frac{\delta}{\rho} \int_0^T e^{-\rho s} x_s^2 \, ds \ge 0, \end{aligned}$$

where we exchange the order of integration in the double integral by Fubini's theorem. □

For the main results in Theorems 9.3 and 9.4, u_i^0 is the optimal tracking based control law for the i-th player given by (9.33) for which s_i and the associated reference tracking trajectory z^* are computed using (9.15)–(9.18) for the large population limit. Thus both s_i and z^* are independent of the population size.

THEOREM 9.3 *Under* (**H1**)–(**H5**), *we have*

$$|J_i(u_i^0, u_{-i}^0) - J_i(u_i^0, z^*)| = O\left(\gamma\bar{\varepsilon}_n + \gamma\bar{\varepsilon}_n' + \frac{\gamma}{\sqrt{n}}\right),$$

as $n \to \infty$, *where* $J_i(u_i^0, z^*)$ *is the individual cost with respect to* z^*, $J(_i(u_i^0, u_{-i}^0)$ *is determined by* (9.32), $\bar{\varepsilon}_n$, $\bar{\varepsilon}_n'$ *and* u_i^0 *are the same as in Lemma 9.4.*

The proof is done by a similar decomposition technique as in proving Theorem 9.4 below and is postponed until after the proof of the latter.

In Theorem 9.4 we need to consider the perturbation in the control of a given agent. We point out that when the control laws change from (u_i^0, u_{-i}^0) to (u_i, u_{-i}^0) for the system of n agents, a change will accordingly take place for each of the n state components since z_k, $k \neq i$, is coupled with $1/nz_i$ even if the set of control laws u_{-i}^0 remains the same. Here we use u_{-i}^0 to denote the row $(u_1^0, \dots, u_n^0)$ with u_i^0 deleted.

THEOREM 9.4 *Under* (**H1**)–(**H5**), *the set of controls* u_i^0, $1 \leq i \leq n$, *for the* n *players is an* ε-*Nash equilibrium with respect to the costs* $J_i(u_i, u_{-i})$, $1 \leq i \leq n$, *i.e.,*

$$J_i(u_i^0, u_{-i}^0) - \varepsilon \leq \inf_{u_i} J_i(u_i, u_{-i}^0) \tag{9.43}$$

$$\leq J_i(u_i^0, u_{-i}^0) \tag{9.44}$$

where $0 < \varepsilon = O(\gamma\bar{\varepsilon}_n + \gamma\bar{\varepsilon}_n' + \gamma/\sqrt{n})$ *(and hence* $\varepsilon \to 0$*) as* $n \to 0$, *and* $u_i \in \mathcal{U}_i$ *is any alternative control which depends on* $(t, z_1, \dots, z_n)$.

Proof. The inequality (9.44) is obviously true. We prove the inequality (9.43). We consider all full state dependent $u_i \in \mathcal{U}_i$ satisfying

$$J_i(u_i, u_{-i}^0) \leq J_i(u_i^0, u_{-i}^0). \tag{9.45}$$

In the remaining part of the proof, (u_i, u_{-i}^0) appearing in each place is assumed to satisfy (9.45). We shall use $C > 0$ to denote a generic constant which is independent of n and the index of the agents, and its value may vary in different places.

We first estimate the RHS of (9.45). Invoking the stability property of the closed-loop of the n agents all adopting u_i^0, $1 \le i \le n$, and using a similar method as in Lemma 9.2, we can show that there exists C (independent of n and i), such that $E\int_0^\infty e^{-\rho t} z_i^2(t)\big|_{(u_i^0,u_{-i}^0)}\,dt < C$ and

$$E\int_0^\infty e^{-\rho t}\left[z_i - \Phi(\frac{1}{n}\sum_{k=1}^n z_k)\right]^2_{(u_i^0,u_{-i}^0)} dt \le C.$$

Furthermore, since $\mathcal{A}$ is compact, for $u_i^0 = -b/r(\Pi_i z_i + s_i)$, we can find C such that $\Pi_i + |s_i| \le C$. It readily follows that the RHS of (9.45) is bounded by C.

Hence for (u_i, u_{-i}^0) satisfying (9.45), we can find a fixed constant C independent of n such that

$$J_i(u_i, u_{-i}^0) = E\int_0^\infty e^{-\rho t}\left\{\left[z_i - \Phi(\frac{1}{n}\sum_{k=1}^n z_k)\right]^2_{(u_i,u_{-i}^0)} + ru_i^2\right\} dt \le C. \quad (9.46)$$

Now, for (u_i, u_{-i}^0) satisfying (9.45) and hence (9.46), we may express z_k, $k \ne i$, $1 \le k \le n$, in terms of their initial conditions $z_k(0)$, $k \ne i$, the Wiener integral, as well as $1/nz_i$ which acts as an input in the closed-loop dynamics of the $n-1$ agents. In addition, similar to establishing (**P1**), we can show that for large n, a mean square stability holds for the $n-1$ agents when $1/nz_i$ is removed, and that the closed-loop (symmetric) gain matrix has a maximum eigenvalue less than $1/2\mu^*$, where $\mu^* < 0$ is determined in (**P1**).

We may write $1/n\sum_{k=1}^n z_k = \Delta + 1/nz_i + 1/n\int_0^t f_n(t-s)z_i(s)\,ds$, where Δ depends only on the initial conditions of z_k, $k \ne i$, and the Wiener processes, $E\Delta^2$ is bounded by a fixed constant independent of n and t, and $\sup_n |f_n(s)| \le ce^{\mu^* s/2}$ with $c > 0$ for $s \ge 0$. Then using Lipschitz continuity of Φ and basic estimates, we obtain from (9.46)

$$E\int_0^\infty e^{-\rho t}\left\{z_i(t) - \Phi\left[\frac{1}{n}z_i(t) + \frac{1}{n}\int_0^t f_n(t-s)z_i(s)\,ds\right]\right\}^2_{(u_i,u_{-i}^0)} dt \le C. \quad (9.47)$$

Fixing $0 < T < \infty$, by tedious but elementary estimates we have

$$E\int_0^T e^{-\rho t}\left\{\left[z_i(t) - \Phi\left(\frac{z_i(t)}{n} + \frac{1}{n}\int_0^t f_n(t-s)z_i(s)\,ds\right)\right]^2 - \frac{z_i^2(t)}{8}\right\}_{(u_i,u_{-i}^0)} dt$$

$$\geq E\int_0^T e^{-\rho t}\left\{\frac{3}{4}z_i^2(t) - 6\gamma^2\left[\frac{z_i(t)}{n} + \frac{1}{n}\int_0^t f_n(t-s)z_i(s)\,ds\right]^2 - 6\Phi^2(0) - \frac{z_i^2(t)}{8}\right\} dt \quad (9.48)$$

$$\geq \int_0^T e^{-\rho t}\left\{\frac{1}{2}z_i^2(t) - \frac{12\gamma^2}{n^2}\left[\int_0^t f_n(t-s)z_i(s)\,ds\right]^2\right\} dt - C$$

$$\geq \int_0^T e^{-\rho t}\left[\frac{1}{2}z_i^2(t) - \frac{12\gamma^2}{n^2}\int_0^t f_n^2(t-s)\,ds\int_0^t z_i^2(s)\,ds\right] dt - C$$

$$\geq E\int_0^T e^{-\rho t}\left[\frac{1}{2}z_i^2(t) - \frac{12\gamma^2 c^2}{n^2|\mu^*|}\int_0^t z_i^2(s)\,ds\right] dt - C.$$

By Lemma 9.5, for all sufficiently large n, we have

$$\int_0^T e^{-\rho t}\left[\frac{1}{2}z_i^2(t) - \frac{12\gamma^2 c^2}{n^2|\mu^*|}\int_0^t z_i^2(s)\,ds\right]_{(u_i,u_{-i}^0)} dt \geq 0. \quad (9.49)$$

Notice that for almost all sample paths, z_i is continuous on $[0,T]$ so that the integral in (9.49) is well defined. Hence by (9.48) and (9.49), we have

$$E\int_0^T e^{-\rho t}\frac{z_i^2(t)}{8}\bigg|_{(u_i,u_{-i}^0)} dt \leq C + E\int_0^\infty e^{-\rho t}$$

$$\times\left[z_i(t) - \Phi\left(\frac{1}{n}z_i(t) + \frac{1}{n}\int_0^t f_n(t-s)z_i(s)ds\right)\right]^2_{(u_i,u_{-i}^0)} dt \leq C \quad (9.50)$$

where the last inequality follows from (9.47) and C is independent of T.

Subsequently, for (u_i, u_{-i}^0) satisfying (9.45), we assert that there exists $C > 0$ independent of n such that

$$E\int_0^\infty e^{-\rho t}z_i^2\big|_{(u_i,u_{-i}^0)} dt + E\int_0^\infty e^{-\rho t}(z_i - z^*)^2\big|_{(u_i,u_{-i}^0)} dt \leq C, \quad (9.51)$$

where z^* is determined from (9.15)–(9.18) as in Lemma 9.4.

We compare $\Phi(1/n \sum_{k=1}^{n} z_k)|_{(u_i,u_{-i}^0)}$ with $\Phi(1/n \sum_{k=1}^{n} z_k)|_{(u_i^0,u_{-i}^0)}$ by use of the $n-1$ dimensional closed-loop dynamics for z_k, $k \neq i$, and after basic estimates using (9.51), we may obtain

$$\begin{aligned}
&E\int_0^\infty e^{-\rho t}\left[z^* - \Phi\left(\frac{1}{n}\sum_{k=1}^{n} z_k\right)\right]^2_{(u_i,u_{-i}^0)} dt \\
&= E\int_0^\infty e^{-\rho t}\left[z^* - \Phi\left(\frac{1}{n}\sum_{k=1}^{n} z_k\right)\Bigg|_{(u_i^0,u_{-i}^0)}\right. \\
&\qquad \left. +\Phi\left(\frac{1}{n}\sum_{k=1}^{n} z_k\right)\Bigg|_{(u_i^0,u_{-i}^0)} - \Phi\left(\frac{1}{n}\sum_{k=1}^{n} z_k\right)\Bigg|_{(u_i,u_{-i}^0)}\right]^2 dt \qquad (9.52)\\
&= E\int_0^\infty e^{-\rho t}\left[z^* - \Phi\left(\frac{1}{n}\sum_{k=1}^{n} z_k\right)\Bigg|_{(u_i^0,u_{-i}^0)}\right]^2 dt + O\left(\frac{\gamma}{n}\right).
\end{aligned}$$

Here and hereafter in the proof, unless otherwise indicated, the state process is always associated with the control (u_i, u_{-i}^0). For notational brevity, in the following we omit the associated control without causing confusion. Now, on the other hand we have

$$\begin{aligned}
&E\int_0^\infty e^{-\rho t}\left\{\left[z_i - \Phi\left(\frac{1}{n}\sum_{k=1}^{n} z_k\right)\right]^2_{(u_i,u_{-i}^0)} + ru_i^2\right\} dt \\
&= E\int_0^\infty e^{-\rho t}\left\{\left[(z_i-z^*)+\left(z^*-\Phi\left(\frac{1}{n}\sum_{k=1}^{n} z_k\right)\right)\right]^2 + ru_i^2\right\} dt \\
&= E\int_0^\infty e^{-\rho t}[(z_i - z^*)^2 + ru_i^2]\, dt \\
&\qquad + E\int_0^\infty e^{-\rho t}\left[z^* - \Phi\left(\frac{1}{n}\sum_{k=1}^{n} z_k\right)\right]^2 dt \qquad (9.53)\\
&\qquad + 2E\int_0^\infty e^{-\rho t}(z_i - z^*)\left[z^* - \Phi\left(\frac{1}{n}\sum_{k=1}^{n} z_k\right)\right] dt \\
&\triangleq I_1 + I_2 + I_3.
\end{aligned}$$

Then we have

$$I_1 = J(u_i, z^*) \geq J_i(u_i^0, z^*) \geq J_i(u_i^0, u_{-i}^0) - O\left(\gamma\bar{\varepsilon}_n + \gamma\bar{\varepsilon}_n' + \frac{\gamma}{\sqrt{n}}\right), \tag{9.54}$$

$$I_2 = O\left(\gamma^2(\bar{\varepsilon}_n + \bar{\varepsilon}_n')^2 + \frac{\gamma^2}{n}\right), \tag{9.55}$$

where (9.54) follows from Theorem 9.3, and (9.55) follows from Lemma 9.4 and (9.52). Moreover, by Schwarz inequality and (9.51) we have

$$\begin{aligned}|I_3| &\leq 2\int_0^\infty e^{-\rho t}[E(z_i - z^*)^2]^{1/2}\left\{E\left[z^* - \Phi\left(\frac{1}{n}\sum_{k=1}^n z_k\right)\right]^2\right\}^{1/2} dt \\ &\leq 2\left[\int_0^\infty e^{-\rho t}E(z_i - z^*)^2 dt\right]^{1/2}\left\{\int_0^\infty e^{-\rho t}E\left[z^* - \Phi\left(\frac{1}{n}\sum_{k=1}^n z_k\right)\right]^2 dt\right\}^{1/2} \\ &= O(\sqrt{I_2}) = O\left(\gamma\bar{\varepsilon}_n + \gamma\bar{\varepsilon}_n' + \frac{\gamma}{\sqrt{n}}\right).\end{aligned} \tag{9.56}$$

Hence it follows that there exists $c > 0$ such that

$$J_i(u_i, u_i^0) \geq J_i(u_i^0, u_{-i}^0) - c\left(\gamma\bar{\varepsilon}_n + \gamma\bar{\varepsilon}_n' + \frac{\gamma}{\sqrt{n}}\right),$$

where c is independent of n. This completes the proof. □

Proof of Theorem 9.3. As in (9.53) we make the decomposition

$$\begin{aligned}&J_i(u_i^0, u_{-i}^0) \\ &= E\int_0^\infty e^{-\rho t}\left\{\left[z_i - \Phi\left(\frac{1}{n}\sum_{k=1}^n z_k\right)\right]^2_{(u_i^0, u_{-i}^0)} + r(u_i^0)^2\right\} dt \\ &= E\int_0^\infty e^{-\rho t}\left\{\left[(z_i - z^*) + \left(z^* - \Phi\left(\frac{1}{n}\sum_{k=1}^n z_k\right)\right)\right]^2_{(u_i^0, u_{-i}^0)} + r(u_i^0)^2\right\} dt \\ &= J_i(u_i^0, z^*) + E\int_0^\infty e^{-\rho t}\left[z^* - \Phi\left(\frac{1}{n}\sum_{k=1}^n z_k\right)\right]^2_{(u_i^0, u_{-i}^0)} dt \\ &\quad + 2E\int_0^\infty e^{-\rho t}(z_i - z^*)\left[z^* - \Phi\left(\frac{1}{n}\sum_{k=1}^n z_k\right)\right]_{(u_i^0, u_{-i}^0)} dt \\ &\triangleq J_i(u_i^0, z^*) + I_2' + I_3'.\end{aligned} \tag{9.57}$$

Finally, similar to (9.55) and (9.56), we apply Schwarz inequality and Lemma 9.4 to obtain

$$|I_2' + I_3'| = O\left(\gamma\bar{\varepsilon}_n + \gamma\bar{\varepsilon}_n' + \frac{\gamma}{\sqrt{n}}\right),$$

and this completes the proof. □

It should be noted that the proof of Theorem 9.3 does not depend on Theorem 9.4.

6. Conclusions and future research

In this paper we study the individual and mass behaviour in large-population weakly coupled dynamic systems with non-uniform agents. In the framework of noncooperative games, we employ a state aggregation technique to develop decentralized control laws for the agents. The resulting set of individual control laws has an ε-Nash equilibrium property, and furthermore, an attraction property of the mass bebaviour is illustrated.

The further investigation of statistical mechanics methods for such weakly coupled systems is of interest. Also, it is of interest to study decentralized optimization in a system configuration where the number of agents changes from time to time. This kind of model is well motivated in many economic and engineering scenarios; see e.g. Liu and Passino (2004); Baccelli, Hong and Liu (2001). In general, the resulting analysis requires appropriately aggregating a more randomized mass effect due to the time varying population, and introducing cost measures for *active agents* as well.

Acknowledgments

This work is partially supported by Australian Research Council and by Natural Science and Engineering Research Council of Canada.

Appendix: Appendix A

Proof of Lemma 9.1. For any $x \in C_b[0,\infty)$, we have

$$\begin{aligned}|(\mathcal{T}x)(t)| \le& \int_{\mathcal{A}}\int_0^t e^{-\beta_1(a)(t-s)}|\alpha x|_\infty\, ds\, dF(a) + \frac{b^2}{r}\int_{\mathcal{A}}\int_0^t\int_s^\infty e^{-\beta_1(a)(t-s)} \\ &\times e^{-\beta_2(a)(\tau-s)}\left[\sup_{\tau\in\mathbb{R}}|\Phi(x(\tau))| + |\alpha x|_\infty \Pi_a\right] d\tau\, ds\, dF(a) \\ \le& |x|_\infty \int_{\mathcal{A}} \frac{|\alpha|}{\beta_1(a)}\, dF(a) + [|\Phi(0)| + \gamma|x|_\infty]\frac{b^2}{r}\int_{\mathcal{A}}\frac{dF(a)}{\beta_1(a)\beta_2(a)} \\ &+ \frac{b^2|x|_\infty}{r}\int_{\mathcal{A}}\frac{|\alpha|\Pi_a}{\beta_1(a)\beta_2(a)}\, dF(a) < \infty.\end{aligned}$$

Hence for a given $x \in C_b[0,\infty)$, by (**H3**) we have $\sup_{t\geq 0}|(\mathcal{T}x)(t)| < \infty$. We now show continuity of $\mathcal{T}x$ on $[0,\infty)$. Assuming $0 \leq t_1 < t_2 < \infty$, we have

$$\begin{aligned}
&(\mathcal{T}x)(t_2) - (\mathcal{T}x)(t_1) \\
&= \int_{\mathcal{A}}\int_0^{t_2}\int_s^{\infty} \frac{b^2}{r} e^{-\beta_1(a)(t_2-s)} e^{-\beta_2(a)(\tau-s)} \Phi\big(x(\tau)\big)\, d\tau\, ds\, dF(a) \\
&\quad - \int_{\mathcal{A}}\int_0^{t_1}\int_s^{\infty} \frac{b^2}{r} e^{-\beta_1(a)(t_1-s)} e^{-\beta_2(a)(\tau-s)} \Phi\big(x(\tau)\big)\, d\tau\, ds\, dF(a) \\
&\quad + \int_{\mathcal{A}}\int_0^{t_2}\int_s^{\infty} \frac{b^2}{r} e^{-\beta_1(a)(t_2-s)} e^{-\beta_2(a)(\tau-s)} (-\alpha)\Pi_a x(\tau)\, d\tau\, ds\, dF(a) \\
&\quad - \int_{\mathcal{A}}\int_0^{t_1}\int_s^{\infty} \frac{b^2}{r} e^{-\beta_1(a)(t_1-s)} e^{\beta_2(a)(\tau-s)} (-\alpha)\Pi_a x(\tau)\, d\tau\, ds\, dF(a) \\
&\quad + \int_{\mathcal{A}}\int_0^{t_2} e^{-\beta_1(a)(t_2-s)} \alpha x(s)\, ds\, dF(a) \\
&\quad - \int_{\mathcal{A}}\int_0^{t_1} e^{-\beta_1(a)(t_1-s)} \alpha x(s)\, ds\, dF(a) \triangleq I_1 - I_2 + I_3 - I_4 + I_5 - I_6
\end{aligned}$$

where the terms I_i, $1 \leq i \leq 6$ are each determined in an obvious manner.

In the following analysis, we will repeatedly use the fact $|e^{-d_1} - e^{-d_2}| \leq e^{-d_1}|d_1 - d_2|$ for $0 \leq d_1 \leq d_2$. We have the estimate

$$\begin{aligned}
I_1 - I_2 &= \frac{b^2}{r}\int_{\mathcal{A}}\int_0^{t_1}\int_s^{\infty} [e^{-\beta_1(a)(t_2-s)} - e^{-\beta_1(a)(t_1-s)}] e^{-\beta_2(a)(\tau-s)} \\
&\qquad\qquad \times \Phi\big(x(\tau)\big)\, d\tau\, ds\, dF(a) \\
&= \frac{b^2}{r}\int_{\mathcal{A}}\int_{t_1}^{t_2}\int_s^{\infty} e^{-\beta_1(a)(t_2-s)} e^{-\beta_2(a)(\tau-s)} \Phi\big(x(\tau)\big)\, d\tau\, ds\, dF(a) \\
&\triangleq \Delta_1 + \Delta_2.
\end{aligned}$$

We have

$$\begin{aligned}
|\Delta_1| &\leq \frac{b^2}{r}[|\Phi(0)| + \gamma|x|_\infty]\int_{\mathcal{A}}\int_0^{t_1}\int_s^{\infty} e^{-\beta_1(a)(t_1-s)} \\
&\qquad \times \beta_1(a)|t_2 - t_1| e^{-\beta_2(a)(\tau-s)} d\tau ds dF(a) \\
&= \frac{b^2}{r}[|\Phi(0)| + \gamma|x|_\infty]\int_{\mathcal{A}}\int_0^{t_1} e^{-\beta_1(a)(t_1-s)}\beta_1(a)|t_2 - t_1|\frac{1}{\beta_2(a)} ds dF(a) \\
&= \frac{b^2}{r}[|\Phi(0)| + \gamma|x|_\infty]|t_2 - t_1|\int_{\mathcal{A}}[1 - e^{-\beta_1(a)t_1}]\frac{1}{\beta_2(a)} dF(a) \\
&\leq \frac{b^2}{r}[|\Phi(0)| + \gamma|x|_\infty]|t_2 - t_1|\int_{\mathcal{A}}\frac{1}{\beta_2(a)} dF(a)
\end{aligned}$$

where $\beta_2(a) \geq \rho/2 + \frac{|b|}{\sqrt{r}}$ for all $a \in \mathbb{R}$, and

$$\begin{aligned}
|\Delta_2| &\leq \frac{b^2}{r}[|\Phi(0)| + \gamma|x|_\infty]\int_{\mathcal{A}}\int_{t_1}^{t_2}\int_s^{\infty} e^{-\beta_1(a)(t_2-s)} e^{-\beta_2(a)(\tau-s)} d\tau ds\, dF(a) \\
&\leq \frac{b^2}{r}[|\Phi(0)| + \gamma|x|_\infty]\int_{\mathcal{A}}\int_{t_1}^{t_2} e^{-\beta_1(a)(t_2-s)}\frac{1}{\beta_2(a)}\, ds\, dF(a)
\end{aligned}$$

$$\le \frac{b^2}{r}[|\Phi(0)| + \gamma|x|_\infty] \int_{\mathcal{A}} [1 - e^{-\beta_1(a)(t_2-t_1)}] \frac{1}{\beta_1(a)\beta_2(a)} dF(a)$$
$$\le \frac{b^2}{r}[|\Phi(0)| + \gamma|x|_\infty]|t_2 - t_1| \int_{\mathcal{A}} \frac{1}{\beta_2(a)} dF(a).$$

Hence,

$$|I_1 - I_2| \le \frac{2b^2}{r}[|\Phi(0)| + \gamma|x|_\infty]|t_2 - t_1| \int_{\mathcal{A}} \frac{1}{\beta_2(a)} dF(a).$$

Similarly we have

$$|I_3 - I_4| \le \frac{2b^2|\alpha x|_\infty}{r}|t_2 - t_1| \int_{\mathcal{A}} \frac{\Pi_a}{\beta_2(a)} dF(a)$$

where the integral is finite since $\Pi_a/\beta_2(a)$ is bounded for $a \in \mathbb{R}$. Furthermore, we have

$$|I_5 - I_6| \le 2|\alpha x|_\infty|t_2 - t_1|.$$

Hence it follows from that

$$\begin{aligned}|(\mathcal{T}x)(t_2) - (\mathcal{T}x)(t_1)| &\le |I_1 - I_2| + |I_3 - I_4| + |I_5 - I_6| \\ &\le \frac{2b^2}{r}[|\Phi(0)| + \gamma|x|_\infty]|t_2 - t_1| \int_{\mathcal{A}} \frac{1}{\beta_2(a)} dF(a) \\ &+ \frac{2b^2|\alpha x|_\infty}{r}|t_2 - t_1| \int_{\mathcal{A}} \frac{\Pi_a}{\beta_2(a)} dF(a) + 2|\alpha x|_\infty|t_2 - t_1|.\end{aligned} \tag{A.1}$$

Then the lemma follows. □

Appendix B

Proof of Proposition 9.7. Let us denote by $\lambda_{\max}(\bar{B}_n)$ the largest eigenvalue of the real symmetric matrix $\bar{B}_n$. Define the set $S = \{x \in \mathbb{R}^n : |x| = (\sum_{i=1}^n x_i^2)^{1/2} = 1\}$. Then we have

$$\lambda_{\max}(\bar{B}_n) = \sup_S x^T \bar{B}_n x \triangleq \sup_S \Lambda(x).$$

It is easy to show that

$$\Lambda(x) = -\sum_{i=1}^n \beta_1(a_i)x_i^2 + \frac{\alpha}{n}\left(\sum_{i=1}^n x_i\right)^2.$$

Assuming the supremum of Λ on S is attained at y, we can obtain the necessary condition for y by the Lagrangian multiplier method and we assert that there exists $\mu \in \mathbb{R}$ such that

$$2\beta_1(a_i)y_i - \frac{2\alpha}{n}\sum_{i=1}^n y_j + 2\mu y_i = 0, \quad 1 \le i \le n, \tag{B.1}$$

$$\sum_{i=1}^n y_i^2 = 1, \tag{B.2}$$

where the first equation is obtained by the necessary condition for the supremum of the function $\Lambda(x) + \mu(\sum_{i=1}^n x_i^2 - 1)$ with $x \in \mathbb{R}^n$.

Let $S^+ = \{x \in S: x_i \geq 0, 1 \leq i \leq n\}$. For any $x \in S$, we denote $\tilde{x} = (|x_1|, \ldots, |x_n|)^T$. Clearly we have $\Lambda(\tilde{x}) \geq \Lambda(x)$ and it is impossible to attain the supremum at x which has both strictly positive and strictly negative entries. Thus we have $\sup_S \Lambda(x) = \sup_{S^+} \Lambda(x)$. Now it suffices to determine the supremum by solving (B.1) and (B.2) under the additional constraint $y_i \geq 0$, $1 \leq i \leq n$, i.e., $y \in S^+$ (then accordingly, $-y$ also attains the supremum by symmetry).

Since $y_i \geq 0$ and $\alpha > 0$, it necessarily follows that $\beta_1(a_i) + \mu > 0$ by (B.1). Furthermore, by (B.1) we may introduce an undetermined constant $c > 0$ such for each $1 \leq i \leq n$,

$$y_i = \frac{c}{\beta_1(a_i) + \mu}. \tag{B.3}$$

Substituting (B.3) into (B.2), we get

$$c^2 \sum_{i=1}^{n} \frac{1}{[\beta_1(a_i) + \mu]^2} = 1 \tag{B.4}$$

which yields

$$0 < c = \left(\sum_{i=1}^{n} \frac{1}{[\beta_1(a_i) + \mu]^2} \right)^{-1/2} \tag{B.5}$$

Combining (B.5) with (B.1) gives

$$(\beta_1(a_i) + \mu) \frac{c}{\beta_1(a_i) + \mu} = \frac{c\alpha}{n} \sum_{i=1}^{n} \frac{1}{\beta_1(a_i) + \mu}.$$

This yields

$$\frac{1}{n} \sum_{i=1}^{n} \frac{\alpha}{\beta_1(a_i) + \mu} = 1. \tag{B.6}$$

Further, by making use of (B.3), (B.6) and then (B.4) we compute

$$\sup_S \Lambda(x) = -\sum_{i=1}^{n} \beta_1(a_i) \frac{c^2}{[\beta_1(a_i) + \mu]^2} + \frac{\alpha}{n} \left(\sum_{i=1}^{n} \frac{c}{\beta_1(a_i) + \mu} \right)^2 = \mu. \tag{B.7}$$

Hence the largest eigenvalue of $\bar{B}_n$ is given by $\lambda_{\max}(\bar{B}_n) = \mu$ which satisfies (B.6). Now for a fixed n, let $G_n(\nu) = 1/n \sum_{i=1}^{n} \alpha/\beta_1(a_i) + \nu$, $\nu \in (-\beta_{1,n}, \infty)$, where $\beta_{1,n} = \inf_{1 \leq i \leq n} \beta_1(a_i)$. Obviously $G_n(\nu)$ is strictly monotone and $\lim_{\nu \to -\beta_{1,n}} G_n(\nu) = \infty$, $G_n(\infty=0$. Therefore there is a unique μ^* satisfying (B.6) on $(-\beta_{1,n}, \infty)$.

Recalling conditions (ii) and (iii), we see this shows there exists a fixed $\mu^* < 0$ which may be taken as satisfying $\mu^* > -\beta^*$, such that for all $n \geq N_0$, $\mu \in [\mu^*, \infty)$,

$$\frac{1}{n} \sum_{i=1}^{n} \frac{\alpha}{\mu + \beta_1(a_i)} < 1$$

which implies that $\mu < \mu^*$ where μ is given by (B.6). This completes the proof. □

Proof of Proposition 9.8. Taking a, $a' \in \mathcal{A}$, by use of (9.20) we obtain

$$\begin{aligned}
|\bar{z}_a(t) - \bar{z}_{a'}(t)| \le & \left| \int_0^t e^{-\beta_1(a)(t-s)} \alpha \bar{z}(s)\, ds - \int_0^t e^{-\beta_1(a')(t-s)} \alpha \bar{z}(s) ds \right| \\
& + \left| \frac{b^2}{r} \int_0^t e^{-\beta_1(a)(t-s)} \int_s^\infty e^{-\beta_2(a)(\tau-s)} \Phi\big(\bar{z}(\tau)\big)\, d\tau\, ds \right. \\
& \left. - \frac{b^2}{r} \int_0^t e^{-\beta_1(a')(t-s)} \int_s^\infty e^{-\beta_2(a')(\tau-s)} \Phi\big(\bar{z}(\tau)\big)\, d\tau\, ds \right| \\
& + \left| \frac{b^2}{r} \int_0^t e^{-\beta_1(a)(t-s)} \int_s^\infty e^{-\beta_2(a)(\tau-s)} \alpha \Pi_a \bar{z}(\tau)\, d\tau\, ds \right. \\
& \left. - \frac{b^2}{r} \int_0^t e^{-\beta_1(a')(t-s)} \int_s^\infty e^{-\beta_2(a')(\tau-s)} \alpha \Pi_{a'} \bar{z}(\tau)\, d\tau\, ds \right| \\
& \triangleq \Delta_1 + \Delta_2 + \Delta_3.
\end{aligned}$$

By direct calculation we have the estimates

$$\begin{aligned}
\Delta_1 &\le \frac{|\alpha \bar{z}|_\infty |\beta_1(a) - \beta_1(a')|}{\min\{\beta_1^2(a), \beta_1^2(a')\}}, \\
\Delta_2 &\le \frac{b^2}{r} \big(|\Phi(0)| + \gamma |\bar{z}|_\infty \big) \\
& \quad \times \left[\frac{|\beta_2(a) - \beta_2(a')|}{\beta_1(a) \min\{\beta_2^2(a), \beta_2^2(a')\}} + \frac{|\beta_1(a) - \beta_1(a')|}{\beta_2(a') \min\{\beta_1^2(a), \beta_1^2(a')\}} \right]
\end{aligned}$$

In order to estimate Δ_3, we write

$$\begin{aligned}
\Delta_3 \le & \left| \frac{b^2}{r} \int_0^t e^{-\beta_1(a)(t-s)} \int_s^\infty e^{-\beta_2(a)(\tau-s)} \alpha \Pi_a \bar{z}(\tau)\, d\tau\, ds \right. \\
& \left. - \frac{b^2}{r} \int_0^t e^{-\beta_1(a')(t-s)} \int_s^\infty e^{-\beta_2(a)(\tau-s)} \alpha \Pi_a \bar{z}(\tau)\, d\tau ds \right| \\
& + \left| \frac{b^2}{r} \int_0^t e^{-\beta_1(a')(t-s)} \int_s^\infty e^{-\beta_2(a)(\tau-s)} \alpha \Pi_a \bar{z}(\tau)\, d\tau\, ds \right. \\
& \left. - \frac{b^2}{r} \int_0^t e^{-\beta_1(a')(t-s)} \int_s^\infty e^{-\beta_2(a')(\tau-s)} \alpha \Pi_{a'} \bar{z}(\tau)\, d\tau\, ds \right| \triangleq \Delta_{31} + \Delta_{32}.
\end{aligned}$$

We have

$$\Delta_{31} \le \frac{b^2}{r} |\alpha \bar{z}|_\infty \frac{\Pi_a}{\beta_2(a)} \frac{|\beta_1(a) - \beta_1(a')|}{\min\{\beta_1^2(a), \beta_1^2(a')\}} \tag{B.8}$$

where it is obvious that $\sup_{a \in \mathbb{R}} \Pi_a / \beta_2(a) < \infty$. Since

$$\begin{aligned}
& \left| \int_s^\infty [e^{-\beta_2(a)(\tau-s)} \Pi_a - e^{-\beta_2(a')(\tau-s)} \Pi_{a'}]\, d\tau \right| \\
& \le \int_s^\infty e^{-\beta_2(a)(\tau-s)} |\Pi_a - \Pi_{a'}|\, d\tau + \int_s^\infty |e^{-\beta_2(a)(\tau-s)} - e^{-\beta_2(a')(\tau-s)}| \Pi_{a'}\, d\tau \\
& \le \frac{|\Pi_a - \Pi_{a'}|}{\beta_2(a)} + \frac{\Pi_{a'} |\beta_2(a) - \beta_2(a')|}{\min\{\beta_2^2(a), \beta_2^2(a')\}},
\end{aligned}$$

it follows that

$$\Delta_{32} \le \frac{b^2 |\alpha \bar{z}|_\infty}{r} \left[\frac{|\Pi_a - \Pi_{a'}|}{\beta_1(a') \beta_2(a)} + \frac{\Pi_{a'} |\beta_2(a) - \beta_2(a')|}{\beta_1(a') \min\{\beta_2^2(a), \beta_2^2(a')\}} \right]$$

where $\sup_{a' \in \mathcal{A}} \Pi_{a'}/|\beta_1(a')| < \infty$.

Since $\beta_2(a) > \beta_1(a) > \hat{\varepsilon} > 0$ for any $a \in \mathcal{A}$, we conclude that there exists a constant C independent of a, a' and t such that

$$|\bar{z}_a(t) - \bar{z}_{a'}(t)| \leq C[|\beta_1(a) - \beta_1(a')| + |\beta_2(a) - \beta_2(a')| + |\Pi_a - \Pi_{a'}|]. \tag{B.9}$$

It is straightforward to further show $\sup_{t \in \mathbb{R}_+, a \in \mathcal{A}} |\bar{z}_a(t)| < \infty$, and then the proposition follows from (B.9) combined with the global Lipschitz continuity for each of $\beta_1(a)$, $\beta_2(a)$, Π_a as a function of a on $\mathbb{R}$. □

References

Altman, E., Basar, T., and Srikant, R. (2002). Nash equilibria for combined flow control and routing in networks: Asymptotic behavior for a large number of users. *IEEE Transactions on Automatic Control*, 47:917–930.

Aubin, J.-P. (1998). *Optima and Equilibria: An Introduction to Nonlinear Analysis*, 2nd ed., Springer, Berlin.

Baccelli, F., Hong, D., and Z. Liu. (2001). Fixed point methods for the simulation of the sharing of a local loop by a large number of interacting TCP connections. *INRIA Technical Report* no. 4154, France.

Bensoussan, A. (1988). *Perturbation Methods in Optimal Control.* Wiley, New York.

Bensoussan, A. (1992). *Stochastic Control of Partially Observable Systems.* Cambridge University Press, Cambridge.

Chow, Y.S. and Teicher, H. (1997). *Probability Theory: Independence, Interchangeability, Martingales*, 3rd ed., Springer-Verlag, New York.

Dziong, Z. and Mason, L.G. (1996). Fair-efficient call admission for broadband networks—A game theoretic framework. *IEEE/ACM Transactions on Networking*, 4:123–136.

Fudenberg, D. and Levine. D.K. (1998). *The Theory of Learning in Games.* MIT Press, Cambridge, MA.

Green, E.J. (1984). Continuum and finite-player noncooperative models of competition. *Econometrica*, 52(4):975–993.

Helbing, D., Farkas, I., and Vicsek, T. (2000). Simulating dynamic features of escape panic. *Nature*, 407:487-490.

Huang, M., Caines, P.E., and Malhamé, R.P. (2003). Individual and mass behaviour in large population stochastic wireless power control problems: centralized and Nash equilibrium solutions. 42nd *IEEE Conference on Decision and Control*, pp. 98–103, Maui, Hawaii.

Huang, M., Caines, P.E., and Malhamé, R.P. (2004). Large-population cost-coupled LQG problems: Generalizations to non-uniform individuals. 43rd *IEEE Conference on Decision and Control*, Atlantis, Paradise Island, Bahamas.

Huang, M., Caines, P.E., and Malhamé, R.P. (2004). Uplink power adjustment in wireless communication systems: A stochastic control analysis. *IEEE Transactions on Automatic Control*, vol. 49, pp. 1693–1708.

Huang, M., Caines, P.E., and Malhamé, R.P. (2004). Large-population cost-coupled LQG problems with non-uniform agents: Individual-mass behaviour and decentralized ε-Nash equilibria. Forthcoming in: *IEEE Transactions on Automatic Control.*

Huang, M., Malhamé, R.P., and Caines, P.E. (2004). On a class of large-scale cost-coupled Markov games with applications to decentralized power control. 43rd *IEEE Conference on Decision and Control*, Atlantis, Paradise Island, Bahamas.

Judd, K.L. (1985). The law of large numbers with a continuum of i.i.d. random variables. *Journal of Economic Theory*, 35:19–35.

Liu, Y. and Passino, K.M. (2004). Stable social foraging swarms in a noisy environment. *IEEE Transactions on Automatic Control*, 49:30–44.

Low, D.J. (2000). Following the crowd. *Nature*, 407:465–466.

Malhamé, R.P. and Chong, C.-Y. (1985). Electric load model synthesis by diffusion approximation of a high-order hybrid-state stochastic system. *IEEE Transactions on Automatic Control*, 30(9):854–860.

Mas-Colell, A., Whinston, M.D., and Green, J.R. (1995). *Microeconomic Theory.* Oxford University Press, New York.

Papavassilopoulos, G. (1982). On the linear-quadratic-Gaussian Nash game with one-step delay observation sharing pattern. *IEEE Transactions on Automatic Control*, 27(5):1065–1071.

Petrovic, B. and Gajic, Z. (1988). The recursive solution of linear quadratic Nash games for weakly interconnected systems. *Journal of Optimization Theory and Applications*, 56(3):463–477.

Phillips, R.G. and Kokotovic, P.V. (1981). A singular perturbation approach to modelling and control of Markov chains. *IEEE Transactions on Automatic Control*, 26:1087–1094.

Sethi, S.P. and Zhang, Q. (1994). *Hierarchical Decision Making in Stochastic Manufacturing Systems*. Birkhäuser, Boston.

Simon, H.A. and Ando, A. (1961). Aggregation of variables in dynamic systems. *Econometrica*, 29:111–138.

Srikant, R. and Basar, T. (1991). Iterative computation of noncooperative equilibria in nonzero-sum differential games with weakly coupled players. *Journal of Optimization Theory and Applications*, 71(1):137–168.

Tanner, H.G., Jadbabaie, A., and Pappas, G.J. (2003). Stable flocking of mobile agents, Part I: Fixed topology. 42nd *IEEE Conf. Decision Control*, pp. 2010–2015, Maui, HI.

Yosida, Y. (1980). *Functional Analysis*, 6th ed. Springer-Verlag, New York.